AF477225

# Art of Writing and Publishing in Agricultural Journals

# Art of Writing and Publishing in Agricultural Journals


**K. Veeranjaneyulu**
University Librarian
ANGR Agricultural University
Hyderabad

**Y. Eswara Prasad**
Professor & University Head
Department of Agricultural Economics
College of Agriculture
Hyderabad

**Azzezuddin Mehmood**
Assistant Professor
Department of English
College of Agriculture
Hyderabad

**Y. Uma Devi**
Assistant Librarian & Head
ANGRAU Regional Library
Agriculture College
Baptla (A.P.)

**A. V. S. S Nagaraju**
Junior Library Assistant
Central Library
ANGR Agricultural University
Hyderabad



**BSP** **BS Publications**
A unit of **BSP Books Pvt., Ltd.**

4-4-309/316, Giriraj Lane, Sultan Bazar,
Hyderabad - 500 095
Phone : 040 - 23445605, 23445688

*Published by :*

**BS Publications**

A unit of **BSP Books Pvt., Ltd.**

4-4-309/316, Giriraj Lane, Sultan Bazar,
Hyderabad - 500 095
Phone : 040 - 23445605, 23445688
e-mail : info@bspbooks.net

**ISBN : 978-93-52300-03-7 (HB)**

**S. Raghu Vardhan Reddy**
**Vice-Chancellor**

**Acharya N.G.Ranga Agricultural University**
Rajendranagar, Hyderabad-500 030.
Phone : Off : 010-24015035
Res : 040-27425305
Grams : 'Agrivarsity'
Fax : 040-24015031
E-mail : angrau_vc@yahoo.com

# FOREWORD

Every year lot of research data is generated from a number of research stations and from postgraduate students research work, but most of the data is confined either to research stations or to the departments in the form of theses. In the context of changing scenario, particularly in the era of globalization, the one who does not expose the work done by him/her in the form of publication of scientific research papers, may not be in a position to compete with the scientists at the national and international levels. In the present-day cut-throat competition, the saying "Publish or Perish" has a lot of relevance. It is in this context that the present publication on "Art of Writing and Publishing in Agricultural Journals" is timely, as it provides valuable information on the art & style of presentation of research work in the form of scientific papers suitable for various scientific journals.

I compliment the efforts made by Dr. K. Veeranjaneyulu, University Librarian, Dr. Y. Eswara Prasad, Professor & University Head, Department of Agricultural Economics, College of Agriculture, Rajendranagar; Sri Azeezuddin Mehmood, Assistant Professor of English, College of Agriculture, Rajendranagar; Mrs. Y. Uma Devi, Assistant Librarian & Head, ANGRAU Regional Library, Bapatla; and Sri A.V.S.S. Nagaraju, Junior Library Assistant, ANGRAU Central Library, Rajendranagar for bringing out such a useful publication. I am sure that this will be very useful reference to researchers, teachers, faculty, postgraduate students and to all the users concerned in getting their research work published in scientific journals.

**(S. RAGHU VARDHAN REDDY)**

# PREFACE

One's genius is known through his/her contribution to knowledge through publications. But unfortunately many learned scholars are restricting much of their innovative research findings to themselves. The reasons for this could be many, ranging from non-availability of publisher to the possession of little information on the finer nuances of publication.

This book enlightens the readers about the lesser-known elements of publication like the list of journals, the process of submitting the manuscript, and the writing style to be adopted while presenting matter in a journal. We have made a good attempt to acquaint the readers with several steps that are to be followed chronologically for getting their articles published in reputed agricultural journals.

We whole heartedly thank our Vice-Chancellor, Dr. S. Raghu Vardhan Reddy for his unstinted moral support and giving us permission to publish this book, which, we feel, will broaden the mental horizon of the readers, attract them and tempt them to have more number of publications.

Read the book, rejoice and, more importantly, use it to increase your publications.

With all good wishes.

*- Editors*

# CONTENTS



## SECTION–I

### Designing an Article

# SECTION – II

## Journal's Profiles

**Chapter 63**


**Chapter 64**


**Chapter 65**


**Chapter 66**


**Chapter 67**


**Chapter 68**


**Chapter 69**


**Chapter 70**


**Chapter 71**


**Chapter 72**


**Chapter 73**


**Chapter 74**


**Chapter 75**


**Chapter 76**

**Chapter 77**



**Chapter 78**



**Chapter 79**



**Chapter 80**



**Chapter 81**



**Chapter 82**



**Chapter 83**



**Chapter 84**



**Chapter 85**



**Chapter 86**



**Chapter 87**



**Chapter 88**



**Chapter 89**



**Chapter 90**

**Chapter 119**



**Chapter 120**



**Chapter 121**



**Chapter 122**



**Chapter 123**



**Chapter 124**



**Chapter 125**



**Chapter 126**



**Chapter 127**



**Chapter 128**



**Chapter 129**



**Chapter 130**



**Chapter 131**



**Chapter 132**

# SECTION – III
## Appendices

**Appendix I**


**Appendix II**


**Appendix III**

**Section – I**

# Designing an Article

# CHAPTER - 1

# Planning and Observation

*"Failing to prepare is preparing to fail."*

Thorough preparation, as such, is a must before writing a scientific paper.

## Type of presentation

The first step towards preparation is to decide the form of your paper, that is whether it is going to be a review article, an original article, or a letter to the editor.

## Selecting the Journal

Select a journal which has a wide audience and which is closely related to your field of study. Ascertain its accessibility, check the members of the editorial board and determine if they are the stalwarts in their fields. Determine the journal's impact factor and try to find out its acceptance rate. The impact factor is calculated by dividing the number of current citations to articles published in the two previous years by the total number of articles published in that period. Have a look at the style and format of the journal; find out the time it takes to publish; and, finally, if it bills the author for pages. For the beginners, it is advisable to go for monthly peer-reviewed journals and when you have got a few publications, you can go for the international journals having high impact factor.

## Planning of Author's Name

Decide on the authorship in the beginning itself if the article is going to be a joint contribution. It would be an intellectual dishonesty if you include the names of persons who have not made any contribution as authors. The framework of "honorary authorship" should be avoided.

*Certain pre-requisites for designing a paper*

Before you start writing a paper, have the following things in your mind.

1. Write as per the format described in the journal, that is your writing style, titles. Formatting style should go according to the journal you are writing for. Also, find out if online submission of the manuscript is allowed or not.

2. Get the appropriate ethical, statistical and copyright clearance for the paper beforehand. It is always better to collect and prepare illustrations, tables, graphs and photographs before you start writing the paper.

# CHAPTER - 2

# Writing

*Writing a paper requires utmost tact and care.*

Certain DO's and DON'Ts while writing a research paper:

1. Do not hesitate to rewrite any number of times since there is no short cut to produce something which consumes the time of others to read.

2. Do not fudge data or mislead, since truth will ultimately come out.

3. Do not duplicate and camouflage the scientific substance through literary flourishes.

4. Do start writing the paper before the experiments are complete since writing is a continuous process and you embellish the quality of your writing with the passage of time.

5. Do not include tables and figures more than necessary.

## Decide your format

There are three basic forms of research articles. Decide before writing the category to which your article belongs:

### Full-length research articles

These articles cover comprehensive research carried out over a length of period. They follow "IMRAD" format – Introduction, Methods, Results and Discussion.

### Short communications

The information is presented as a single section with tables and figures.

### Rapid communications

Present "hot" findings in a brief communicational format.

In a nutshell your paper should contain the following.

| Experimental process | Section of paper |
| --- | --- |
| What did I do (brief) | Abstract |
| What is the problem | Introduction |
| How did I solve the problem | Methods and Materials |
| What did I find out | Results |
| What does it mean | Discussion |
| Who helped me out | Acknowledgements (Optional) |
| Whose work did I refer to | References |
| Extra Information | Tables and Figures |

## LET US EXAMINE THESE STEPS IN DETAIL

### Title

The readability of your book to a large extent depends on the cover and title. Select a short, simple and eye-catching title which goes well with its contents.

### Abstract

Your chances of getting your paper published depend on the Abstract. In a capsule, it should cover the entire information like the aim of the study, the methodology used, results obtained and conclusions drawn. An abstract between 150-250 words is considered to be "good". Devote a considerable amount of time for preparing an abstract.

### Introduction

Should not be too long as seen in most of the cases. Explain the significance of the study by spelling out hypothesis. Also, explain the reason for undertaking the research.

### Methods and Materials

Explain the following clearly in this part:

- Type of study/approach you have undertaken – descriptive, ethnographic, narrative/holistic with details like location, date and time of carrying out experiments.
- Sampling design and methods used for data collection.
- Statistical procedures used for analyzing data.

### Style

The preferred style is third person passive voice. Avoid writing in first person. Instead, use the words "The Researcher" or "The Investigator".

## Results

Avoid using terms like majority, most of the, in some cases, significantly greater than, more or less, could be due to, etc. Explain the results with confidence and have faith in yourself. Do not hesitate to give negative results or unexpected findings if any. Ignore percentages or statistical figures when the number is small.

## Discussion

It is usually a good idea to begin this section by giving the answer to the question you set out with. State the main findings clearly through flow diagrams and tables and, if possible, compare this data with other published information. Examine the ramification in detail. Do not, however, introduce new results in this part. Reserve this for your "major findings" unit.

## Acknowledgements

There are no hard and fast rules for acknowledging somebody's help, be it that of a person, an institution or an organization. The investigator, as such, is at liberty to write this part in his own style. This unit is also not compulsory but a word of thanks will really go a long way in paying intellectual dividends to all those who helped you in carrying out your work.

## References

The purpose of this section is to provide the full citation for articles referred to in the text. A complete reference includes all the authors' names, the titles of the articles, the journals' name, the volume number, page numbers and the year of publication. A wide range of styles is used for citing references in the text and bibliography. Check the journals' "instructions to authors" for information about the content and formatting references.

At the end of the paper, a list of references or bibliography is provided in alphabetical order. In some cases citations appear in the text as numbers, usually a superscript, which then refer to a particular item in the reference list. It is compulsory for the author to check all the references thoroughly before citing them since simply copying them is likely to be error-prone. There are no rules to be followed in letter and spirit with regard to the number of references. Some journals nevertheless impose a limit to conserve space. A rule of thumb is not more than 6 references for a particular point and not more than 100 references per paper.

*Guidelines of manuscript design and basic rules of writing a research article*

| Sections | Main functions | Preferred style | Basic rules |
|---|---|---|---|
| Title | Indicates the entire matter, catches the reader's attention | Short, simple, clear and unambiguous | Avoid high flown language  make it catchy and avoid redundancy<br>Spend some time on finalizing the title |
| Abstract | Explains the entire story in a capsulated form | Past tense and passive voice – no citations, tables, graphs etc. | Explain what was done and what was the outcome without introducing the theme |
| Introduction | Introduces the topic, explains the terminology, the focus of the paper, the research objectives, hypothesis and limitations of the study | Simple present tense, for referring to established knowledge | Do not prolong this section by inserting redundancies<br>Explain the significance of the study step by step in a logical move |
| Review of related literature | Cites the related research undertaken in the area concerned | Past tense | Use state-of-the-art references.<br>Avoid irrelevancy |
| Methodology | Explains the 5 W's, namely who, what, when, where and why and H namely How of your  study | Past tense | Explain everything in detail.  Use authoritative tone and simple language |
| Results | Gives summary results obtained through statistical tests in written as well as graphic forms | Past tense | Emphasise the most significant findings.<br>Distinguish your work from that of others |
| Conclusions and Recommendations | States the important research findings and future implications | Past tense for conclusions – Researcher's style for Re-comendations | Do not recapitulate results but make strong statements.<br>Do not hide unexpected results |

# CHAPTER - 3

# Editing and Revising

*Editing and revising involves the following steps in the order :*

1. ***Make major alterations:*** Fill in the gaps, change the order of contents to present the material in an acceptable form.

2. ***Fine-tune the style:*** Refine the text and make it free from grammatical and spelling errors.

3. ***Get feedback:*** Get a feedback of your document from your peers. Make modifications as per necessity and requirement.

4. ***Format the document:*** Make your document attractive and easy to read.

Self-revising involves revision and modification of matter starting from a global perspective, coming down to paragraph content and finally down to sentence-level editing.

## Global

Check whether there is cohesive, logical and sequential presentation of ideas from one topic to another.

## Paragraph

Check the order of paragraphs to find out if there is a smoother transition between them.

## Line editing

Eliminate superfluos and colloquial phrases; avoid redundancy; read the paper aloud; and check and recheck grammatical and spelling errors.

## Checklist Before the Final Submission of Your Document

1. Check all your sources are cited correctly.

2. Check the number of tables and figures.

3. Ensure that the authors' names are spelt out correctly.

4. Check the line spacing.

5. Check you have given your document for proofreading at least twice to two different persons.

6. Re-read the document for the last time before giving it for publication.

# CHAPTER - 4

# Submitting Manuscript and Follow-ups

As soon as your manuscript is ready, start the process of sending it for publication. Remember that you will have to go through the following procedure before getting your document published.

## Identify the mode of submission

Prefer online submission if this option is available as it has its own advantages over the normal postal submission.

## Prepare a covering letter

The covering letter should be simple, short and effective addressing the editor by name as it has its own impact. Do not forget to write the title of the article/ book, authors' name and date.

## Submit the manuscript to the Editor

Follow the instructions to authors scrupulously. Some journals ask for a 'pre-review", i.e., a letter indicating the content of the article to determine its worth and credibility. List out the possible reviewers in this case.

## Handling the reviewer's comments

Acceptance of manuscripts on first submission is rare. Do not get discouraged if this happens to you. Modify your manuscript with tact if the editor so desires. Answer every concern of the reviewer and indicate where the changes were made. Approach the revision using global, paragraph and line editing strategy as discussed earlier. While resubmitting, mention in your cover letter that this a revised version. Do not submit your manuscript to more than one journal at a time.

## Check the proofs

Your manuscript, if accepted for publication, will be sent to you for proofreading. The proofs may be sent via email or hard copy. Check the proofs with utmost care and patience.

## Celebrate

Thomas Alva Edison says that success is 99% perspiration and 1% inspiration. After putting immeasurable amount of hard work and perspiration, you have now ultimately reached your goal, i.e., publication. Enjoy and rejoice the occasion.

RECIPE FOR A SUCCESSFUL PUBLICATION. Avoid Plot till your manuscript is published.

Postponement

Laziness

Oversleeping

Television.

# CHAPTER - 5

# Avoid Rejection

*Be thorough with the following in order to avoid rejection.*

**Be well conversant with the editorial process**

Your article will be reviewed by the editor, associate editor and reviewer. They will evaluate your document in the following areas:

A. Significance and relevance of the topic in the present-day scenario.

B. Literature review.

C. Sequential and logical presentation of ideas.

D. Conclusions/Recommendations that have emerged out of your study.

E. Clarity and consistency of writing.

Few manuscripts are published as originally submitted. Most journals use some variation of the following set of categories:

1. Publish (with least revision).

2. Publish (with some revision).

3. Ask for major revisions before considering publication.

4. Reject.

Do not feel disheartened if the editor asks you for any of these modifications. Resubmit your manuscript as per the requirements of the publisher.

## Avoid Pitfalls

The most important reasons for rejecting a research article are:

| Area | Reason for Rejection |
|---|---|
| Topic | Irrelevant and restricted. |
| Newness | Repetition of the work done earlier · |
| Focus | No cohesion between the topic, objectives and conclusions. |
| Methodology | Weak, not clearly spelt out. Data with insignificant sample number and flaws. |
| Style | Detached and unfocussed. |

There are a few rules for improving the probability of acceptance of your research article.

| What to do | How to do |
|---|---|
| Take a reader's view | Imagine yourself to be the reader while you write. |
| Simplify | Be yourself.  Use short sentences and simple examples to explain complex methodology. |
| Avoid | Redundancy, repetition and over-explanation of familiar techniques and terminology. |
| Make strong statements | Use "We/I concluded" instead of "It may be concluded". |
| Be self-critical | Be open to unexpected conclusions and acknowledge the work of others. |

# CHAPTER - 6

# Scientific Review Writing

Reviews play an important role in scientific communication and understanding. They also give the policy-makers and also the researchers, a vivid insight into the ever-expanding horizon of emerging knowledge. Review writing, however, is a task which is quite intensive and uniquely demanding. Given below are some points to be kept in mind while writing a review paper.

- Your review paper must be a smooth and integrated synthesis of literature with a systematic and orderly cohesion of ideas. It should not sound like a mere regurgitation of facts.

- Take sufficient time since the quality of your paper depends on extensive library searches. Take a month or so for writing a 10-20-page review paper.

- Jot down whatever comes to your mind immediately without bothering anything. These could be of utmost significance while preparing your final manuscript.

- Maintain index cards, notes, word processors to make your job easy.

- Decide the format headings and subheadings before setting pen to paper. This will markedly improve the overall quality of your final product.

- Use the draft system. Write the first draft and come to it again after two to three days. Revise and improvise it, if necessary.

- Write with authority. Avoid statements like, I think, It, seems, etc.,

**Your review paper should have the following sections:**

*Title*

Short, simple and relevant.

*Abstract*

Should be written after the body of the paper is completed.

**Introduction**

Better to write it after the body of the paper is completed. Should be short and concise and try to bring readers from different backgrounds to match up with the speed of the objective of your paper. It should also explain the significance of the issue which is being reviewed.

## Body

Figure out the perspectives at the beginning. Build your ideas logically by giving outlines and headings. Each paragraph should have clear, well-thought-out points and should contain only the information needed and nothing more.

## Conclusion

Do not separate it from the body if it is not too long. Restate your objectives and list out all the major conclusions.

# CHAPTER - 7

# Open Access Journals

## Introduction

The desire for publishing the results of research, particularly in science and technology is an age-old tradition. Thus, over the years, the number of journals have increased both in volume and variety. This, on the one hand, has led to the exponential growth of scholarly literature. On the other hand, it has put severe hindrance on their accessibility. The libraries, particularly in developing countries, are vexed with the problems of escalating costs and shrinking funds to provide access to literature and act as a true partner of research. Though the scientists and scholars are communicating the fruits of their research through scholarly journals, all are not available to many, if not to all. There is a growing concern now to make the public-funded research available to wider cross-sections of people, so that its benefits reach out to the society in general. To augment the accessibility of information, an array of initiatives have been taken by libraries though resource initiatives could only bring the limited access of literature to the researchers. The access to printed literature in libraries is slowly becoming non-attractive.

The development of technology (Internet) has brought enormous opportunities to bring the results of primary research to all through digital communication anywhere and at any time. The impact of convergence of tradition and technology brings the facility of accessing information conveniently and instantaneously. Lawrence says, "Scientists now have almost instant access to large and rapidly increasing amount of information that was previously gathered through trips to the library and inter-library loan. Although it is now possible to have free access to significant amount of literature on the Web, still a significant amount of research is not available freely. The reasons may be many – the non-availability of high-impact journals on the Web freely, the researchers' preference for high impact (peer-reviewed) journal, the impediment of subscription price, etc. It is here that the importance of open access comes where the question of availability is more on the individuals and institutions in contrast to the earlier model of dissemination through formal publishing channels.

Open Access (OA) is the free online availability of digital content. It is best-known and most feasible for peer-reviewed, scientific and scholarly journal articles, which scholars publish without expectation of payment.  Open Access publishing, where the author (usually the author's research funder or institution) pays the publication costs, has been proposed as an alternative to a subscription-based cost-recovery model.  The growth of this alternative cost-recovery model has been slow so far.

Open access to scientific journal articles means online access without charge to readers or libraries.  Committing to open access means dispensing with the financial, technical and legal barriers that are designed to limit access to scientific research articles to paying customers.  It means that, for the sake of accelerating research and sharing knowledge, publishers  will recoup their costs from other sources.

Open access to the scientific journal literature would be hard to defend if its obvious advantages require sacrificing any of the obvious advantages of traditional journals. No such sacrifice, however, is necessary. Open access to scientific journal literature is compatible with all the major advantages of traditional journals, which are  innumerable in number.

Open access journals provide access to full-text contents of scholarly, peer-reviewed journals. There are two types of open access journals – the one, available in electronic version only and the other, available in both electronic as well as print versions, viz., *Current Science.* In the first type, the journals are published at regular intervals on the Internet that do not have any print-on-paper counterpart.  In the second type, the journals are published in print-on-paper format and distributed to the subscribers.  The same contents of print-on-paper are available to the scholars free of charge in electronic form.

## Objectives of Open Access Initiative

Based on the experience and comprehensive literature survey, the Open Access Initiative has the following objectives:

1. To remove access barriers to research
2. To assist institutions in solving the problems of price hike in journals' subscription
3. To help the researches to solve permission problem
4. To ensure the widest possible dissemination of research work
5. To bridge the knowledge gap
6. To provide low-cost solutions through Institute Repositories
7. To make openly available the research work to the academic community, irrespective of region and financial resources
8. To assist the academic community in sharing their research work and experiences
9. To maximize impact of research by maximizing access to research.

## Peer-Reviewing

Researchers could put their own articles on their home pages through peer-review, but that is not the kind of open access advocated by the Public Library of Science, the Budapest Open Access Initiative or BioMed Central (the publishers of *Journal of Biology*).  All the major open-access initiatives agree that peer-review is essential to scientific journals, whether these journals are online or in print, free of charge or 'priced'.  Open access removes the barrier of price, not the filter of quality control.

## Preservation

So far, paper is the only commonly used medium that we know can preserve tests for hundreds of years. There are many creative methods emerging for storing digital texts electronically with at least the security of paper.  The only problem is that it will take hundreds of years to monitor the outcome of present-day experiments. But we don't have to chose between insecure storage and retreat from the digital revolution. The short cut to preservation is to print digital texts on paper. Individual researchers can make printouts for their own use, and journal publishers can print entire issues, either for routine sale or specifically for deposit in long-term archives. Preservation in the digital era will be as good as paper, just as it was before the digital era.

## Intellectual Property

Open access is compatible with copyright as long as the holder of the copyright consents to open access. The fact that most copyright holders want to restrict access to paying customers has created the illusion that all copyright holders want this, or that copyright requires payment. This is not the case. Copyright law gives the rightsholder the authority to decide – but most rightsholders are profit-seekers whose interest lies in controlling access, distribution, and copying. But in their role as authors of journals and articles, scientists are not profit-seekers and their interest lies in dissemination to the widest possible audience. For this purpose, it doesn't matter whether scientists retain copyright of their own articles or transfer the copyright to an open-access journal or repository. Copyright assures authors that authorized copies will not mangle or misattribute their work. And the fact that the holder of the copyright consents to free access sharply separates this kind of open access from what might be called 'Napster for science'.

## Profit

Open-access publishing is compatible with revenue, and even profit, just as it is compatible with a non-profit business model. For example, BioMed Central is a for-profit publisher. Publishers adopt open access not to make a charitable donation or political statement, but to provide free online access to a body of literature, accelerate research in that field, create opportunities for sophisticated indexing and searching, help readers by making new work easier to find and retrieve, and help authors by enlarging their audience and increasing their impact.

If these benefits are expensive to produce, they would nevertheless be worth paying for – but it turns out that open access can cost much less than traditional forms of dissemination. For journals that dispense with print, with subscription management, and with software to block online access to non-subscribers, open access can cost significantly less than traditional publication, creating the compelling combination of increased distribution and reduced cost. The revenue would thus violate the barrier-free nature of open access. Instead of charging readers or their sponsors for access, BioMed Central charges authors or their sponsors a fee for dissemination. Its revenue consists of these dissemination fees plus proceeds from the sale of add-ons and auxiliary services.

## Promotion of Need-Based Scientific Products or Services

An open-access journal gives readers access to the essential literature without charge. But this is compatible with selling an enhanced edition, or other products and services, to the same community of readers. A scientific journal might sell 'add-ons' and auxiliary services such as current awareness, reference linking, customization ('My Journal'), or a print edition. Revenue from these add-ons may offset, or even exceed, the cost of providing open access to essential literature.

## Open Access Journals in India

Open access is free online access, and is perfectly compatible with other kinds of access to the same content. A publisher of an open-access journal might lose money by producing a print edition of the same content, and this is one reason why some publishers might not create a print edition. But a publisher might decide to sell a print edition that can generate at least as much revenue as is needed to cover its costs. Promotion of product or services can generate even more revenue and publishers need no longer see the print edition of a journal as the economic centerpiece of the enterprise. Open access is compatible with printing copies for the purpose of long-term preservation, and is very much compatible with the users' printing individual articles through their browsers.

It is, therefore, concluded that open access journals have got almost all the advantages of traditional journals. Most of the big publishers have entered this sector. The best example being the Springerlink and Elsevier, who are providing their journals in the open-access form along with the traditional one.

# Section – II

# Journals' Profiles

# CHAPTER - 8

# African Journal of Food, Agriculture, Nutrition and Development

## Author Guidelines

The *African Journal of Food Agriculture, Nutrition and Development* (AJFAND) is open to both African and non-African contributors. Besides academic research, the journal will provide an avenue for sharing information on national- level food and nutrition programs. It is anticipated that three issues will be produced every year, with each issue combining both research and programs aspects. The journal will be produced in English. In the medium term, abstracts will be produced in both English and French. French translations of the articles will be available as soon as more stable funding is assured.

## Submissions

Manuscripts will be accepted for consideration on condition that they are submitted exclusively to *African Journal of Food Agriculture, Nutrition and Development* (AJFAND). This restriction does not apply to abstracts on press reports published in connection with scientific meetings. All materials submitted for publication must be typed in double line space on numbered pages and should conform to the uniform requirements for manuscripts submitted to biomedical journals (1). A word count excluding abstracts and references must be given in all submissions. Manuscripts must be submitted in triplicate including a diskette preferably in word format.

Address all submissions to:

The Editor-in-Chief

*African Journal of Food Agriculture, Nutrition and Development* (AJFAND)

P. O. Box 29086

Nairobi, Kenya

Tel: +254 2 249799 Fax: +254 2 249799

E-mail: oniango@iconnect.co.ke

## AUTHORSHIP

It is a requirement that all authors must give signed consent to publication. They should provide their names, qualifications, designations, current address including fax and e-mail numbers for purposes of correspondence. A covering letter signed by all authors should identify the corresponding author responsible concerning the manuscript. Credit for authorship requires substantial contributions to: (a) conception and design, or analysis and interpretation of data; (b) the drafting of the article or critical revision for important intellectual content and (c) final approval of the version to be published.

### Format

*Original articles and Reviews:* Should be between three and four thousand words with a maximum of eight tables or other illustrations. Original articles should review data from original research. Reviews must be critical analyses of the subjects reviewed giving the current and a balanced view of all the issues, for instance controversies. Reviews should preferably be contributed by authorities and experts in the respective field. The message carried in both reviews and original articles must be clear and of significance. An abstract not exceeding 250 words must be included. Copies of related papers already published and any non-standard questionnaires used should be submitted. This requirement is important where details of study methods are published elsewhere or when the manuscript is part of a series e.g. part 11 of a series where part 1 has been published elsewhere. References should not exceed a maximum of thirty. *Short communications:* Should not be more than 1000 words and one illustration. This should focus on any interesting findings/news an author wishes to convey to the readers of AJFNS. The abstract should be in prose form.

### Letters to the Editor

Must not be more than 500 words; contain one illustration with not more than five references. The abstract should in prose form.

### Key words

A maximum of five key words should be added at the bottom of the abstract in reviews and original articles.

### References

This must be numbered consecutively as they are cited. All authors must be listed when they are not more than three. When they are three or more the first three are listed followed by "et al". The title, source, and journal abbreviations should conform to those in index medicus, year, volume and inclusive page numbers.

Review articles should have not more than 40 references and original articles not more than 30. Personal communications, unpublished data and manuscripts "in preparation" or "submitted for publication" are unacceptable. If necessary such material may be incorporated in the appropriate place in the text. The following are sample references:

1. *Periodical*

   Bunce G.E., Kinoshita J., Horwitz J. Nutritional factors in cataract. Ann. Rev. Nutr. 1990; 10: 233 - 254.

2. *Chapter in a book*

   Spencer N. Poverty and child health in less developed countries. In: Poverty and child health; Oxford, UK and New York, NY: Radcliffe Medical Press, 1996: 74-94.

3. *Book*

   Parry J. Introduction to Ophthalmology. Oxford University Press, Oxford. Ed 3. 1989 Pp 188 - 194.

## Illustration Figures

Should be professionally designed and numbered in Arabic numerals. Symbols and numbering should be clear and large enough to remain legible after reduction to fit the width of a single column. The back of each figure should include the sequence number, the name of the author and proper orientation (e.g. "Top"). Legends for illustrations should be on a separate sheet and should not appear on the illustration. Definitions of any abbreviations that appear on the figure should be provided.

## Tables

Should be provided each on a separate page, double spaced and a title provided for each. Tables should be numbered in Arabic numerals.

## Copyright

Published material in the AJFNS is covered by copyright. Authors transfer all rights to the journal upon publication. Permission must be granted by the Editor for use of any published material.

## Reference

1. Uniform requirements for manuscripts submitted to biomedical journals. International Committee of Medical Journal Editors. Ann. Intern. Med. 1997; 126: 36-47.

## Manuscript Submission Checklist

Three hard copies

A diskette preferably in word format

Abstract of approximately 250 words in reviews and original articles

References, figures and tables cited consecutively in text

Separate illustrations from text appropriately labelled

Category of article specified

Covering letter, including appended signatures of all authors

Name and address of corresponding author

## Submission Preparation Checklist (All items required)

The submission has not been previously published nor is it before another journal for consideration; or an explanation has been provided in Comments to the Editor.

## Copyright Notice

Copyright for articles published in this journal is retained by the journal.

## Privacy Statement

The names and email addresses entered in this journal site will be used exclusively for the stated purposes of this journal and will not be made available for any other purpose or to any other party.

# CHAPTER - 9

# Agribusiness

## Guidelines to Author

*Agribusiness : An Internal Journal* publishes research that improve our understanding of how food systems work, how they are evolving, and how public and/or private actions affect the performance of the global agro-industrial complex. The Journal focuses on the application of economic analysis to the organization and performance of firms and markets in industrial food systems. Subject matter areas include supply and demand analysis, industrial organization analysis, price and trade analysis, marketing, finance, and public policy analysis. International, cross-country comparative, and within-country studies are welcome. To facilitate research the journal's Forum section, on an intermittent basis, offers commentary and reports on business policy issues.

Major articles should be no longer than 30 double-spaced typewritten pages. Notes or comments should be no longer than 8-10 manuscript pages. For the sake of readability, limit the amount of mathematical modeling in articles. If extensive mathematical modeling is important to the topic, put it in an appendix to the article.

Submission by e-mail is preferred; however, surface, air, or express mail are also acceptable. Each submission must indicate the name, address, telephone number, facsimile number, and e-mail, address of the author to whom all correspondence is to be addressed. If the manuscript has been presented, published, or submitted for publication elsewhere, please inform the Editor. Our primary objective is to publish technical material not otherwise available, but we do occasionally publish papers of unusual merit that have appeared or that will appear elsewhere when there is full justification for this.

Prospective authors should submit one copy of the complete manuscript via e-mail, or three paper copies by regular mail to: Ronald W.Cotterill, Food Marketing Policy Center, Agricultural and Resource Economics, Unit 4021, 1376 Storrs Road, Storrs, CT 06269-4021; E-mail: fmpc@uconn.edu; Telephone: 860-486-2742; Fax:860-486-246d1. Please see below for more detailed disk submission instructions.

*Forma :* Manuscripts should contain the following:

Title; names and complete affiliations of all co-authors, including mailing addresses, phone numbers, facsimile numbers, and e-mail addresses. Please provide an informative 150-word abstract at the beginning of your manuscript, and identify not more than five keywords. The abstract should provide an overview of your paper and not a statement of conclusions only. Since abstracts of accepted papers are included in the EconLit citation system, please classify your papers in at least one, and not more than three, EconLit subject matter areas. The EconLit classification of subject matter areas is available at www.econlit.org/subjectdescriptors.html. Put the EconLit alphanumeric subject codes at the end of your abstract.

If applicable, please provide contract grant sponsor(s) and contract grant number(s). Regular manuscripts should be no longer than 30 typewritten, double-spaced pages, plus references. Notes should be no longer than 8-10 type-written double-spaced pages, including references. Authors should write succinctly and should note that a significant element of the review process will be its length relative to its content and the clarity of writing. Manuscripts should be subdivided into sections and subsections, numbered with Arabic numerals as in 1.1 or 2.3, to aid readability. For information about references, tables, list of figure captions, and figure specification, please see below. Authors of accepted papers will be asked to supply a technical biography of one short paragraph. The biography should be in the form used for biographies in professional journals. It should contain name, degrees and dates earned, current position, a few current research interests, phone number, and e-mail address.

*Instructions for Keyboarding:* Manuscripts must be double-spaced on a single side of standard 8.5 by 11 in.(21.5 by 28cm) white paper with one-inch margins. Material intended for footnotes should be inserted in the text as parenthetical material whenever possible. All mathematical symbols, equations, formulas, Greek and/or unusual symbols should be typed. Variables should be set in italic type, and vectors should be set as boldface. Identifiers (nonvariables) and abbreviations that are not variables should be roman.

*Reference:* References are styled according to the fifth edition of the Publication Manual of the American Psychological Association.

Within the text, the author's name and year of publication should accompany direct quotations or discussions of published material. The format for citations within the text is either "Smith (1997) and Jones and Black (1996)" or "(Smith, 1997; Jones & Black, 1996)." Full references should be listed alphabetically in the last section of the paper titled "References."

A sampling of the most common entries in reference lists appears below. Please note that for journal articles, issue numbers are not included unless each issue in the volume begins with page one. Hence, it is Agribusiness, 20,

155-166, not Agribusiness, 20(2), 155-166.  Entries not exemplified below are modeled in the *Publication Manual*.  Please note that the use of italics has been eliminated for the printed publication (this is a departure from the *Publication Manual style*).

*Book :* Senauer, B., Asp, E., &Kinsey, J. (1991).  Food trends and the changing consumer.  St.Paul, MN: Eagan Press.

*Journal :* Tahani, N. (2004).  Valuing credit derivatives using Gaussian quadrature: A stochastic volatility framework.  The Journal of Futures Markets, 24, 3-35.

### Article in edited book :
Brown, E.D., & Sechrest, L.(1980).  Experiments in cross-cultural research. In H.C. Triandis & J.W.Berry (Eds.), Handbook of cross-cultural psychology (Vol.2, pp.297-318).  Boston: Allyn and Bacon, Inc.

### Unpublished paper presented at a meeting :
Black, L., &Loveday,G.(1998,February).  The development of sign language in hearing children.  Paper presented at the annual meeting of the Professional Linguistics Society, Munich, Germany.

### Published proceedings of meetings and symposia :
Shang, H., &chen, C.(1990).  Air quality improvement efforts in three urban areas in eastern China.  In S. Shea (Ed), Eastern Symposium on Air Pollution: Vol.27 (pp.155-172).  Hong Kong: Tech University.

### Unpublished doctoral dissertation :
Smith, A. (2001).  Analyses of nonunion American companies in the late 1990s. Unpublished doctoral dissertation, Georgetown University, Washington, DC.

References should refer only to material listed within the text. Do not abbreviate journal names.  Authors should review and verify references before manuscripts are submitted for consideration, because they alone are responsible for accuracy and completeness.  Material that is not retrievable, such as papers presented at meetings and symposia, unpublished works, and personal communications should be limited to material essential to the article.

Anthologies and collections must include names of editors and pages on which the reference appears. Books in a series must include series title and number/volume if applicable. Because of the large quantity of conference proceedings available, it is critical to give as much information as possible when citing references from proceedings.  Please include the complete title of the meeting, symposium, etc. (do *not abbreviate titles),* and the city and dates of the meeting. If a proceeding has been published, please provide the editor's names, publisher, city, and year of publication, and pages on which the article appears.

## Wiley's Journal Styles Can Be Created in End Note

End Note is a software product that we recommend to our journal authors to help simplify and streamline the research process. Using EndNote's bibliographic management tools, you can search bibliographic databases, build and organize your reference collection, and then instantly output your bibliography in any Wiley journal style.

***Download the reference style for this Journal :*** If you already use EndNote, you can download the reference style for this journal. Visit www.interscience .wiley.com/jendnotes.

***How to Order :*** To learn more about EndNote, or to purchase your own copy, visit www.endnote.com.

***Technical Support :*** If you need assistance using EndNote, e-mail endnote@isiresearchsoft.com, or visit www.endnote.com/support.

***Tables :*** Tables should not be embedded in the text. Within the main text of the article, please place a note in a separate line indicating where you would like each table to be inserted (e.g. "Insert Table 1 here"). Each table should be cited in the text. Tables are best grouped together in a separate electronic file and clearly labeled with your manuscript number or last name, an underscore, and the word "tables" (e.g. 2314_tables). All tables should be numbered consecutively with Arabic numerals and should include an explanatory heading. With the hard copy of the manuscript, group together the hard copies of tables at the end, each new table starting on a fresh page. The fifth edition of the *Publication Manual of the American Psychological Association* has helpful guidelines for both tables and figures.

***Figures (Gray Scale Art & Line Art).*** As with tables (see above), figure should not be embedded in the text. Please place a note in the text indicating approximate place figure should go, cite the figure in the text, and provide a list of figure captions at the end of the manuscript, after the references. With the hard copy of the manuscript, group together the hard copies of figures at the end, each starting on a fresh page. Figures must be numbered consecutively with Arabic numerals.

For best quality printing, submit gray scale figures (e.g. careen shots, photos, or charts requiring shades of gray) that are high resolution (above 300 dpi). Figures pasted directly from the Web are low resolution (72dpi) Bitmapped line art (made only of black & white lines—often simple charts or graphs) is best submitted at higher resolutions yielding 600-1200 dpi.

Figures are best submitted in TIFF or EPS (with preview) formats. If these formats are not possible, use JPEG. Please do not submit proprietary graphics formats such as Corel Draw or Adobe Illustrator. Authors who wish to submit figures in Microsoft Word should place them in a separate document from the article text file, following the naming convention suggested above for tables.

Figures submitted only in hard copy are acceptable if they have been printed with a high-quality laser printer. Authors are cautioned to provide lettering of graphs and figure labels that is large, clear, and open so that letters and numbers do not become illegible when reduced. Likewise, authors are cautioned that very thin lines and other fine details in figures may not successfully reproduce. Original figures should be drawn with these precautions in mind.

***Final Submissions :*** Authors are to submit the final, accepted version of their manuscript on disk to the Editor. Three hard copies of the manuscript should accompany the disk.

***Copyright Information:*** Normally only original papers will be accepted. No article can be published unless accompanied by a signed copyright transfer agreement (CTA), which serves as a transfer of copyright from author to publisher. ACTA may be obtained from http://www3.interscience.wiley.com/cgi-bin/jabout/35917/For Authors.html.Copyright of published papers will be vested in the publisher. It is the author's responsibility to obtain written permission to reproduce material (often, figures) from another publication. Please keep in mind that obtaining permissions can sometimes take several weeks. A form for this purpose is sent with the manuscript acceptance letter, or can be obtained from http://www3.interscience.wiley.com/cgi-bin/jabout/35917/ForAuthors.html.

***Author Page Proofs:*** After the manuscript is sent to the publisher, it will be copy edited and typeset. After being typeset, page proofs will be e-mailed in PDF format to the designated corresponding author. Authors have 48 hours after receiving page proofs to return them with any corrections. Authors may be charged for alterations to the proofs beyond those needed to correct typesetting errors. **Authors are advised to keep a copy of the original manuscript to refer to when checking the page proofs.** The manuscript and artwork sent to the publisher will not be returned unless a request is made when the manuscript is originally submitted. Authors with no e-mail access will be mailed their page proofs.

***Copies of the Published Article:*** A complimentary copy of the journal for each co-author will be mailed to the corresponding author after publication. Additional copies of an article can be ordered in bulk (100 copies minimum). A reprint order form will be supplied with the page proofs for bulk orders. Reprints are shipped approximately six weeks after publication of the issue in which the article appears.

***Other Correspondence :*** To obtain contact information for subscriptions, advertising, and other matters, go to http://www3.interscinece.wiley.com/cgi-bin/jabout/ 35917/ Contact.html. Alternatively, contact the Publisher, Professional & Trade Division, Wiley Periodicals, 111 River Street, Hoboken, NJ 07030-5774.

***Disk Submission Instructions:*** *Please submit your final, revised manuscript on disk as well as hard copy. The hard copy must match the disk.*

Please return the disk submission slip on the next page with your manuscript and labeled disk(s).

*Storage medium.* 3-1/2" high-density disk, Zip disk, or CD-ROM in Windows, Macintosh, or Linux format. At authors' requests, disks will be returned after publication.

*Software and format.* Microsoft Word is preferred. Other word processing programs and LaTeX are less preferable. If you use software other than Microsoft Word, please provide a plain text files in PDF format, Adobe PageMaker, Quark Xpress, or other desktop publishing software. If you prepared your manuscript with desktop publishing software, export the text to a word processing format. Refrain from complex formatting; the publisher will style your manuscript according to the Journal design specifications. Please make sure your word processing program's "fast save" feature is turned off. Please do not deliver files that contain hidden text: for example, do not use your word processor's automated features to create footnotes or reference lists.

*File names.* Files can be named with your manuscript number or last name, an underscore, and a very brief description of the file's contents (e.g., "2314_text" or "2314_tables 1 to 4").

*Labels.* Label all disks with the journal name, author name, file name, and the word processing program and version used. For figures or artwork, label disks with journal name, author name, file names, formats, and compression schemes (if any) used. Hard copy output must accompany all files.

Please scan your disks for viruses before you send them, and keep a copy of what you send in a safe place in case any of the files need to be replaced.

## Photocopy and return with labeled diskette(s)

| Corresponding author's name | E-mail address: | Telephone: |
|---|---|---|

| Manuscript number: | Type of computer: | Program(s)& version(s)used: |
|---|---|---|

| Miscellaneous: |
|---|

**I certify that the material on the enclosed diskette(s) is identical in both word and content to the printed copy herewith enclosed.**

**Signature:**_______________________ **Date:**_______________________

Production questions can be directed to Margaret Ziomkowski, Senior Production Editor, Phone: 201-748-8824; E-mail: mziomkow@wukey.com.

## Disk Submission Instructions

The following guidelines are intended to be used when preparing your manuscript for submission to a Wiley journal. A hard copy of your manuscript should always accompany the electronic version. Please ensure that the electronic version of the manuscript matches the printed one; they must be identical. Please return the disk submission slip below with your manuscript and labeled disk(s).

## Word Processing

We prefer to receive manuscripts prepared in Microsoft Word 6.0 or higher, although manuscripts prepared with other word processors are usually acceptable. Figure should be prepared separately – not embedded in your Word file. Please make sure to include an electronic version of your manuscript with the printed version.

## TeX/LaTeX Guidelines

Some publications accept submissions in TeX and/or LaTeX. Further information about this is available on the journal's home page. LaTeX-class files are usually available for journals in the areas of mathematics and engineering. If you use any nonstandard document classes, please include them with the files you send us. As with manuscripts prepared in Word, we prefer to receive illustrations in EPS or TIFF format, one file per figure.

## Other Software

If you use other word processing software, please provide us with a plain text file of your manuscript, as we may not be able to cope with all proprietary files. Most applications support the export of plain text files.

## Illustrations

Figures should be prepared separately in a suitable graphics application and saved as Encapsulated PostScript (EPS) or Tagged Image File Format (TIFF) files, one file per figure. If these formats are not possible, use JPEG instead. These can be submitted on a floppy disk, an Iomega Zip disk, or a CD-ROM.

*Please do not:*

- Use desktop publishing applications such as PageMaker or QuarkXPress.
- Use proprietary graphics formats such as Corel Draw or Adobe Illustrator.
- Send files that contain hidden text. For example, do not use your word processor's automated features to create footnotes or reference lists.

## Electronic Manuscript Submission

Some journals allow authors to submit their manuscripts through various electronic means, such as disk (floppy disk, Iomega Zip disk, CD-ROM, etc.), the Internet (FTP), or email attachment. The journal's editor can give your details on what methods are available, and how to prepare and submit the necessary files. Details also appear on the 'Instructions to Authors' page of each journal's website.

For files submitted electronically, the required hard copy is usually replaced by a PostScript or PDF version of the article in order to facilitate the review process; not every reviewer has access to every kind of software. The source files will be needed if the article is accepted.

***Please do not send:***

Magneto-optical disks, SyQuest cartridges, 5 ¼"floppy disks, or LS-120 disks.

## Before You Send Your Files

Name your document file using your surname and the standard three-letter extension that identifies the file format (for example, "wright.doc" or "deacon.txt").

Label the disk clearly with the journal name and the file name (for example J Comp Neuro-smith.doc). Avoid vague file names such as "paper.doc" or "article.txt". Also label with the word processing program and version used.

Please scan your disks for viruses before you send them, and keep a copy of what you send in a safe place in case any of the files need to be replaced.

**For more detailed instructions on general journal manuscript submission guidelines, please visit our website at www.wiley.com and use the Resources for Authors link. For your convenience you can also go directly to the link at http://journalauthors.wiley.com.**

**Detach and return with labeled disk(s)**

Corresponding  
Author:_____________

E-mail  
Address:_____________

Telephone:_____________

Manuscript  
No.:_____________

Type of  
Computer:_____________

Program(s) & Version(s)  
Used:_____________

Miscellaneous:_____________

**I certify that the material on the enclosed disk(s) is identical in both word and content to the printed copy herewith enclosed.**

Signature:_____________          Date:_____________

# CHAPTER - 10

# Agricultural and Forest Meteorology

**An International Journal**
**Former title: Agricultural Meteorology**

## Guide for Authors Online Submission

All submissions are handled online at http://ees.elsevier.com/agrformet/. Once you have logged on as author using your JOURNAL username and password you will be guided through the creation and uploading of your files. The system automatically converts source files to a single Adobe Acrobat PDF version of the article, which is used in the peer-review process. Please note that even though manuscript source files are converted to PDF at submission for the review process, these source files are needed for further processing after acceptance. All correspondence, including notification of the Editor's decision and requests for revision, takes place by e-mail and via the Author's homepage only. Therefore users need to keep their contact coordinates on the registration page up-to-date with the "UPDATE MY INFORMATION" option.

There is no submission fee or paper charges.

## Legal and Copyright

Submission of an article implies that the work described has not been published previously (except in the form of an abstract or as part of a published lecture or academic thesis), that it is not under consideration for publication elsewhere, that its publication is approved by all Authors and tacitly or explicitly by the responsible authorities where the work was carried out, and that, if accepted, it will not be published elsewhere in the same form, in English or in any other language, without the written consent of the Publisher.

Upon acceptance of an article, authors will be asked to sign a ?Journal Publishing Agreement?? (for more information on this and copyright see http://www.elsevier.com/copyright). Acceptance of the agreement will ensure the widest possible dissemination of information. An e-mail (or letter) will be sent to the corresponding author confirming receipt of the manuscript together with a 'Journal Publishing Agreement? form or a link to the online version of this agreement.

If excerpts from other copyrighted works are included, the Author(s) must obtain written permission from the copyright owners and credit the source(s) in the article. Elsevier has forms for use by Authors in these cases: contact Elsevier's Rights Department, Oxford, UK: phone (+44) 1865 843830, fax (+44) 1865 853333, e-mail permissions@elsevier.com. Requests may also be completed on-line via http://www.elsevier.com/locate/permissions

## Word processors

Save the file in the native format of the word processor used. The text should be in single-column format. Keep the layout of the text as simple as possible. Most formatting codes will be removed and replaced on processing the article. In particular, do not use the word processor's options to justify text or to hyphenate words. However, do use bold face, italics, subscripts, superscripts etc. Do not embed "graphically designed" equations or tables, but prepare these using the word processor's facility. When preparing tables, if you are using a table grid, use only one grid for each individual table and not a grid for each row. If no grid is used, use tabs, not spaces, to align columns. Do not import the figures into the text file but, instead, indicate their approximate locations directly in the electronic text. To avoid unnecessary errors you are strongly advised to use the "spellchecker" function of your word processor.

## Article

*Language*. Articles must be written in good English

*Title*. Concise and informative. Avoid abbreviations and formulae.

*Line numbering*. Manuscripts should be prepared with numbered lines, with wide margins and double spacing throughout, i.e. also for abstracts, footnotes and references. Every page of the manuscript, including the title page, references, tables, etc. should be numbered. However, in the text no reference should be made to page numbers; if necessary, one may refer to sections. Underline words that should be in italics, and do not underline any other words. Avoid excessive use of italics to emphasize part of the text.

*Author names and affiliations*. Where the family name may be ambiguous (e.g., a double name), please indicate this clearly using appropriate script (capital cases as first letter of authors' first and surnames followed by lower cases). The Present the Authors' affiliation addresses (where the actual work was done) below the names. Indicate all affiliations with a lower-case superscript letter immediately after the Author's name and in front of the appropriate address. Provide the full postal address of each affiliation, including the country name, and, if available, the e-mail address of each Author.

*Corresponding Author*. Clearly indicate who is willing to handle correspondence at all stages of refereeing and publication, also post-publication. Ensure that telephone and fax numbers (with country and area code) are provided in addition to the e-mail address and the complete postal address.

*Abstract*. A concise abstract should briefly state the purpose of the research and the main results. An abstract is often presented separate from the article, so it must be able to stand alone.

*Keywords*. Authors should provide 4 - 6 keywords.

*Illustrations*. A guide on electronic artwork is available on

http://authors.elsevier.com/artwork

## References

(a) References in the text consist of the surname of the author(s), followed by the year of publication in parentheses. All references cited in the text should be given in the reference list and vice versa.

(b) The reference list should be in alphabetical order and on sheets separate from the text.

Citing and listing of Web references. As a minimum, the full URL should be given. Any further information, if known (Author names, dates, reference to a source publication, etc.), should also be given. Web references can be listed separately (e.g., after the reference list) under a different heading if desired, or can be included in the reference list.

## Reprints

Twenty-five reprints of each paper are supplied free of charge to the corresponding author; additional reprints are available at cost if they are ordered when the proof is returned.

## Submission checklist

- One Author designated as corresponding Author:
- E-mail address
- Full postal address
- Telephone and fax numbers
- All necessary files have been uploaded
- Keywords
- All figure captions
- All tables (including title, description, footnotes)
- Manuscript has been "spellchecked"
- References are in the correct format for this journal
- All references mentioned in the Reference list are cited in the text, and vice versa
- Permission has been obtained for use of copyrighted material from other sources (including the Web)

For any further information please contact the Author Support Department at authorsupport@elsevier.com.

# CHAPTER - 11

# Agricultural Economics – Research Review

## GUIDELINES FOR SUBMISSION OF PAPERS/ABSTRACTS

1. Research articles and research notes based on basic and applied research on economic aspects of Agriculture and Rural Development, abstracts of **M.Sc./Ph.D.** theses and short communications may be accepted for publication.

2. Authors should be members of the Association, although it will be the discretion of the Editor to consider papers submitted by even non-members.

3. Three copies of manuscript typed in double spacer should be sent to the *Editor, Agricultural Economics Research Review*, Division of Agricultural Economics, IARI, New Delhi 110012 and soft copy at **aera @ rediffmail.com.** All articles must include an abstract in about 100-150 words.

4. The length of papers should not be more than 20 (double space) typed pages including tables, diagrams and appendices. Abstracts of **M.Sc.** and **Ph.D.** theses should not exceed 2 and 4 typed pages, respectively.

5. Name(s) and affiliation(s) of the author(s) with e-mail address (es) should be provided on a separate page along with the title of the article.

6. Only essential mathematical notations may be used. All statistical formulae should be neatly typed. Footnotes should be numbered consecutively in plain Arabic superscripts.

7. Only cited works should be included in reference list. The citation of references should be in the following order: author(s) name(s); year; title of article; name of journal; volume number and pages. Please follow the style of citations as in latest issue of this journal. Papers not submitted in standard format as suggested above will not be considered for publication.

8. Papers submitted for publication should be exclusively written for this journal and should not have been published or sent for publication elsewhere.

9. No free reprints will be supplied. Reprints may be ordered in multiples of 15 reprints, at the time of submitting the papers for publication. The rates for a unit of 15 reprints will be Rs.300.

- Students must submit a letter from the Head of their Institution certifying their bonafides.

***Address for correspondence and sending remittances:***

**The Secretary**
**Agricultural Economics Research**
**Association (India)**
**National Centre for Agricultural**
**Economics and Policy Research**
**Post Box 11305**
**Pusa Campus, New Delhi – 110012**
**Ph.: 011-25843036**
**Email: aera@rediffmail.com**

**Joint Secretary**
**Agricultural Economics Research**
**Association (India)**
**Division of Agricultural Economics**
**Indian Agricultural Research Institute**
**Pusa, New Delhi 110012**
**Ph.: 011-25847501, 011-25842951**
**Email: npsingh@iari.res.in**

# CHAPTER - 12

# Agricultural Situations in India

## NOTE TO CONTRIBUTORS

Non-controversial articles on various facets of Indian Agriculture are accepted for publication in the Journal. The Journal intends to provide a forum for scholarly work and also to promote technical competence for research in agricultural and allied subjects. The articles generally not exceeding five thousand words, may be sent in duplicate, typed in double space on one side of fullscap paper addressed to the Economic and Statistical Adviser, Directorate of Economics and Statistics, Room No. 152, Krishi Bhawan, New Delhi-110001 alongwith a declaration by author(s) that the article has neither been published nor submitted for publication elsewhere. All references and footnotes may be given only at the end of the articles. Floppy is also required.

Although authors are solely responsible for the factual accuracy and the opinion expressed in their signed articles, yet the Editorial Board of the Directorate, reserves the right to edit, amend and delete any portion of the article with a view to make it more presentable and to reject any article, if not found suitable. The articles which are not found suitable will not be returned unless accompanied by a self-addressed and stamped envelope. No correspondence will be entertained on the articles rejected by the Editorial Board.

_The Directorate of Economics and Statistics_ pays a nominal honorarium upto (i) Rs. 1,000 per article of atleast 1,500 words in general case and (ii) upto Rs. 2,000 per article of not less than 1,500 words included in Special Number of the Journal. In addition, ten reprints of each article will be provided free of cost to the first named author of main article.

# CHAPTER - 13

# Agricultural Systems

## GUIDE FOR AUTHORS

*Agricultural Systems* is an international journal that deals with interactions – among the components of agricultural systems, among hierarchical levels of agricultural systems, between agricultural and other land use systems, and between agricultural systems and their natural and social environments. In particular, its aim is to encourage integration of knowledge among those disciplines that underpin agriculture. Many contributions will therefore be multi- or inter-disciplinary. Papers generally focus on either methodological approaches to understanding and managing interactions within or among agricultural systems, or the application of holistic or quantitative systems approaches to a range of problems within agricultural systems and their interactions with other systems. Because of the nature of the readership of Agricultural Systems, the contents of papers should be easily accessible (properly introduced, presented and discussed) to readers from a wide range of disciplines.

The scope includes the development and application of systems methodology, including system modelling, simulation and optimisation; ecoregional analysis of agriculture and land use; studies on natural resource issues related to agriculture; impact and scenario analyses related to topics such as GMOs, multifunctional land use and global change; and the development and application of decision and discussion support systems; approaches to analysing and improving farming systems; technology transfer in tropical and temperate agriculture; and the relationship between agricultural development issues and policy.

The journal publishes original scientific papers, short communications, review articles and book reviews. Review articles and book reviews should only be submitted after consultation with the Editors.

Please bookmark this page as: http://www.elsevier.com/locate/agsy

For more information/suggestions/comments please contact

AuthorSupport@elsevier.com

## Types of contribution

The journal welcomes submissions that address interactions among the components of agricultural systems, among hierarchical levels of agricultural systems, between agricultural and other land use systems, and between agricultural systems and their natural and social environments. Submissions can also deal with methodological approaches to understanding and managing interactions within or among agricultural systems, or with the application of holistic or quantitative systems approaches to a range of problems within agricultural systems and their interactions with other systems. Because of the interdisciplinary and international readership of Agricultural Systems, the contents of papers should be easily accessible (properly introduced, presented and discussed) to readers from a wide range of disciplines. The journal publishes original scientific papers, short communications, review articles and book reviews. Review articles and book reviews should be submitted only after consultation with the editors.

## Online Submission of manuscripts

Submission of an article implies that the work described has not been published previously (except in the form of an abstract or as part of a published lecture or academic thesis), that it is not under consideration for publication elsewhere, that its publication is approved by all authors and tacitly or explicitly by the responsible authorities where the work was carried out, and that, if accepted, it will not be published elsewhere in the same form, in English or in any other language, without the written consent of the Publisher.

Upon acceptance of an article, authors will be asked to transfer copyright (for more information on copyright see http://www.elsevier.com/authorsrights. This transfer will ensure the widest possible dissemination of information. A letter will be sent to the corresponding author confirming receipt of the manuscript. A form facilitating transfer of copyright will be provided.

If excerpts from other copyrighted works are included, the author(s) must obtain written permission from the copyright owners and credit the source(s) in the article. Elsevier has preprinted forms for use by authors in these cases: contact Elsevier's Rights Department, Oxford, UK; phone: (+44) 1865 843830, fax: (+44) 1865 853333, e-mail: permissions@elsevier.com. Requests may also be completed on-line via the Elsevier homepage (http://www.elsevier.com/authors).

Submission to this journal proceeds totally on-line. Use the following guidelines to prepare your article. Via the following site you will be guided stepwise through the creation and uploading of the various files (http://ees.elsevier.com/agsy or http://www.elsevier.com/locate/guidepublication). Once the uploading is done, our system automatically generates an electronic (PDF) proof, which is then used for reviewing. It is crucial that all graphical elements be uploaded in separate files, so that the PDF is suitable for reviewing. Authors can upload their article as a LaTex or Microsoft (MS) Word. All correspondence, including notification of the Editor's decision and requests for revisions, will be by e-mail.

## Electronic format requirements for accepted articles

We accept most wordprocessing formats, but Word and LaTeX are preferred. Always keep a backup copy of the electronic file for reference and safety. Save your files using the default extension of the program used.

## Word processor documents

It is important that the file be saved in the native format of the wordprocessor used. The text should be in single-column format. Keep the layout of the text as simple as possible. Most formatting codes will be removed and replaced on processing the article. In particular, do not use the word processor's options to justify text or to hyphenate words. However, do use bold face, italics, subscripts, superscripts etc. Do not embed 'graphically designed' equations or tables, but prepare these using the word processor's facility. When preparing tables, if you are using a table grid, use only one grid for each individual table and not a grid for each row. If no grid is used, use tabs, not spaces, to align columns. The electronic text should be prepared in a way very similar to that of conventional manuscripts (see also http://www.elsevier.com/artworkinstructions). Do not import the figures into the text file but, instead, indicate their approximate locations directly in the electronic text and on the manuscript. See also the above mentioned website. To avoid unnecessary errors you are strongly advised to use the 'spellchecker' function of your wordprocessor.

## Preparation of manuscripts

1.  Manuscripts should be written in English. Authors whose native language is not English are strongly advised to have their manuscripts checked by an English-speaking colleague prior to submission.

    ***English language help service:*** Upon request, Elsevier will direct authors to an agent who can check and improve the English of their paper (before submission). Please contact authorsupport@elsevier.com for further information or visit http://www.elsevier.com/locate/languagepolishing.

2.  Manuscripts should be prepared with numbered lines, with wide margins and double spacing throughout, i.e. also for abstracts, footnotes and references. **Every page of the manuscript, including the title page, references, tables, etc. should be numbered.** However, in the text no reference should be made to page numbers; if necessary, one may refer to sections. Underline words that should be in italics, and do not underline any other words. Avoid excessive use of italics to emphasize part of the text. **Authors are requested to submit, with their manuscripts, the names and addresses of four potential referees.**

3.  Manuscripts in general should be organized in the following manner:
    - Title (should be clear, descriptive and not too long)
    - Name(s) of author(s)
    - Complete contact details
    - Telephone and E-mail of the corresponding author
    - Present address(es) of author(s) if applicable
    - Abstract
    - Key words (indexing terms), normally 3-6 items
    - Introduction
    - Material studied, area descriptions, methods, techniques
    - Results
    - Discussion
    - Conclusion
    - Acknowledgements and any additional information concerning research grants, etc.
    - References
    - Tables
    - Figure captions

4.  In typing the manuscript, titles and subtitles should not be run within the text. They should be typed on a separate line, without indentation. Use lower-case letter type.

5.  Elsevier reserves the privilege of returning to the author for revision accepted manuscripts and illustrations which are not in the proper form given in this guide.

## Abstracts

The abstract should be clear, descriptive and not longer than 400 words, reporting concisely on the purpose of the paper. Four or five keywords should also be included.

## Formulae

1.  Subscripts and superscripts should be clear.

2.  Give the meaning of all symbols immediately after the equation in which they are first used.

3.  For simple fractions use the solidus (/) instead of a horizontal line.

4.  Equations should be numbered serially at the right-hand side in parentheses. In general only equations explicitly referred to in the text need be numbered.

5.  The use of fractional powers instead of root signs is recommended. Also powers of e are often more conveniently denoted by exp.

6. Levels of statistical significance which can be mentioned without further explanation are * $P<0.05$, ** $P<0.01$ and *** $P<0.001$.

7. In chemical formulae, valence of ions should be given, as, e.g. $Ca^{2+}$ not as $Ca^{++}$.

8. Isotope numbers should precede the symbols, e.g. $^{18}O$.

9. The repeated writing of chemical formulae in the text is to be avoided where reasonably possible; instead, the name of the compound should be given in full. Exceptions may be made in the case of a very long name occurring very frequently or in the case of a compound being described as the end product of a gravimetric determination (e.g. phosphate as $P_2O_5$).

## Units and abbreviations

In principle SI units should be used except where they conflict with current practise or are confusing. Other equivalent units may be given in parentheses. Units and their abbreviations should be those approved by ISO (International Standard 1000:1992. SI units and recommendations for the use of their multiples and of certain other units). Abbreviate units of measure only when used with numerals.

## Nomenclature

1. Authors and editors are, by general agreement, obliged to accept the rules governing biological nomenclature, as laid down in the *International Code of Botanical Nomenclature*, the *International Code of Nomenclature of Bacteria*, and the *International Code of Zoological Nomenclature*.

2. All biota (crops, plants, insects, birds, mammals, etc.) should be identified by their scientific names when the English term is first used, with the exception of common domestic animals.

3. All biocides and other organic compounds must be identified by their Geneva names when first used in the text. Active ingredients of all formulations should likewise be identified.

4. For chemical nomenclature, the conventions of the *International Union of Pure and Applied Chemistry* and the official recommendations of the *IUPAC_IUB Combined Commission on Biochemical Nomenclature* should be followed.

### *Supplementary data*

Elsevier now accepts electronic supplementary material to support and enhance your scientific research. Supplementary files offer the author additional possibilities to publish supporting applications, movies, animation sequences, high-resolution images, background datasets, sound clips and more. Supplementary files supplied will be published online alongside the electronic version of your article in Elsevier web products, including ScienceDirect: http:/

/www.sciencedirect.com. In order to ensure that your submitted material is directly usable, please ensure that data is provided in one of our recommended file formats. Authors should submit the material in electronic format together with the article and supply a concise and descriptive caption for each file. For more detailed instructions please visit \http://www.elsevier.com/authors.

## Tables

1. Authors should take notice of the limitations set by the size and lay-out of the journal. Large tables should be avoided. Reversing columns and rows will often reduce the dimensions of a table.

2. If many data are to be presented, an attempt should be made to divide them over two or more tables.

3. Drawn tables, from which prints need to be made, should not be folded.

4. Tables should be numbered according to their sequence in the text. The text should include references to all tables.

5. Each table should be typewritten on a separate page of the manuscript. Tables should never be included in the text.

6. Each table should have a brief and self-explanatory title.

7. Column headings should be brief, but sufficiently explanatory. Standard abbreviations of units of measurement should be added between parentheses.

8. Vertical lines should not be used to separate columns. Leave some extra space between the columns instead.

9. Any explanation essential to the understanding of the table should be given as a footnote at the bottom of the table.

10. Wherever possible, columns should represent individual variables or variables with common units, and rows should represent observations.

11. Present data with no more digits than justified by the accuracy of their measurement or simulation, and no more digits than needed for the purpose of the table. Using fewer digits usually enhances readability of tables.

## Preparation of electronic illustrations

Submitting your artwork in an electronic format helps us to produce your work to the best possible standards, ensuring accuracy, clarity and a high level of detail.

### *General points*

- Always upload high-quality e-files of your artwork.
- Make sure you use uniform lettering and sizing of your original artwork.
- Save text in illustrations as "graphics" or enclose the font.
- Only use the following fonts in your illustrations: Arial, Courier, Helvetica, Times, Symbol.

- Number the illustrations according to their sequence in the text.
- Use a logical naming convention for your artwork files.
- Provide all illustrations as separate files.
- Provide captions to illustrations separately.
- Produce images near to the desired size of the printed version.
- Illustrations should be numbered according to their sequence in the text. References should be made in the text to each illustration.
- Illustrations should be designed with the format of the page of the journal in mind. Illustrations should be of such a size as to allow a reduction of 50%.
- Make sure that the size of the lettering is big enough to allow a reduction of 50% without becoming illegible. The final font size in printing should be about 6-8pt. The lettering should be in English. Use the same kind of lettering throughout and follow the style of the journal.
- If a scale should be given, use bar scales on all illustrations instead of numerical scales that must be changed with reduction.

A detailed guide on electronic artwork is available on our website:

http://www.elsevier.com/artworkinstructions

You are urged to visit this site; some excerpts from the detailed information are given here.

## Colour illustrations

Please make sure that artwork files are in an acceptable format (TIFF, EPS, or MS Office files) and with the correct resolution. Polaroid colour prints are *not* suitable. If, together with your accepted article, you submit usable colour figures then Elsevier will ensure, at no additional charge, that these figures will appear in colour on the Web (e.g., ScienceDirect and other sites) regardless of whether or not these illustrations are reproduces in colour in the printed version. For colour reproduction in print, you will receive information regarding the costs from Elsevier after receipt of your accepted article. Please indicate your preference for colour print or on the Web only. For further information on the preparation of electronic artwork, please see http://www.elsevier.com/artworkinstructions.

Please note : Because of technical complications which can arise by converting colour figures to 'grey scale' (for the printed version should you not opt for colour in print) please submit in addition usable black and white prints corresponding to all the colour illustrations.

## Supplementary files

Preparation of supplementary data. Elsevier now accepts electronic supplementary material (e-components) to support and enhance your scientific research. Supplementary files offer the Author additional possibilities to publish supporting applications, movies, animation sequences, high-resolution images, background datasets, sound clips and more. Supplementary files supplied will be published online alongside the electronic version of your article in Elsevier Web products, including ScienceDirect: http://www.sciencedirect.com. In order to ensure that your submitted material is directly usable, please ensure that data is provided in one of our recommended file formats. Authors should submit the material in electronic format together with the article and supply a concise and descriptive caption for each file. For more detailed instructions please visit our artwork instruction pages at http://www.elsevier.com/artworkinstructions. This journal offers electronic submission services and supplementary data files can be uploaded via http://ees.elsevier.com/agsy.

## References

1.  All publications cited in the text should be presented in a list of references following the text the manuscript. The manuscript should be carefully checked to ensure that the spelling of author's names and dates are exactly the same in the text as in the reference list.

2.  In the text refer to the author's name (without initial) and year of publication, followed – if necessary – by a short reference to appropriate pages. Examples: "Since Peterson (1988) has shown that..." "This is in agreement with results obtained later (Kramer,1989, pp. 12-16)".

3.  If reference is made in the text to a publication written by more than two authors the name of the first author should be used followed by "*et al.*" This indication, however, should never be used in the list of references. In this list names of first author and co-authors should be mentioned.

4.  References cited together in the text should be arranged chronologically. The list of references should be arranged alphabetically on author's names, and chronologically per author. If an author's name in the list is also mentioned with co-authors the following order should be used: publications of the single author, arranged according to publication dates – publications of the same author with one co-author – publications of the author with more than one co-author. Publications by the same author(s) in the same year should be listed as 1974a, 1974b, etc.

5.  Use the following system for arranging your references:

    (a)  *For periodicals*

    Tietema, A., Riemer, L., Verstraten, J.M., van der Maas, M.P., van Wijk, A.J., van Voorthuyzen, I.,1992. Nitrogen cycling in acid forest soils subject to increased atmospheric nitrogen input. For. Ecol. Manage. 57, 29-44.

    (b)  *For edited symposia, special issues, etc. published in a periodical*

    Rice, K., 1992. Theory and conceptual issues. In: Gall, G.A.E., Staton, M. (Eds.), Integrating Conversation Biology and Agricultural Production. Agric. Ecosyst. Environ. 42, 9-26.

(c)    *For books*

Gaugh, Jr., H.G., 1992. Statistical Analysis of Regional Yield Trials. Elsevier, Amsterdam.

(d)    *For multi-author books*

Baker, Jr., 1993. Insects. In: De Hertogh, A., Le Nard, M. (Eds.), The Physiology of Flower Bulbs. Elsevier, Amsterdam, pp. 101-153.

6. In the case of publications in any language other than English, the original title is to be retained.However, the titles of publications in non-Latin alphabets should be transliterated, and a notation such as "(in Russian)" or "(in Greek, with English abstract)" should be added.

7. Work accepted for publication but not yet published should be referred to as "in press".

8. References concerning unpublished data and "personal communications" should not be cited in the reference list but may be mentioned in the text.

## Footnotes

Footnotes are not generally acceptable in the main body of an Agricultural Systems manuscript. Any information that is essential to understanding should be incorporated into the text. Footnotes can be used within tables as discussed previously.

## Copyright

1. An author, when quoting from someone else's work or when considering reproducing an illustration or table from a book or journal article, should make sure that he is not infringing a copyright.

2. Although in general an author may quote from other published works, he should obtain permission from the holder of the copyright if he wishes to make substantial extracts or to reproduce tables, plates, or other illustrations. If the copyright-holder is not the author of the quoted or reproduced material, it is recommended that the permission of the author should also be sought.

3. Material in unpublished letters and manuscripts is also protected and must not be published unless permission has been obtained.

4. A suitable acknowledgment of any borrowed material must always be made.

## Proofs

One set of page proofs in PDF format will be sent by e-mail to the corresponding author (if we do not have an e-mail address then paper proofs will be sent by post). Elsevier now sends PDF proofs which can be annotated; for this you will need to download Adobe Reader version 7 available free from http://www.adobe.com/products/acrobat/readstep2.html. Instructions on how to annotate PDF files will accompany the proofs. The exact system requirements are given at the Adobe site: http://www.adobe.com/products/acrobat/acrrsystemreqs.html#70win. If you do not wish to use the PDF annotations function, you may list the corrections (including replies to the Query Form) and return to Elsevier in an e-mail. Please list your corrections quoting line number. If, for any reason, this is not possible, then mark the corrections and any other comments (including replies to the Query Form) on a printout of your proof and return by fax, or scan the pages and e-mail, or by post.

Please use this proof only for checking the typesetting, editing, completeness and correctness of the text, tables and figures. Significant changes to the article as accepted for publication will only be considered at this stage with permission from the Editor. We will do everything possible to get your article published quickly and accurately. Therefore, it is important to ensure that all of your corrections are sent back to us in one communication: please check carefully before replying, as inclusion of any subsequent corrections cannot be guaranteed. Proofreading is solely your responsibility. Note that Elsevier may proceed with the publication of your article if no response is received.

## Offprints

The corresponding author, at no cost, will be provided with a PDF file of the article via e-mail or, alternatively, 25 free paper offprints. The PDF file is a watermarked version of the published article and includes a cover sheet with the journal cover image and a disclaimer outlining the terms and conditions of use. Additional paper offprints can be ordered by the authors. An order form with prices will be sent to the corresponding author.

### *Agricultural Systems* has no page charges

For further information please see the journal home page at:

http://www.elsevier.com/locate/agsy

# CHAPTER - 14

# Agricultural Water Management

## Types of contribution

1. Original research papers

2. Review articles

3. Short communications

4. (Guest) editorials

5. Book reviews

*Original research papers* should report the results of original research. The material should not have been previously published elsewhere, except in a preliminary form.

*Review articles* should cover subjects falling within the scope of the journal which are of active current interest. They may be submitted or invited.

*A Short communication* is a concise but complete description of a limited investigation, which will not be included in a later paper. Short Communications should be as completely documented, both by reference to the literature and description of the experimental procedures employed, as a regular paper. They should not occupy more than 6 printed pages (about 12 manuscript pages, including figures, tables and references).

*Book reviews* will be included in the journal on a range of relevant books which are not more than 2 years old. Book reviews will be solicited by the Book Review Editor. Unsolicited reviews will not usually be accepted, but suggestions for appropriate books for review may be sent to the Book Review Editor:

T. Howell

Water Management Research Lab.

USDA-ARS

P.O. Drawer 10

Bushland, TX 79012

USA.

## SUBMISSION OF MANUSCRIPTS

Submission of an article is understood to imply that the article is original and is not being considered for publication elsewhere. Submission also implies that all authors have approved the paper for release and are in agreement with its content. Upon acceptance of the article by the journal, the author(s) will be asked to sign a "Journal Publishing Agreement" (for more information on this and copyright see http://www.elsevier.com/copyright). Acceptance of the agreement will ensure the widest possible dissemination of information. Agricultural Water Management uses an online, electronic submission system. By accessing the website http://ees.elsevier.com/agwat you will be guided stepwise through the creation and uploading of the various files. When submitting a manuscript to the Elsevier Editorial System, authors need to provide an electronic version of their manuscript. For this purpose original source files, not PDF files, are preferred. The author should specify a category designation for the manuscript (full length article, review article, short communication, etc.), choose a set of classifications from the prescribed list provided online and select an editor. Authors may send queries concerning the submission process, manuscript status, or journal procedures to the Journal Manager. Once the uploading is complete, the system automatically generates an electronic (PDF) proof, which is then used for reviewing. All correspondence, including the Editor's decision and request for revisions, will be communicated by e-mail.

**Authors are requested to provide the names of four potential referees in their covering letter.**

## PREPARATION OF MANUSCRIPTS

1.  Manuscripts should be written in English. Authors whose native language is not English are strongly advised to have their manuscripts checked by an English-speaking colleague prior to submission. **English language help service**: Upon request, Elsevier will direct authors to an agent who can check and improve the English of their paper (before submission). Please contact authorsupport@elsevier.com for further information

2.  Manuscripts should be typewritten, typed on one side of the paper (with numbered lines), with wide margins and double spacing throughout, i.e. also for abstracts, footnotes and references. Every page of the manuscript, including the title page, references, tables, etc. should be numbered. However, in the text no reference should be made to page numbers; if necessary, one may refer to sections. Avoid excessive usage of italics to emphasize part of the text.

3.  Manuscripts in general should be organized in the following order: Title (should be clear, descriptive and concise)

    Name(s) of author(s)

    Complete postal address(es) of affiliations

    Full telephone, E-mail and fax number of the corresponding author

Present address(es) of author(s) if applicable

Complete correspondence address to which the proofs should be sent

Abstract

Key words (indexing terms), normally 3-6 items

Introduction

Material studied, area descriptions, methods, techniques

Results

Discussion

Conclusion

Acknowledgements and any additional information concerning research grants, etc.

References

Tables

Figure captions

4.  In typing the manuscript, titles and subtitles should not be run within the text. They should be typed on a separate line, without indentation. Use lower-case lettertype.

5.  SI units should be used.

6.  If a special instruction to the copy editor or typesetter is written on the copy it should be encircled. The typesetter will then know that the enclosed matter is not to be set in type. When a typewritten character may have more than one meaning (e.g. the lower case letter l may be confused with the numeral), a note should be inserted in a circle in the margin to make the meaning clear to the typesetter. If Greek letters or uncommon symbols are used in the manuscript, they should be written very clearly, and if necessary a note such as "Greek lower-case chi" should be put in the margin and encircled.

7.  Elsevier reserves the privilege of returning to the author for revision accepted manuscripts and illustrations which are not in the proper form given in this guide.

## Abstract

The abstract should be clear,descriptive and not longer than 400 words.

## Tables

1.  Authors should take notice of the limitations set by the size and lay-out of the journal. Large tables should be avoided. Reversing columns and rows will often reduce the dimensions of a table.

2.  If many data are to be presented, an attempt should be made to divide them over two or more tables.

3. Tables should be numbered according to their sequence in the text. The text should include references to all tables.

4. Each table should be typewritten on a separate page of the manuscript. Tables should never be included in the text.

5. Each table should have a brief and self-explanatory title.

6. Column headings should be brief, but sufficiently explanatory. Standard abbreviations of units of measurement should be added between parentheses.

7. Vertical lines should not be used to separate columns. Leave some extra space between the columns instead.

8. Any explanation essential to the understanding of the table should be given as a footnote at the bottom of the table.

## Illustrations

Preparation of electronic illustrations. Submitting your artwork in an electronic format helps us to produce your work to the best possible standards, ensuring accuracy, clarity and a high level of detail. General points

- Always supply high-quality printouts of your artwork, in case conversion of the electronic artwork is problematic
- lettering and sizing of your original artwork.
- Save text in illustrations as "graphics" or enclose the font.
- Only use the following fonts in your illustrations: Arial, Courier, Helvetica, Times and Symbol.
- Number the illustrations according to their sequence in the text.
- Use a logical naming convention for your artwork files, and supply a separate listing of the files and the software used.
- Provide all illustrations as separate files and as hardcopy printouts on separate sheets.
- Provide captions to illustrations separately.
- Produce images near to the desired size of the printed version
- Submit colour illustrations as original photographs, high-quality computer prints or transparencies, close to the size expected in publication, or as 35 mm slides. Polaroid colour prints are not suitable. If, together with your accepted article, you submit usable colour figures then Elsevier will ensure, at no additional charge, that these figures will appear in colour on the web (e.g. Science Direct and other sites) regardless of whether or not these illustrations are reproduced in colour in the printed version. For colour reproduction in print, you will receive information regarding the costs from Elsevier after receipt of your accepted article. For further information on the preparation of electronic artwork, please visit http://authors.elsevier.com/artwork.

- Please note : Because of technical complications which can arise by converting colour figures to 'grey scale' (for the printed version should you not opt for colour in print) please submit in addition usable black and white prints corresponding to all the colour illustrations.

This journal offers electronic submission services and graphic files can be uploaded.

## Non-electronic illustrations

Provide all illustrations as high-quality printouts, suitable for reproduction (which may include reduction) without retouching. Number illustrations consecutively in the order in which they are referred to in the text. They should accompany the manuscript, but should not be included within the text. Clearly mark all illustrations on the back (or - in case of line drawings - on the lower front side) with the figure number and the author's name and, in cases of ambiguity, the correct orientation. Mark the appropriate position of a figure in the article.

## Captions

Ensure that each illustration has a caption. Supply captions on a separate sheet, not attached to the figure. A caption should comprise a brief title (not on the figure itself) and a description of the illustration. Keep text in the illustrations themselves to a minimum but explain all symbols and abbreviations used.

## Supplementary data

Elsevier now accepts electronic supplementary material to support and enhance your scientific research. Supplementary files offer the author additional possibilities to publish supporting applications, movies, animation sequences, high-resolution images, background datasets, sound clips and more. Supplementary files supplied will be published online alongside the electronic version of your article in Elsevier web products, including ScienceDirect: http://www.sciencedirect.com. In order to ensure that your submitted material is directly usable, please ensure that data is provided in one of our recommended file formats. Authors should submit the material in electronic format together with the article and supply a concise and descriptive caption for each file. More detailed instructions can be obtained at http://authors.elsevier.com.

## Colour charges

Authors will be charged for including colour illustrations in the printed version at the following rates and are encouraged only to consider colour if necessary for clarity or comprehension: 1st page: Euro 350. Every 2nd page: Euro 175 (Prices per October 2003).

## References

1. All publications cited in the text should be presented in a list of references following the text of the manuscript. The manuscript should be carefully

checked to ensure that the spelling of author's names and dates are exactly the same in the text as in the reference list.

2. In the text refer to the author's name (without initial) and year of publication, followed – if necessary– by a short reference to appropriate pages. Examples: "Since Peterson (1988) has shown that..." "This is in agreement with results obtained later (Kramer, 1993, pp. 12-16)".

3. If reference is made in the text to a publication written by more than two authors the name of the first author should be used followed by "et al.". This indication, however, should never be used in the list of references. In this list names of first author and co-authors should be mentioned.

4. References cited together in the text should be arranged chronologically. The list of references should be arranged alphabetically on authors' names, and chronologically per author. If an author's name in the list is also mentioned with co-authors, the following order should be used: publications of the single author, arranged according to publication dates –publications of the same author with one co-author – publications of the author with more than one co-author. Publications by the same author(s) in the same year should be listed as 1993a, 1993b, etc.

5. Use the following system for arranging your references:

    (a)    *For periodicals*

    Tillman, R.W., Scotter, D.R., Clothier, B.E., White, R.E., 1991. Solute transport during intermittent water flow in a field soil and some implications for irrigation and fertilizer application. Agric. Water Manage. 20, 119-133.

    (b)    *For edited symposia, special issues, etc. published in a periodical*

    Iwata, M., Hirano, T., Hasegawa, S., 1982. Behavior and plasma sodium regulation of chum salmon fry during transition into seawater. In: Bern, H.A., Mahnken, C.V.W. (Eds.), Salmonid Smoltification. Proceedings of a Symposium, 29 June-1 July 1981, at La Jolla, CA, U.S.A. Aquaculture 28, 133-142.

    (c)    *For books*

    Bartik, M., Piskaè, A. (Eds.), 1981. Veterinary Toxicology, Developments in Animal and Veterinary Sciences, 7. Elsevier, Amsterdam.

    (d)    *For multi-author books*

    Vermeer, J.G., Joosten, J.H.J., 1992. Conservation and management of bog and fen reserves in the Netherlands. In: Verhoeven, J.T.A. (Ed.), Fens and Bogs in The Netherlands: Vegetation, History, Nutrient Dynamics and Conservation. Geobotany, 18. Kluwer Academic Publishers, Dordrecht, pp. 433-478.

(e)    *For unpublished reports, departmental notes, etc*

Dickson, J.W., Henshall, J.K., O'Sullivan, M.F., Soane, B.D., 1979. Compaction effects under commercial and experimental cage wheels in comparison with rubber types on loose soil. Scot. Inst. Agric. Eng., Dep. Note SIN/261 (unpubl.) 9 pp.

6. Abbreviate the titles of periodicals mentioned in the list of references according to International List of Periodical Title Word Abbreviations.

7. In the case of publications in any language other than English, the original title is to be retained.However, the titles of publications in non-Latin alphabets should be transliterated, and a notation such as "(in Russian)" or "(in Greek, with English abstract)" should be added. 8. In referring to a personal communication the two words are followed by the year, e.g. "(J. McNary, personal communication, 1990)".

## Formulae

1. Formulae should be typewritten, if possible. Leave ample space around the formulae.

2. Subscripts and superscripts should be clear.

3. Greek letters and other non-Latin or handwritten symbols should be explained in the margin where they are first used. Take special care to show clearly the difference between zero (0) and the letter O, and between one (1) and the letter l.

4. Give the meaning of all symbols immediately after the equation in which they are first used.

5. For simple fractions use the solidus (/) instead of a horizontal line.

6. Equations should be numbered serially at the right-hand side in parentheses. In general only equations explicitly referred to in the text need be numbered.

7. The use of fractional powers instead of root signs is recommended. Also powers of e are often more conveniently denoted by exp.

8. Levels of statistical significance which can be mentioned without further explanation are $^* P<0.05$, $^{**} P<0.01$ and $^{***} P<0.001$.

9. In chemical formulae,valence of ions should be given as e.g. $Ca^{2+}$ and not as $Ca^{++}$.

10. Isotope numbers should precede the symbols, e.g. $^{18}O$.

11. The repeated writing of chemical formulae in the text is to be avoided where reasonably possible; instead, the name of the compound should be given in full. Exceptions may be made in the case of a very long name occurring very frequently or in the case of a compound being described as the end product of a gravimetric determination (e.g. phosphate as $P_2O_5$).

## Footnotes

1. Footnotes should only be used if absolutely essential. In most cases it should be possible to incorporate the information in normal text.

2. If used, they should be numbered in the text, indicated by superscript numbers, and kept as short as possible.

## Nomenclature

1. Authors and editors are, by general agreement, obliged to accept the rules governing biological nomenclature, as laid down in the *International Code of Botanical Nomenclature*, the *International Code of Nomenclature of Bacteria*, and the *International Code of Zoological Nomenclature*.

2. All biota (crops, plants, insects, birds,mammals, etc.) should be identified by their scientific names when the English term is first used, with the exception of common domestic animals.

3. All biocides and other organic compounds must be identified by their Geneva names when first used in the text. Active ingredients of all formulations should be likewise identified.

4. For chemical nomenclature, the conventions of the *International Union of Pure and Applied Chemistry* and the official recommendations of the *IUPAC-IUB Combined Commission on Biochemical Nomenclature* should be followed.

## Copyright

1. An author, when quoting from someone else's work or when considering reproducing an illustration or table from a book or journal article, should make sure that he/she is not infringing a copyright.

2. Although in general an author may quote from other published works,he/she should obtain permission from the holder of the copyright if he wishes to make substantial extracts or to reproduce tables, plates, or other illustrations. If the copyright-holder is not the author of the quoted or reproduced material, it is recommended that the permission of the author should also be sought.

3. Material in unpublished letters and manuscripts is also protected and must not be published unless permission has been obtained.

4. A suitable acknowledgement of any borrowed material must always be made.

## Proofs

One set of proofs will be sent to the corresponding author as given on the title page of the manuscript. Only typesetter's errors may be corrected; no changes in, or additions to, the edited manuscript will be allowed.

Elsevier will do everything possible to get your article corrected and published as quickly and accurately as possible. Therefore, it is important to ensure that all of your corrections are sent back to us in one communication. Subsequent corrections will not be possible, so please ensure your first sending is complete.

## Offprints

1. The corresponding author, at no cost, will be provided with a PDF file of the article via e-mail or, alternatively, 25 free paper offprints will be supplied (100 for Review articles). The PDF file is a watermarked version of the published article and includes a cover sheet with the journal cover image and a disclaimer outlining the terms and conditions of use.

2. Additional paper offprints can be ordered on an offprint order form, which is included with the proofs.

3. UNESCO coupons are acceptable in payment of extra paper offprints.

*Agricultural Water Management* **has no page charges.**

# CHAPTER - 15

# Agricultural Ecosystem and Environment

## GUIDE FOR AUTHORS

Agriculture, Ecosystems and Environment deals with the interface between agriculture and the environment. Preference is given to papers that develop and apply interdisciplinarity, bridge scientific disciplines, integrate scientific analyses derived from different perspectives of agroecosystem sustainability, and are put in as wide an international or comparative context as possible. It is addressed to scientists in agriculture, food production, agroforestry, ecology, environment, earth and resource management, and administrators and policy-makers in these fields.

The journal regularly covers topics such as: ecology of agricultural production methods; influence of agricultural production methods on the environment, including soil, water and air quality, and use of energy and non-renewable resources; agroecosystem management, functioning, health, and complexity, including agro-biodiversity and response of multi-species ecosystems to environmental stress; the effect of pollutants on agriculture; agro-landscape values and changes, landscape indicators and sustainable land use; farming system changes and dynamics; integrated pest management and crop protection; and problems of agroecosystems from a biological, physical, economic, and socio-cultural standpoint.

## Types of contribution

1.  *Original papers* (Regular Papers) should report the results of original research. The material should not have been previously published elsewhere, except in a preliminary form.

2.  *Reviews* should cover a part of the subject of active current interest. They may be submitted or invited.

3.  A *Short Communication* is a concise, but complete, description of a limited investigation, which will not be included in a later paper. Short Communications should be as completely documented, both by reference

to the literature and description of the experimental procedures employed, as a regular paper. They should not occupy more than 6 printed pages (about 12 manuscript pages, including figures, etc.).

4. The section *Views and Ideas* offers comment or useful critique on material published in the journal or on relevant issues. Contributions to this section should not occupy more than 2 printed pages (about 4 manuscript pages)

5. *Book Reviews* will be included in the journal on a range of relevant books which are not more than 2 years old. Book reviews will be solicited by the Book Review Editor. Unsolicited reviews will not usually be accepted, but suggestions for appropriate books for review may be sent to the Book Review Editor:'

Edward Gregorich

Agriculture Canada

Neatby Bldg.

Central Experimental Farm

Ottowa

Ontario K1A 0P6

Canada

Please bookmark this page as: http://www.elsevier.com/locate/agee

For more information/suggestions/comments please contact AuthorSupport@elsevier.com

## Online Submission of manuscripts

Submission of an article implies that the work described has not been published previously (except in the form of an abstract or as part of a published lecture or academic thesis), that it is not under consideration for publication elsewhere, that its publication is approved by all authors and tacitly or explicitly by the responsible authorities where the work was carried out, and that, if accepted, it will not be published elsewhere in the same form, in English or in any other language, without the written consent of the Publisher.

Upon acceptance of an article, authors will be asked to transfer copyright (for more information on copyright see http://wwwauthors.elsevier.com/copyright. This transfer will ensure the widest possible dissemination of information. A letter will be sent to the corresponding author confirming receipt of the manuscript. A form facilitating transfer of copyright will be provided.

If excerpts from other copyrighted works are included, the author(s) must obtain written permission from the copyright owners and credit the source(s) in the article. Elsevier has preprinted forms for use by authors in these cases: contact Elsevier's Rights Department, Oxford, UK; phone: (+44) 1865 843830, fax: (+44) 1865 853333, e-mail: permissions@elsevier.com. Requests may also be completed on-line via the Elsevier homepage (http://elsevier.com/locate/permissions).

Papers for consideration should be submitted to: Elsevier Editorial System

Submission to this journal proceeds totally on-line. Use the following guidelines to prepare your article, via the journal's homepage (http://www.elsevier.com/locate/agee) you will be guided stepwise through the creation and uploading of the various files. Once the uploading is done, our system automatically generates an electronic (PDF) proof, which is then used for reviewing. It is crucial that all graphical elements be uploaded in separate files, so that the PDF is suitable for reviewing. Authors can upload their article as a LaTex, Microsoft (MS) Word, WordPerfect, or PostScript files. All correspondence, including notification of the Editor's decision and requests for revisions, will be by e-mail.

## Electronic format requirements for accepted articles

We accept most wordprocessing formats, but Word, WordPerfect or LaTeX is preferred. Always keep a backup copy of the electronic file for reference and safety. Save your files using the default extension of the program used.

## Word processor documents

It is important that the file be saved in the native format of the word processor used. The text should be in single-column format. Keep the layout of the text as simple as possible. Most formatting codes will be removed and replaced on processing the article. In particular, do not use the word processor's options to justify text or to hyphenate words. However, do use bold face, italics, subscripts, superscripts etc. Do not embed 'graphically designed' equations or tables, but prepare these using the word processor's facility. When preparing tables, if you are using a table grid, use only one grid for each individual table and not a grid for each row. If no grid is used, use tabs, not spaces, to align columns. The electronic text should be prepared in a way very similar to that of conventional manuscripts (see also Elsevier's Quickguide (www.elsevier.com/locate/guidepublication). Do not import the figures into the text file but, instead, indicate their approximate locations directly in the electronic text and on the manuscript. See also the section on Preparation of electronic illustrations.

To avoid unnecessary errors you are strongly advised to use the 'spellchecker' function of your wordprocessor.

## Preparation of manuscripts

1.  Manuscripts should be written in English. Authors whose native language is not English are strongly advised to have their manuscripts checked by an English-speaking colleague prior to submission.

    ***English language help service :*** Upon request, Elsevier will direct authors to an agent who can check and improve the English of their paper (before submission). Please contact authorsupport@elsevier.com for further information.

2.  Manuscripts should be prepared with numbered lines, with wide margins and double spacing throughout, i.e. also for abstracts, footnotes and references.**Every page of the manuscript, including the title page, references, tables, etc. should be numbered. Authors are requested to submit, with their manuscripts, the names and addresses of four potential referees.** However, in the text no reference should be made to page numbers; if necessary, one may refer to sections. Underline words that should be in italics, and do not underline any other words. Avoid excessive use of italics to emphasize part of the text.

3.  Manuscripts in general should be organized in the following order:
    - Title (should be clear, descriptive and not too long)
    - Name(s) of author(s)
    - Complete postal address(es) of affiliations
    - Full telephone, Fax. no. and E-mail of the corresponding author
    - Present address(es) of author(s) if applicable
    - Complete correspondence address to which the proofs should be sent
    - Abstract
    - Key words (indexing terms), normally 3-6 items
    - Introduction
    - Material studied, area descriptions, methods, techniques
    - Results
    - Discussion
    - Conclusion
    - Acknowledgements and any additional information concerning research grants, etc.
    - References
    - Tables
    - Figure captions

4.  In typing the manuscript, titles and subtitles should not be run within the text. They should be typed on a separate line, without indentation. Use lower-case lettertype.

5.  Elsevier reserves the privilege of returning to the author for revision accepted manuscripts and illustrations which are not in the proper form given in this guide.

## Abstracts

The abstract should be clear, descriptive and not longer than 400 words.

## Formulae

1. Subscripts and superscripts should be clear.

2. Give the meaning of all symbols immediately after the equation in which they are first used.

3. For simple fractions use the solidus (/) instead of a horizontal line.

4. Equations should be numbered serially at the right-hand side in parentheses. In general only equations explicitly referred to in the text need be numbered.

5. The use of fractional powers instead of root signs is recommended. Also powers of e are often more conveniently denoted by exp.

6. Levels of statistical significance which can be mentioned without further explanation are * $P<0.05$, ** $P<0.01$ and *** $P<0.001$.

7. In chemical formulae,valence of ions should be given, as, e.g. $Ca^{2+}$ not as $Ca^{++}$.

8. Isotope numbers should precede the symbols, e.g. $^{18}O$.

9. The repeated writing of chemical formulae in the text is to be avoided where reasonably possible; instead, the name of the compound should be given in full. Exceptions may be made in the case of a very long name occurring very frequently or in the case of a compound being described as the end product of a gravimetric determination (e.g. phosphate as $P_2O_5$).

## Units and abbreviations

In principle SI units should be used except where they conflict with current practise or are confusing. Other equivalent units may be given in parentheses. Units and their abbreviations should be those approved by ISO (International Standard 1000:1992. SI units and recommendations for the use of their multiples and of certain other units). Abbreviate units of measure only when used with numerals.

## Nomenclature

1. Authors and editors are, by general agreement, obliged to accept the rules governing biological nomenclature, as laid down in the *International Code of Botanical Nomenclature, the International Code of Nomenclature of Bacteria, and the International Code of Zoological Nomenclature,*.

2. All biotica (crops, plants, insects, birds, mammals, etc.) should be identified by their scientific names when the English term is first used, with the exception of common domestic animals.

3. All biocides and other organic compounds must be identified by their Geneva names when first used in the text. Active ingredients of all formulations should be likewise identified.

4.  For chemical nomenclature, the conventions of the International Union of Pure and Applied Chemistry and the official recommendations of the IUPAC IUB Combined Commission on Biochemical Nomenclature should be followed.

## Tables

1.  Authors should take notice of the limitations set by the size and lay-out of the journal. Large tables should be avoided. Reversing columns and rows will often reduce the dimensions of a table.

2.  If many data are to be presented, an attempt should be made to divide them over two or more tables.

3.  Drawn tables, from which prints need to be made, should not be folded.

4.  Tables should be numbered according to their sequence in the text. The text should include references to all tables.

5.  Each table should be typewritten on a separate page of the manuscript. Tables should never be included in the text.

6.  Each table should have a brief and self-explanatory title.

7.  Column headings should be brief, but sufficiently explanatory. Standard abbreviations of units of measurement should be added between parentheses.

8.  Vertical lines should not be used to separate columns. Leave some extra space between the columns instead.

9.  Any explanation essential to the understanding of the table should be given as a footnote at the bottom of the table.

10. Wherever possible, columns should represent individual variables or variables with common units, and rows should represent observations.

11. Present data with no more digits than justified by the accuracy of their measurement or simulation, and no more digits than needed for the purpose of the table. Using fewer digits usually enhances readability of tables.

### *Preparation of electronic illustrations*

Submitting your artwork in an electronic format helps us to produce your work to the best possible standards, ensuring accuracy, clarity and a high level of detail.

### *General points*

- Always supply high-quality printouts of your artwork, in case conversion of the electronic artwork is problematic.
- Make sure you use uniform lettering and sizing of your original artwork.
- Save text in illustrations as "graphics" or enclose the font.
- Only use the following fonts in your illustrations: Arial, Courier, Helvetica, Times, Symbol.

- Number the illustrations according to their sequence in the text.
- Use a logical naming convention for your artwork files, and supply a separate listing of the files and the software used.
- Provide all illustrations as separate files and as hardcopy printouts on separate sheets.
- Provide captions to illustrations separately.
- Produce images near to the desired size of the printed version.
- Illustration should be numbered according to their sequence in the text. References should be made in the text to each illustration.
- Illustrations should be designed with the format of the page of the journal in mind. Illustrations should be of such a size as to allow a reduction of 50%.
- Make sure that the size of the lettering is big enough to allow a reduction of 50% without becoming illegible.
- If a scale should be given, use bar scales on all illustrations instead of numerical scales that must be changed with reduction.

A detailed guide on electronic artwork is available on our website:

http://www.elsevier.com.com/artworkinstruction

**You are urged to visit this site; some excerpts from the detailed information are given here.**

### Colour illustrations

Please make sure that artwork files are in an acceptable format (TIFF, EPS or MSOffice) and with the correct resolution. Polaroid colour prints are *not* suitable. If, together with your accepted article, you submit usable colour figures then Elsevier will ensure, at no additional charge, that these figures will appear in colour on the Web (e.g. ScienceDirect and other sites) regardless of whether or not these illustrations are reproduces in colour in the printed version. For colour reproduction in print, you will receive information regarding the costs from Elsevier after receipt of your accepted article. Please indicate your preference for colour print or on the Web only. For further information on the preparation of electronic artwork, please see http://www.elsevier.com.com/artworkinstruction.

Please note: Because of technical complications which can arise by converting colour figures to 'grey scale' (for the printed version should you not opt for colour in print) please submit in addition usable black and white prints corresponding to all the colour illustrations.

### Non-electronic illustrations

For illustrations that are unable to be uploaded electronically hard copies will be accepted.

Please contact the journal office at: agee@elsevier.com

Provide all illustrations as high-quality printouts, suitable for reproduction (which may include reduction) without retouching. Number illustrations consecutively in the order in which they are referred to in the text. They should accompany the manuscript, but should not be included within the text. Clearly mark all illustrations on the back (or - in case of line drawings - on the lower front side) with the figure number and the author's name and, in cases of ambiguity, the correct orientation.

Mark the appropriate position of a figure in the article.

Note that photocopies of photographs are not acceptable

## Supplementary files

Supplementary files offer the author additional possibilities to publish supporting applications, movies, animation sequences, high- resolution images, background datasets, sound clips and more. Supplementary files supplied will be publshed online alongside the electronic version of your article in Elsevier web products, including ScienceDirect:http://www.sciencedirect.com. In order to ensure that your submitted material is directly usable, please ensure that data is provided in one of our recommended file formats. Authors should submit the material in electronic format together with the article and supply a concise and descriptive caption for each file. For more detailed instructions please visit the artwork instruction pages at http://www.elsevier.com.com/artworkinstruction.

## References

1.  All publications cited in the text should be presented in a list of references following the text of the manuscript. The manuscript should be carefully checked to ensure that the spelling of author's names and dates are exactly the same in the text as in the reference list.

2.  In the text refer to the author's name (without initial) and year of publication, followed - if necessary - by a short reference to appropriate pages. Examples: "Since Peterson (1988) has shown that..." "This is in agreement with results obtained later (Kramer,1989, pp. 12-16)".

3.  If reference is made in the text to a publication written by more than two authors the name of the first author should be used followed by "*et al.*" This indication, however, should never be used in the list of references. In this list names of first author and co-authors should be mentioned.

4.  References cited together in the text should be arranged chronologically. The list of references should be arranged alphabetically on author's names, and chronologically per author. If an author's name in the list is also mentioned with co-authors the following order should be used: publications of the single author, arranged according to publication dates - publications of the same author with one co-author - publications of the author with

more than one co-author. Publications by the same author(s) in the same year should be listed as 1974a, 1974b, etc.

5.  Use the following system for arranging your references:

    (a)  *For periodicals*

    Tietema, A., Riemer, L., Verstraten, J.M., van der Maas, M.P., van Wijk, A.J., van Voorthuyzen, I.,1992. Nitrogen cycling in acid forest soils subject to increased atmospheric nitrogen input. For. Ecol. Manage. 57, 29-44.

    (b)  *For edited symposia, special issues, etc. published in a periodical*

    Rice, K., 1992. Theory and conceptual issues. In: Gall, G.A.E., Staton, M. (Eds.), Integrating Conversation Biology and Agricultural Production. Agric. Ecosyst. Environ. 42, 9-26.

    (c)  *For books*

    Gaugh, Jr., H.G., 1992. Statistical Analysis of Regional Yield Trials. Elsevier, Amsterdam.

    (d)  *For multi-author books*

    Baker, Jr., 1993. Insects. In: De Hertogh, A., Le Nard, M. (Eds.), The Physiology of Flower Bulbs. Elsevier, Amsterdam, pp. 101-153.

6.  In the case of publications in any language other than English, the original title is to be retained.However, the titles of publications in non-Latin alphabets should be transliterated, and a notation such as "(in Russian)" or "(in Greek, with English abstract)" should be added.

7.  Work accepted for publication but not yet published should be referred to as "in press".

8.  References concerning unpublished data and "personal communications" should not be cited in the reference list but may be mentioned in The text.

## Copyright

1.  An author, when quoting from someone else's work or when considering reproducing an illustration or table from a book or journal article, should make sure that he is not infringing a copyright.

2.  Although in general an author may quote from other published works, he should obtain permission from the holder of the copyright if he wishes to make substantial extracts or to reproduce tables, plates, or other illustrations. If the copyright-holder is not the author of the quoted or reproduced material, it is recommended that the permission of the author should also be sought.

3.  Material in unpublished letters and manuscripts is also protected and must not be published unless permission has been obtained.

4.  A suitable acknowledgment of any borrowed material must always be made.

## Proofs

When your manuscript is received by the Publisher it is considered to be in its final form. Proofs are not be regarded as 'drafts'.

One set of proofs in PDF format will be sent to the corresponding author, to be checked for typesetting/ editing. No changes in, or additions to, the accepted (and subsequently edited) manuscript will be allowed at this stage. Proofreading is solely your responsibility.

The Publisher reserves the right to proceed with publication if corrections are not communicated. Return corrections within 3 working days of receipt of the proofs. Should there be no corrections, please confirm this.

Elsevier will do everything possible to get your article corrected and published as quickly and accurately as possible. In order to do this we need your help. When you receive the (PDF) proof of your article for correction, it is important to ensure that all of your corrections are sent back to us in one communication. Subsequent corrections will not be possible, so please ensure your first sending is complete. Note that this does not mean you have any less time to make your corrections, just that only one set of corrections will be accepted.

## Offprints

1. Twenty-five offprints will be supplied free of charge.
2. One hundred free offprints will be supplied to the first author of a review article.
3. Additional offprints can be ordered on an offprint order form, which is included with the proofs.
4. UNESCO coupons are acceptable in payment of extra offprints.

***Agriculture, Ecosystems &Environment* has no page charges**

Information about Agriculture, Ecosystems &Environment is available on the World Wide Web at the following address: http://www.elsevier.com/locate/agee.

# CHAPTER - 16

# Agronomy Journal

## INSTRUCTIONS TO AUTHORS

Articles must be original reports of research not simultaneously submitted to or previously published in any other scientific or technical journal and must make a significant contribution to the advancement of knowledge or towards a better understanding of existing agronomic concepts. The study reported should be applicable to a sizable geographic area or an area of ecological or economic significance and of potential interest to a significant number of scientists. Membership is not a requirement for publishing in *Agronomy Journal* (AJ). Consult the ASA-CSSA-SSSA (2004) style manual and recent issues of AJ for guidance. For questions not answered in the style manual (www.asa-cssa-ssa.org/publoications/style/), consult Susan Ernst, Managing Editor (sernst@agronomy.org), or Matt Nilsson, Associate Production Editor (mnilsson@agronomy.org).

## Scope

After critical review and approval by the editorial board. AJ publishes articles reporting research findings in soil-plant relationships; crop science; soil science; biometry; crop, soil, pasture, and range management; crop, forage, and pasture production and utilization; turfgrass; agroclimatology; agronomic models; integrated pest management; integrated agricultural systems; and various aspects of entomology, wee science, animal science, plant pathology, and agricultural economics as applied to production agriculture. Notes are published about apparatus, observations, and experimental techniques. Observations usually are limited to studies and reports of unrepeatable phenomena or other unique circumstances. Review and interpretation papers are also published, subject to standard review. Contributions to the Forum section deal with current agronomic issues and questions, in brief in thought provoking form. Such papers are reviewed by the Editor in consultation with the editorial board.

## Statistical Methods

Report enough details of your experimental design so that the results can be judged for validity and previous experiments may serve as a basis for the design of future experiments.

Means separation procedures are frequently misused. Such misuse may result in incorrect scientific conclusions. Pairwise multiple comparison tests (LSD) should be used only when the treatment structure is not well understood (e.g. studies to compare cultivars). Authors should be aware of the limitations of multiple comparison tests when little information exists on the structure of the treatments (Carmer and Walker, 1985; Chew 1980; Little, 1978; Nelson and Rawlings, 1983; Petersen, 1977; see also Chew, 1976; Miller, 1981). When treatments have a logical structure, orthogonal contrasts among treatments should be used.

## Validation of Field Results

Field experiments that are sensitive to environmental interactions and in which the crop environment is not rigidly controlled or monitored, such as studies on crop yield and yield components, usually should be repeated (over time or space, or both) to demonstrate that similar results can, or cannot, be obtained in another environmental regime.

## Symposia Series

Manuscripts resulting from symposia having appropriate subject matter will be considered for publication as a compilation in a single issue of AJ. Set/s of manuscripts considered may originate from ASA, CSSA, or SSSA sponsored symposia or from appropriate subject matter symposia sponsored by other organizations. Symposia organizers desiring to publish a compilation of manuscripts in AJ must solicit the Editor with the following prospectus materials: (i) title, location, and date of the symposium; (ii) the organization affiliated with the symposium; (iii) names, addresses, telephone numbers, and email addresses of the solicitors; (iv) a short abstract (= 250 words) outlining the overall purpose of the symposium and reasons justifying why the manuscripts should be published as a compilation; and (v) titles and abstracts, written according to the *Publications Handbook and Style Manual,* for each paper to be considered for publication. Prospectus materials may be submitted to the Editor during any time of the year. Symposia papers are subject to the usual page and production charges for the journal.

## Style

- Use a comma before the final time in a list of three or more items. For example : "Cores were kept inside plastic liners, capped,  and stored on ice.."

- Define all abbreviations at first mentioned in the abstract or text and again in the tables and figures. Once an abbreviation is used, it should be used throughout the entire article, except at the beginning of a sentence.

- The Latin binomial or trinomial and authority must be shown for all plants, insects, pathogens, and animals at first listing. For example:  "In this experiment, 15.5-h corn (Zea mays L.) fields were studied…"

- Both the common and chemical name of pesticides  must be given when first  mentioned. For example : "Atrazine (2-chloro-4-ethylamino-6isopropylamino-1,3,5-triazine) was most   persistent..."

- Identify soils at the series and family level, or at least the Great Group when mentioned first. For soils outside the USA, give both the local identification and the U.S. equivalent.  Up-to-date U.S. soil descriptions may be checked online (http://soils.usda.gov/technical/classification/osd/index.html).

- SI units must be used in all manuscripts. Non-SI units may be added in parentheses.   Spell out numbers one through nine, except when used with units. For decimal    quantities <1, place a zero before the decimal point. Use spaces instead of commas  for the decimal  separator. In text, the space is not necessary for four-digit numbers (e.g., 73 722, but 73722).

- Use the 24-h time system, with four digits for hours and minutes (e.g., 1430 h for  2:30 p.m.). Report dates with the day first, then the month and the year last. Abbreviate months with more than four letters (e.g., 14 May 1997, 7 June 1983, 10 Aug, 1996, or 26 Sept.1974).

## Manuscript Preparation [*Style Manual ch.* 1]

All accepted manuscript files will be edited in Microsoft Word. Therefore, authors should compose manuscripts in Word. Corel WordPerfect is also acceptable although authors should be aware that errors are occasionally introduced during the conversion process. Authors should avoid using work processing features such as automated bulleting and numbering, head and subhead formatting, internal linking, or styles. Avoid using more than one font and font size. Limited use of italics, bold, superscripts, and subscripts is acceptable. Manuscript text and tables should be double spaced, with line numbering.

*Agronomy Journal* has a double-blind review in that the reviewers do not know name of the authors and the authors do not know the name of the reviewers. Therefore, authors should prepare the manuscript with no  information about the author (e.g., no byline, addresses/affiliations, acknowledgments, etc.; these will be added after a manuscript has been accepted –see Accepted Manuscripts). Take care to label tables and figures with reference to the paper's title, not theames of the author. Any identification in headers or footers should be similarly anonymous. As a last consideration, authorship may be unintentionally revealed through such software features as document summaries. If this is a concern consult your local software experts. When authors submit a manuscript electronically via Manuscript Tracker (see Submitting Manuscripts), they will be asked to enter contact information into the system database, and the Editor and Technical Editors will have access to this information so that they can contact the authors about the outcome of the review. The typical sequence for a paper is title, abstract, a list of abbreviations, introduction (without any heading), materials and methods, results, discussion, summary or conclusions, references, figure captions (without any heading), and finally tables. Some papers may have a theory section, a few have no materials and methods; the results and discussion section may be combined, and the summary may be incorporated into the discussion. If an appendix is needed, it comes before the references.

## Title *[Style Manual ch. 1, p. 7]*

A good title briefly identifies the subject and indicates the purpose of the study or the major findings. Use common name of crops where possible, and avoid abbreviations. The recommended length is 12 words.

## Abstract *[Style Manual ch. 1, p.8-9]*

Each paper must have an abstract, a single self-contained paragraph of 250 words or less for papers and 150 words or less for notes. State the rationale, objectives, methods, results, and their meaning or scope of application. Be specific. Identify the crops or organisms involved, as well as soil type, chemicals, or other details important to interpretation of the results. Do not cite figures, tables, or references. Avoid equations.

## Tables *[Style Manual ch. 1, p. 11; ch. 5, p. 41-45]*

- Start each table on a new page.

- Always use your word processor's (Word or WordPerfect) table feature. That is, the table that you create should have defined cells. DO NOT create table by using the space bar and/or tab keys.

- Do not use the enter key within the body of the table. Instead, separate data horizontally with a new row.

- Do not insert blank columns or rows. If you want extra spacing between columns or rows, indicate this on the hard copy you send to headquarters after your manuscript is accepted.

- Asterisks or letters next to values indicating statistical significance should appear in the same cell as the value, not an adjacent cell (i.e., they should not have their own column).

- Use the following symbols for footnotes in the order shown:†, ‡, §, ¶, # ††, ‡‡, etc. The symbols *, **, and *** are always used to show 0.05, 0.01, and 0.0001 probability levels, respectively, and are not used for other footnotes. Footnote symbols should not be set in superscript type, and all footnotes should be set on separate lines.

- Spell out abbreviations on first mention in tables, even if they have already been defined in the text. The reader should be able to understand the table content without referring back to the text.

- Tables in the journals are set entirely in boldface type, so bold cannot be used to highlight individual values. Italic type and underlining are acceptable alternatives, but shading is not allowed.

- In the body of a table, only the first word in a row should be capitalized (the exception would be proper nouns, which should always be capitalized).

## Figures [*Style Manual ch. 1, p. 11; ch. 5, p. 41, 45-49*]

Figures in accepted manuscripts are prepared for publication by scaning printed copies of the figures. Therefore, although authors will be asked to submit an electronic file for the review process via Manuscripts Tracker (see Submitting Manuscripts), authors should prepare all figures keeping in mind that ultimately they will need to provide high quality printed copies (see Accepted Manuscripts).

To maintain clear contrast, use line patterns instead of shading and avoid thin, light lines. As feasible, plan for reduction to one-column width (84 mm, or = 3.25 inches). The original should be one-third to one-half larger than the intended final size. You can test the reduction quality on a copier. Keep relative sizes in mind when adding symbols, letters, and numbers.

Label each figure along the top or bottom margin with its number and the title of the article – but not the author's name, to preserve anonymity. If there could be any doubt at all, indicate orientation (which way is up).

*Agronomy Journal* will publish four-color images. Authors may publish color illustration such as photos, figures, or maps in papers, but at their own expense. The cost is $1000 per journal page

## Reference [*Style Manual ch. 1 p. 11-20*]

The reference section is typically limited to published literature and unpublished but available reports, abstracts, theses, and dissertations. Alphabetize the list by the surnames of the first authors and then by the second and third authors. Cite unpublished data, personal communications, and reports not available to the public in the text only (in parentheses; state the year).

For journal articles, give the authors, year, complete article title, abbreviated journal title, volume number, and inclusive pages. For book chapters, give the authors, year, chapter title, pages, book editor (if any), complete book title, publisher, and place of publication. For proceedings, give the place and date of the conference also. For electronic references, see the style manual.

## Examples

*Journal article*
Simth, D.T., D.L. Johnson, and J.K. Thomas, 2001. Phosphorus losses in irrigation runoff. J. Environ. Qual. 30:2569-2580.

*Books*
Lindsay, W.L., 1979. Chemical equilibria in soils, John Wiley & Sons, New York.

*Chapter in a book*
Nelson, D.W., and L.E. Sommers. 1982. Total carbon, organic carbon, and organic matter. p. 539-579. *In* A.L., Page, R.H. Miller, and D.R. Keeney (ed.) Methods of soil analysis, Part 2. 2$^{nd}$ ed. Agron. Monogr. 9. ASA and SSSA, Madison, WI.

***Chapter in proceedings***
Power, J.F., and V.O. Biederbeck, 1991. Role of cover crops in integrated crop production systems. p. 167-174. *In* W.L. Hargrove (ed.). Cover crops for clean water. Proc. Int. Conf., Jackson, TN. 9-11 Apr. 1991. Soil and Water Conserv. Soc., Ankeny, IA.

## Submitting Manuscripts
Authors should submit manuscripts electronically via our Manuscript Tracker system (www.manuscripttracker.com/aj/). Manuscript Tracker allows authors to track the status of their manuscript through the review process. To ensure a double-blind review, authors should not include any identifying information in the manuscript (e.g., no byline, addresses/affiliations, acknowledgments, etc,; these will be added after a manuscript has been accepted – see Manuscript Preparation and Accepted Manuscripts). Detailed instructions for using Manuscript Tracker can be found on the website above.

## Potential Reviewers
When authors submit manuscripts through the Manuscript Tracker system, they will be prompted to provide a list of potential reviewers. These reviewes must not have a conflict of interest involving the authors or paper, and the editorial board has the right not to use any reviewes suggested by authors.

## Changes in Authors Byline
From time to time, names of authors are added or deleted from a manuscript between the time of submission and publication. In situations such as this, the ethical and responsible manner of handling this type of change is for the lead author to advise the author about the material that is being added or deleted and to notify, in writing, the Editor and Managing Editor of the journal.

## Publishing Supplemental Information
Video, color, animation, data sets, and other information that is expensive or difficult to publish on paper can be published in the online version of the journal at no additional charge. Authors who wish to publish supplemental information online should include it with the original submission and notify the Editor immediately. Arrangements will be made so the supplemental information can be included in the review process.

## Accepted Manuscripts
Once a manuscript has been accepted, authors should prepare a cover page that includes the title, byline (all authors, with an asterisk identifying the corresponding author), author-paper documentation, and acknowledgements (if any). The author-paper documentation *[Style Manual ch. 1, p. 7-8]* should be a single paragraph that lists all authors with their complete current addresses. It should include any applicable documentation for the paper, such as the institutional article number, contributing institutions, or a brief statement of financial support. It should end with "Received- *Corresponding author" and the author's email

address. The corresponding author is the one who works with the editors during review and production and acts as the primary contact for reprints and information requests after publication.

Authors should attach the cover page to a printed copy of the final version of the manuscript (text, tables, and figures) and send it to headquarters, along with a word processing file for the text and tables (Word preferred). The printed copy and word processing file must match exactly in all parts of the manuscript. Do not submit files for the figures; do submit high quality copies (glossy prints preferred). Authors are also invited to submit color photos or computer graphics for possible use on the cover at no added expense. An ideal cover combines scientific meaning and visual beauty. It should be related to the subject of your paper. Please send slides, prints, or electronic TIF or EPS files with at least 300 dpi in the finished size. See current issues for examples, size, and shape.

Send the printed copy, disk/CD-ROM with the manuscript file, and cover submission (if any) to: Matt Nilsson, Associate Production Editor, *Agronomy Journal,* American Society of Agronomy, 677 South Segoe Road, Madison, WI, USA 53711 (mnilsson @ agronomy.org).

## Publication Charges

Authors who are members of ASA, CSSA, or SSSA will pay a flatrate production charge of $450.00; non members will pay $700.00. The cost for non-color figures is $10.00 per figure, less $15.00 (amount contributed by ASA-CSSA-SSSA). Color figures will cost $1000 per journal page. Authors alterations after a paper has been typeset are $5.00 per line.

## References


1. ASA, CSSA, and SSSA. 2004. Publications handbook and style manual [Online]. Available at www.asa-cssa-sssa.org/publicaions/style/. ASA, CSSA, and SSSA, Madison. WI.

2. Carmer, S.G., and W.M. Walker. 1985. Pairwise multiple comparisons of treatment means in agronomic research. J. Agron. Educ. 14:19-26.

3. Chew, V. 1976. Comparing treatment means: A compendium. Hort Science 11:348-357.

4. Chew, V. 1980. Testing differenes among means: Correct interpretation and some alternatives. HortScience 15:467-470.

5. Little, T.M. 1978. If Galileo published in HortScience. HortScience 13:504-506.

6. Miller, R.G., Jr. 1981. Simultaneous statistical inference. Springer-Verlag, New York.

7. Nelson, L.A., and J.O. Rawlings. 1983. Ten common misuses of statistics in agronomic research and reporting. J. Agron. Educ. 12:100-105.

8. Petersen, R.G. 1977. Use and misuse of multiple comparison procedures. Agrony. J. 69:205-208.

# CHAPTER - 17

# Agronomy Research

## INSTRUCTIONS TO AUTHORS

*Agronomy Research* is a biannually published peer-reviewed international Journal intended for publication of broad-spectrum original articles, reviews and short communications in actual problems of modern agriculture including crop and animal science, genetics, economics, technical aspects, agriculture and environmental relations etc. The journal also publishes special issues with conference papers that present new findings of important advances in research. Agricultural scientists throughout the world are cordially invited to submit original articles for inclusion in its pages.

**Papers must be in English (British spelling). English is revised by a languagereviewer, but** authors are strongly urged to have the papers reviewed linguistically prior to submitting. **Contributions should be sent electronically. Papers are considered by referees before acceptance.**

Papers should strictly follow instructions.

## Structure

Title, Authors (names), Authors' place of work with full address (each on a separate line), Abstract (up to 250 words), Key words, Introduction, Materials and methods, Results and discussion, Conclusions, Acknowledgements, References.

## Example

**A study of synergistic effect of greater wax moth *Galleria mellonella***

**(Lepidoptera: Pyralidae)**

M. Kuusk1 and K. Lepp2

1    Institute of Agricultural and Environmental Sciences, Estonian University of Life Sciences, Kreutzwaldi 64, EE51014 Tartu, Estonia; e-mail:

2.   Institute of …

## Abstract

In laboratory pupal…

## Key words

*Galleria mellonella*, **respirography, synergistic interaction**

## Introduction

In many countries rural inhabitants for insect control use various local plants (Smith & Jones, 1996; Brown et al., 1997; Adams, 1998).

## Page size and font

- The file should be prepared using **Microsoft Word 97** or a later version
- Set page size to **B5 Envelope (17,6 × 25 cm),** all margins at 2 cm
- Use **single line** spacing and justify the text
- Use font **Times New Roman**, size 11; for Abstract, Key words, References and tables use **10** points
- **Use** tabs 0,8
- Do not use page numbering
- Use *italics* for Latin biological names and for statistical terms (*t*-test, $n = 193$, $P > 0.05$)
- Use single ('......') instead of double quotation marks ("......")

## Tables

- All tables and figures must be referred to in the text (Table 1; Tables 1, 2)
- For tables use font Times New Roman, regular, 10 points
- Use **TAB** and not space bar between columns
- Do not use vertical lines as dividers, only **horizontal** lines are allowed
- Primary column and row headings should start with an initial capital, secondary headings without initial capital

## Figures

- Use only black and white for figures
- **Use font** Arial **within the figures**
- Legend below the figure must not be in a frame of the figure
- All figures must be referred to in the text (Fig. 1; Fig. 1, a, b; Figs 1, 3; Figs 1–3)

## References

- Within the text

  In case of **two** authors use **&.** In case of more than two authors, reduce to first author "et al."

  Smith & Jones (1996); (Smith & Jones, 1996)

  Brown et al. (1997); (Brown et al., 1997)

Adams (1998); (Adams, 1998)

When referring to more than one publication, arrange them using the following keys: 1. year of publication (ascending), 2. alphabetical order for the same year of publication: (Smith & Jones, 1996; Brown et al. 1997; Adams, 1998; Smith, 1998)

- For whole books

Name(s) and initials of the author(s), year of publishing, title of the book (*in italic*), publisher, town of publishing, number of pages.

Tritton, D. Y. 1988. *Physical Fluid Dynamics*. Clarendon Press, Oxford, 350 pp.

Shiyatov, S. G. 1986. *Dendrochronology of the Upper Timberline in the Urals*. Nauka,

Moskva, 350 pp. (in Russian).

- For journals articles

Name(s) and initials of the author(s), year of publishing, title of the article, abbreviated journal title (*in italic*), volume (*in bold*), page numbers.

Titles of papers published in languages other than English, German, French, Italian, Spanish, and Portuguese should be replaced by an English translation, with an explanatory note at the end, e.g. (in Russian, English abstr.).

Karube, I. & Tamiyra, M. Y. 1987. Biosensors for environmental control. Pure Appl. Chem. **59**, 545–554.

Frey, R. 1958. Zur Kenntnis der Diptera brachycera p.p. der Kapverdischen Inseln. Commentat.Biol. **18**(4), 1–61.

Danielyan, S.G. & Nabaldiyan, K.M. 1971. The causal agents of meloids in bees. Veterinariya **8**, 64–65 (in Russian).

- For articles in collections:

Name(s) and initials of the author(s), year of publishing, title of the article, name(s) and initials of the editor(s) (preceded by **In**), title of the collection (*in italic)*, publisher, town of publishing, page numbers: Yurtsev, B.A., Tolmachev, A.I. & Rebristaya, O.V. 1978. The floristic delimitation and subdivisions of the Arctic. In Yurtsev, B. A. (ed.): *The Arctic Floristic Region*. Nauka, Leningrad, pp. 9–104 (in Russian).

- For conference proceedings:

Name(s) and initials of the author(s), year of publishing, proceedings title, name(s) and initials of the editor(s) (preceded by **In**), proceedings name (***in italic)***, publisher, town of publishing, page numbers:

Ritchie, M.E. & Olff, H. 1999. Herbivore diversity and plant dynamics: compensatory and additive effects. In Olff, H., Brown, V.K. & Drent,

R.H. (eds): *Herbivores between plants and predators. The 38th Symposium of the British Ecological Societ.* Blackwell Science, Oxford, UK, pp. 175–204.

## Please note

Use '.' ( not ',') : $0.6 \pm 0.2$

Use a 'comma' for thousands - 1,230.4 (one thousand two hundred and thirty and four tenths)

Without space: 5°C, 5% (not 5 °C, 5 %)

Use '–' (not '-') and without space: pp. 27–36, 1998–2000, 4–6 min, 3–5 kg

Spaces: 5 h, 5 kg, 5 m, $C : D = 0.6 \pm 0.2$

Use 'kg ha-1' (not 'kg/ha')

Use ' ° ' : 5 °C (not 5oC)

## Mailing address

Agronomy Research

Estonian Grassland Society

St. Teaduse 4, 75501 Saku, Estonia

E-mail: rein.viiralt@emu.ee

*Agronomy Research* is available online at http://www.eau.ee/~agronomy/

# CHAPTER - 18

# Agropedology

## NOTE TO AUTHORS

### Journal policy and notes

Papers will be considered for publication if they make an original contribution to the knowledge in any branch of soil science, notably soil genesis, soil morphology and classification; soil physics and hydrology; soil chemistry and mineralogy; soil and water management and conservation. The journal particularly welcomes papers that promote understanding of soils in Indian sub-continent.

Submission of a paper is taken to mean that the results have not been published and are not being considered for publication elsewhere.

We also consider critical review articles that indicate fruitful areas of further research and are original and innovative. Reviews should not exceed 15 printed pages in length (30 pages A4 double-spaced type).

All papers are referred. **No free prints are provided.**

Original illustrations will be returned after publication if this is requested at the time of submission.

The Journal assumes that all authors of a multi-authored papers agree to its submission. It takes no responsibility for fraud or inaccuracy on the part of the contributors.

### Guidelines for the preparation of manuscripts

### General presentation

The Work should be presented in concise and clear English. The Introduction should not exceed what is necessary to indicate the reason for the work and its essential background. Sufficient experimental details should be given to enable the work to be repeated. The Discussion should explain the significance of the results.

### Manuscripts

Submit two clearly legible copies. Type them with double-spacing throughout, making the left-hand margin at least 3 cm wide, on good quality paper. Place tables, figures, and captions to figures after the text, and number all pages of the manuscript consecutively. Refer to each figure and table in the text.

When the manuscript has been referred, the author will be requested to submit a computer disk bearing the final version of the manuscript, as well as a hard copy of the same manuscript for editing

When you submit a manuscript, please enclose a separate sheet containing your telephone number, facsimile number, and email address, as well as your postal address; we many need to contact you urgently.

Submission of accepted manuscripts in a computer readable form.

Authors must submit accepted manuscripts on disc or by email prepared in IBM-PC word processors, stating the software's name and version and listing file names. MicroSoft Word and WordPerfect are preferred formats.

*Title :* This should be concise and appropriately informative and should contain all keywords necessary to facilitate retrieval by searching techniques. An abridged title suitable for use as a running head at the top of the printed page and not exceeding 50 letter spaces should also be supplied.

*Abstract :* The abstract (preferably less than 200 words) should not just recapitulate the results but should state concisely the scope of the work and give the principal findings. It should be complete enough for direct use by abstracting services. Acronyms and references should be avoided.

*Keywords :* Up to six key words not contained in the title may be listed beneath the abstract to assist searching techniques.

*References* are cited chronologically in the text. All references in the text must be listed at the end of the paper, with the names of authors arranged alphabetically; all entries in this list must correspond to references in the text. In the text, the names of two co-authors are linked by "and"; for three or more the first author's name is followed by'et al'. No editorial responsibility can be taken for the accuracy of the references; authors are requested to check these with special care. Titles must be included for  all references. Papers that have not been accepted for publication may not be included in the list of references and must be cited as "unpublished data' or as "personal communication'; the use of citations is discouraged.

### Give titles of books and journals in full

### Examples :

#### Journal article

Buyanovsky, G.A., Aslam, M., and Wagner, G.H. (1994). Carbon turnover in soil physical fractions. Soil Science of America Journal 58, 1167-1173.

## Chapter in book

Oades, J.M. (1998). An introduction to organic matter in mineral soil. In 'Minerals in Soil Environments'. (Eds. J. B. Dixon and S.B. Weed) pp.92-159. (Soil Science Society of America : Wisconsin). of headings to rows and vertical columns should be capitalized. The symbol for the unit of measurement should be placed in parentheses of the column heading. Footnotes should be reserved for specific items in columns. Horizontal rules should be inserted only above and below column headings and at the foot of the table. Vertical rules must not be used. Each table must be referred in the text. Short tables can frequently be incorporated into the text as a sentence or as a brief untitled tabulation llustrations.

Authors should submit lettered line drawings and lettered and mounted photographs which comply with the instructions below. Unsatisfactory figures will be returned for correction.

*Line drawings :* Original line drawings may be produced using computer graphics with laser printing, or drawn with black ink on white board or on drawing or tracing paper, and with regard to the size of the printed page (12.5 by 20 cm). Explanations of symbols used should be given in the caption to the figure. Lettering of the graphs should be kept to minimum. Grid marks should point inwords; legends to axes should state the quantity being measured and be followed by the appropriate SI units in parentheses.

*Photographs :* Photographs must be of the highest quality with a full range of tones and of good contrast. Lettering should be in contrast with its back ground. The size should be such that the final height after reduction is 1.5-2 mm. A scale bar must be inserted on each photomicrograph and electron micrograph. Important features to which attention has been drawn in the text should be indicated. Colour photographs will be accepted if they are essential but the cost of production must be borne by the author.

## Checklist for preparation of manuscripts

Submit manuscripts in hard copy to start with. When the manuscript is returned to the authors with the referee's reports for revision, the editor will make notes on presentation, layout and general matters concerning style, and these should be incorporated in the revised paper. A disc will be requested once the paper is accepted in principle.

The version on the disk should be an exact copy of the hard copy. Please give the name of the file and the word-processing package used.

1.  Type manuscript double-speaced throughout, including references, figure captions and tables.

2.  Type the title and all headings aligned left, with only the first letter of the first word and of any proper name capitalized.

3. Main heading (Introduction, Materials and methods, Results and discussion Acknowledgements, References ) are set in bold roman (not italic) type, Minor headings are set in the light italic type.

4. Use the conventions "From.....to', "between ....and', "range x-y),

5. Check that all references mentioned in the text are in the References, and vice versa.

6. List references in the text in chronological order, separated by semi-colons. Do not use a comma between the author's name and the date. List references in the References in alphabetical order.

7. Give full journal and book titles in the References.

8. Use Arabic numerals in the text but in headings spell out numbers less than 10,  type space between and its unit(e.g. 3 mm).

9. Check that the stippling and/or symbols are legible at the size likely to be used in the published article.

10. Type tables with title as a separate paragraph.

11. Indicate approximate positions of figures and tables on the manuscript.

12. Check that figures are numbered in the order in which they are discussed in the text.

13. Suggest a running head for the paper of not more than 50 characters (including spaces).

14. Return the requested number of revised manuscripts; also, return the original manuscript annotated by the editor.

15. Provide an email and Postal address for the corresponding author.

## Correspondence

When you submit a manuscript, please supply us with your telephone number, e-mail as well as your postal address; we many need to contact you urgently.

## Address for submissions

**The Honorary Secretary**

Indian Society of Soil Survey and Land Use Planning,

NBSS&LUP, Amravati Road,

Nagpur – 440 010, India

# CHAPTER - 19

# Allelopathy Journal International

## INSTRUCTIONS TO AUTHORS

### Submission of Manuscript

**Manuscript (Ms.) must be the report of original research and neither simultaneously submitted to, nor previously published in any other scientific Journal. Ms. Should be written in active voice, clear, concise, grammatical English.** Contributors. Whose native language is not English, are strongly recommended to get their Ms. Checked by native English speaker or a colleague with adequate experience in written English. **The Ms. Submission implies that consent of all authors and permission of institute has been obtained.**

Please submit Electronic Ms. (Text in MS Word, Figs in MS Excel files) by E. Mail to any Regional Editors as under:

### Regional Editors

Prof. J.V. Lovett, Lovett Associates Pty Ltd., PO Box 400, Hall ACT 2618, Australia. E.Mail: lovettassociates@bigpond.com

Prof. R.D. Williams, USDA-ARS, Langston University, Langston, OK 73050, U.S.A. E.Mail: rdwms@mail.luresext.edu

Prof. M.J. Reigosa, Facultad Ciencias, Universidade de Vigo, Ap 874, 36200, Vigo, Spain. E.Mail: merigosa@uvigo.es

Ms. From China, Korea, Japan : Prof. C. Kong, Institute of Applied Ecology, Chinese Academy of Science, Wenhua Road, Shenyang – 110016, CHINA, E.Mail: knogch@mail.edu.cn

**Ms. only from India: Prof. S.S. Narwal**, 8/515, Haruam Agri University, Hisar-125 004, India. Allelopathy1974@yahoo.com

### Reviews

Prof. L.S. Dias, Centre of Ecology, University of Evora, Ap 94, 7002-554, Evora, Portugal. E.Mail: Isdias@uevora.pt

## Page Charges

There are no page charges for publishing the Ms.

## Time of Publication

The Accepted Ms. is published within 3-months after acceptance.

## Scope

Original Papers, Reviews, Short Communications, Reports of Conferences/ Meetings and their Announcements, Book Reviews on all aspects of allelopathy and related areas in both aquatic and terrestrial ecosystems are invited.

## Nomenclature

The **Latin binomial or trinomial names** (must be in italics) and the authority must be written for plants/insects/microbes, etc. and full chemical names for compounds must be written when used first time. Crop cultivars (not experimental line) must be identified by a single quotation mark when mentioned in the text (e.g. cv. 'Coral'), thereafter, the common or generic names may be used. Don't use acronyms/abbreviations without its full form, give its full form, when mentioned for the first time.

## Units of Measurement and Abbreviations

Metric Units should be used. Abbreviations should be explained, when they first appear in Text. For mineral contents the elements(N.P.K, etc.) should be used. **Isotopes** should be indicated as $^{14}C$, $^{32}P$. etc., **Ions** should be mentioned as $H^+$ and $Mg^{2+}$, **etc. for molar concentration** M should be used.

## Reporting Time and Dates

Use the 24-h time system with four digits, the first two for hours and last two for minutes (e.g. 1430 h for 2.30 PM). Give dates in full e.g. 10 February, 1992.

## Manuscript Preparation

### Research Paper

The Manuscript (Ms.) should be divided into following sections: *Abstract, keywords, Introduction, Materials and Methods, Results and Discussion, Acknowledgement, References.*

### Scripts

For faster publication, please critically follow these instructions and the format of Journal. Type the Ms in double space in 12 Font size (for Text), 10 Font Size (for References/Bibliography) and 9 Font size (for Tables, their Titles/Foot notes etc.) on one side of A4 paper with 4.0 cm margin on the left hand side and 2.0 cm on right hand side. All pages should be numbered consecutively in Arabic numerals on the top right hand corner. All abbreviations used should be full explained at first mention. Spelling should confirm to British English Oxford

Dictionary. The Editors/Publisher do not accept responsibility for loss of, or damage to Ms, hence, authors must retain copies of all materials submitted.

## Title page

This page should contain following information in order of (a) **Running title**, not exceeding 50 characters including letters and spaces, (b) **Full title of paper, it** should be informative, contain maximum number of key words and should not exceed 70 characters, (c) **Authors name** and full mailing address at which the research was conducted and the present address, if different from the place of research, (d) Phone and FAX numbers **and E. Mail address of Correspondence Author**, (e) **Abstract** of 150-200 words, (f) **Key words** not exceeding 10 in alphabetical order and (g) **Introduction,** if space is left. If there are more than one authors, use* to indicate the name of **Correspondence** Author as footnote.

## Author's address

Multiple authors with different addresses must indicate their respective address separately by superscript number e.g. S.S. NARWAL and G.S. DHALIWAL[1]

Department of Agronomy, CCS Haryana Agricultural University, Hisar – 125 004, India.

Phone (Office):...,

FAX (Office):......

phone Home)/Mobile No:........,

E.Mail:......... [1]Department of Entomology,

Punjab Agricultural University, Ludhiana – 141 004, India.

## Important

Add Phone (Office, Home/Mobile), FAX No. with ISD/STD Codes and E. Mail address of Correspondence Author.

## Abstract

It should briefly (150-200 words) describe the experiments(s) details including the year, place of study, the purpose of research and main results.

## Key words

Add upto 10 keywords in alphabetical order and separate them by commas.

## Text

The paper may be written in the past tense and divided into sections e.g., Introduction, Materials and Methods, Results and Discussion and References. For paragraph, leave 10 spaces at the start of line. Avoid too many headings and sub-headings.

## Introduction

Why did you conduct the study? Present only essential background, include a concise statement of objectives, avoid detailed review, findings or conclusions.

## Materials and Methods

What did you use and do in this study? Present full details of the techniques, the treatments and the methods of statistical analysis used in the past tense. For field experiments, give site location (Latitude, longitude, annual rainfall and mean height above Sea level), date, month and year of start and completion, chemical properties of soil (pH, soil fertility status).

For plant samples, please add phenological phase at the time of Sample collection. For Lab studies, indicate temp C and duration of light/dark in Growth Chamber/Incubator. For preparing extracts/leachates, specify if you have used fresh or dried sample and the duration of soaking.

# CHAPTER - 20

# American Journal of Agricultural Economics

## SUBMITTING MANUSCRIPTS TO THE AJAE
## SUBMISSION GUIDELINES

### Conditions

Submission of a paper will be held to imply that (a) the material in the manuscript has not been published, is not being published or considered for publication elsewhere, and will not be submitted for publication elsewhere unless rejected by the journal editor or withdrawn by the author(s); (b) the material in the manuscript, so far as the author(s) knows, does not infringe upon other published material covered by copyright; (c) the author's (s') employer, if any, either does not assert an ownership interest in the manuscript or is willing to convey such interest to the American Agricultural Economics Association (AAEA); and (d) submission of the manuscript gives the AAEA exclusive right to publish, to copyright, and to allow or deny reproduction of it, in whole or in part. If the applicability of point (a) is unclear, the author(s) must provide an explanation in the cover letter.

Data and documentation. Authors are expected to document their data sources, models, and estimation procedures as thoroughly as possible, and to make the data used available to others for replication purposes. If an exception to this rule is desired, reasons should be given and this should be explicitly noted in the cover letter.

### What and where to submit

WE ONLY ACCEPT MANUSCRIPTS IN PDF FORMAT. For the initial submission, please send the manuscript, model documentation and all supporting materials through the AJAE's electronic submission system. Subsequent editorial correspondence should be with the editor in charge of the manuscript. The AJAE does not ordinarily consider for publication manuscripts exceeding thirty double-spaced pages (everything included).

**Formatting Information - For articles appearing in AJAE volume 88 (year 2006) or later**

- Please make sure all manuscripts adhere to the style and formatting instructions below.

- Once the manuscript is formatted correctly, please read the electronic submission instructions.

## Formatting Instructions

- CHANGE (February 2004) - Authors' identification and title page. To protect their anonymity in the review process AUTHORS SHOULD NOT IDENTIFY THEMSELVES ON THE TITLE PAGE OR IN ANY HEADERS. A SEPARATE TITLE PAGE MUST BE SENT AS AN ATTACHMENT TO THE EDITORS via the online submission management system and should include: (a) title; (b) author(s) names; and (c) name, address, phone and fax numbers, and email address of the author serving as the contact person.

- Text preparation. Please assure that at least one sentence of text occurs between any two headings (in particular, a section heading should be followed by at least some text preceding any subsection heading). Double-space all material, including footnotes, references, and tables, on 8-1/2- by 11-inch standard-weight white paper. Use 1-1/4-inch margins and 12-point Times Roman or a similar type style. All headings and subheadings are flush left. Provide short headings for each section and subsection. Do not number sections or subsections. Section headings are denoted in bold and subsection headings, in italics. Do not indent the first paragraph after any heading. Do not use a heading prior to the first paragraph of the article (e.g., no heading for "introduction"), and do not indent the first paragraph of the article. All material should be double-spaced and single sided. Follow The Chicago Manual of Style, by the University of Chicago Press, and recent issues of AJAE for style. NOTE: When referring to your paper, use the word "article."

- Style. Follow A Manual of Style by University of Chicago Press and recent issues of the AJAE.

- Data and documentation. Authors are expected to document their data sources, models, and estimation procedures as thoroughly as possible, and to make the data used available to others for replication purposes. If an exception to this rule is desired, reasons should be given and this should be explicitly noted in the cover letter.

- Mathematical notation. Use only essential mathematical notation. Avoid using the same character for both superscripts and subscripts or using capital letters for such, and whenever possible avoid overbars, tildes, carets, and other modifications of standard type. Asterisks, primes, and small English letter superscripts are suitable.

- Mathematical typesetting. Use standard type to the maximum extent possible (for example, use Symbol font for Greek characters in simple notation). Refrain from use of embellished letters (dots, bars, tildes, carets). Run equations into text if at all possible (rather than displaying). Simplify notation to avoid costly typesetting. All displayed equations (i.e., equations not run into regular text) should be numbered consecutively, placed in parentheses (not square brackets) and left-justified on the same line as the equal or inequality sign.

- Footnotes to text. Number footnotes consecutively throughout the article, not page by page. Type all footnotes, double-spaced, on a separate page following the article. Footnotes should be only explanatory and not for citations or for directing the reader to a particular work. Such information can be incorporated into the text. (Note: for accepted manuscripts, a leading unnumbered footnote providing authors' affiliations and acknowledgements will be taken from the title page.)

  Instructions - How to footnote for the AJAE

- References and citations. Place References, alphabetized by author, in a list at the end of the paper, double spaced (without extra blank lines between references). Provide issue number whenever possible and always for journals that do not number pages sequentially through complete volumes (e.g., for Journal of Economic Perspectives). Format reference with hanging indentation (first line flush left, second and subsequent lines indented). See AJAE Reference Guide for other critical details. Only cited works may be included in the reference list. All citations should appear in the text and contain the name of author and year, with page numbers when necessary; text citations should omit any comma or other punctuation between the author's name and the year of publication. Citations can be inserted parenthetically, e.g. (Doe 1998, p. 5). If the author's name is used as part of the sentence, include year of publication parenthetically, with page numbers if necessary; e.g., Doe, Smith and Jones (2002) show that . . . . . Use et al. only with four or more authors. For text citations listing more than one source, separate sources by a semi-colon: (Doe 1998; Smith and Jones 2000; Smith, Jones, and Erp 2003; Thomas et al. 2004). Do not use et al. in the reference section.

## AJAE Reference Guide (PDF)

- Tables and Figures. Place each table and figure on a separate page at the end of the paper. Double-space all material and omit vertical rules in tables. Each table and figure must have a legend. Place legends for tables at the top of the table, flush left, and bold. Table and figure titles should be fully descriptive. Omit a period at the end of the legend. Capitalize the first letter in each word. Example:

**Table 1. Summary Statistics for Pennsylvania Wheat Crops, 1998-2000**

- Place legends for figures at the bottom of the figure, flush left, and bold. Capitalize only the first letter of the first word. Example:

**Figure 1. Annual net sales of Mississippi catfish farmers, 2002-2003**

- Use lowercase for the words "table" and "figure" in the text unless, of course, they appear at the beginning of a sentence.

- ***Footnotes to tables :*** Use lower case English letters to attach footnotes to specific items within the table, and place the footnotes below the bottom line of the table in (un-indented) paragraph form. For general explanatory notes, use the heading "Note:" and continue on the same line with the first word of the note, in paragraph form. The "Note:" paragraph may define the use of asterisk (e.g., * or **) to denote statistical significance levels. For example, say "Asterisk (*) and double asterisk (**) denote variables significant at 5% and 10% respectively." See a recent issue after January 2006 of the journal for other examples.

- ***Footnotes to figures :*** Normally figures do not carry footnotes to specific items within the figure. General explanatory information may be included in a paragraph bearing the heading of "Note:" and placed below the figure legend. See a recent issue of the journal for examples.

- The *AJAE* has arranged with AgEconSearch to post supplemental appendices for published articles on their web site for reliable availability to readers. While authors may still maintain their own web-based supplements, any materials to which a published article would have previously referred as "available on request" will now be posted to the AgEconSearch database for easy online access by prospective readers. Thus the article to be published in the *AJAE* should cite a paper posted on AgEconSearch rather than simply telling readers that supplementary materials are available upon request. Please see the instructions on how to prepare the supplementary appendices.

## How to Submit electronically instructions

- The AJAE's New Electronic Editorial Management System (1.5 MB) (updated 8/3/04) A "how to" presentation on how to submit a publication electronically.

- Once your manuscript is formatted correctly (see Manuscript Formatting), you are ready to submit your manuscript electronically for review.

- Please make sure you review the following instructions/notes for electronic submissions:

- **Authors' identification and title page.** To protect their anonymity in the review process AUTHORS SHOULD NOT IDENTIFY THEMSELVES ON THE TITLE PAGE OR IN ANY HEADERS. A *SEPARATE* TITLE

PAGE MUST BE SENT AS AN ATTACHMENT TO THE EDITORS via the online submission management system and should include: (a) title; (b) author(s) names; (c) name, address, phone and fax numbers, and email address of the author serving as the contact person; (d) date of submission of the manuscript.

- *Instructions :* How to Convert Your Manuscript to PDF Format. In order for your manuscript to be handled correctly by the electronic submission system, it must be in PDF format AND it must conform to AJAE instructions for creating PDFs.

- *Electronic Submission Information :* This form allows you to upload a manuscript, cover letter, or other files or supporting documentation for your submission to American Journal of Agricultural Economics. This information will be transmitted *securely* and *anonymity will be maintained* as your submission is transmitted to referees. Moreover, Adobe(R) Acrobat(R) PDF files will be automatically "cleansed" so that no identifying features will appear in the internal document metadata. Thus submissions will be as anonymous as if sent by hard copy. However you are responsible for making sure that the names of authors or any other identifying information is purged from your PDF file prior to submission.

- Submission and verification proceeds in 5 easy steps. It will take only a few minutes of your time.

  Web site: Electronic Submission Form and Author Account Management

- *IMPORTANT NOTE :* Please make sure you review your uploaded file(s) to ensure that all maths, figures, tables, etc.,. come through cleanly before approving the final upload.

# CHAPTER - 21

# Annals of Plant Physiology

## GUIDELINES TO AUTHORS FOR PREPARATON OF MANUSCRIPT

The Annals of Plant Physiology is abstracted in all abstracting journals of CAB International U.K. Send Manuscript in duplicate to the Editor-in-Chief APP, Saint Tukaram Hospital Chowk, 01, Kamala Nagar, Akola-444 004 (M.A) India.

1. Manuscript should be typed with 1.5 spacing on an elite typewriter on good quality bond paper measuring $23 \times 29$ cm. Margin should be 3 cm. on left side and 2 cm. on other sides. MS must be dark and clear. Use space judiciously.

2. The title should be typed 5 cm below the top of the paper. Leave two single spaces between the title and the name of the authors. Leave two single spaces till you reach ABSTRACT. Foot note should cover Designation and present place of work, if different.

3. The manuscript should be forwarded in duplicate. First copy should be sharp, neat, clean and clear and should be made on good black ribbon on computer. Final copy should show no trace of erasures and creases. Forward the MS alongwith self addressed envelop ($27 \times 12$ cm) with sufficient postage affixed to return back for revision. Typing should be absolutely without errors. MS without this may not be considered. Xerox copies are unacceptable. Send computerized disc (CD, floppy 1.44 MB) along with manuscript.

4. Typed material should not exceed more than 6 to 7 pages including illustrations and tables and short communication of 3 pages only.

5. Format of the paper should be in following order:

   Title (ALL CAPITALS), Name of authors (usual like S.B.Lall) Name of Institute, ABSTRACT (Single Space), Key words, Designation and Current Addresses of the authors, (Foot note on first page), INTRODUCTION, MATERIALS AND METHODS, RESUTLS AND DISCUSSION, ACKNOWLEDGEMENT and REFERENCES. All captions should be in the centre and prominent.

6. Reference should be indicated by subscript in the text and should be in alphabetical order

Style of the references should be Names, year, title, Name of the journal (abbreviated in international standard), volume and page numbers. For books mention Edition, year and Publisher.

7.  All pages should be measured lightly with blue pencil only on lower rightmost corner of   the paper.  Manuscript may be returned if it does not satisfy the requirements. Please avoid unnecessary correspondence.

8.  Tables should be typed in single space as part of the text and should be inserted in the text close to the point of reference. Use full width of the paper for caption and matter.

9.  Drawings and figures should be prepared with black Indian ink excluding margin ($23 \times 15$ cm) and letters should be bold enough to permit reduction by 25 per cent of its original. Type captions below the figures. Photographs should be on glossy paper. Authors must pay for block charges and for additional pages if any.

10.  To avoid creasing, manuscript should be placed between heavy cardboards and securely bound before mailing. Stapling and punching should be avoided.

---

Reprints shall be supplied only on extra cost to the senior author. Put up your requirements to the Editor-in-Chief immediately.

# CHAPTER - 22

# Applied Soil Ecology

## GUIDE FOR AUTHORS

*Applied Soil Ecology* addresses the role of soil organisms and their interactions in relation to: agricultural productivity, nutrient cycling and other soil processes, the maintenance of soil structure and fertility, the impact of human activities and xenobiotics on soil ecosystems and bio(techno)logical control of soil-inhabiting pests, diseases and weeds. Such issues are the basis of sustainable agricultural and forestry systems and the long-term conservation of soils in both the temperate and tropical regions.

The disciplines covered include the following, and preference will be given to articles which are interdisciplinary and integrate two or more of these disciplines.

- soil microbiology and microbial ecology
- soil invertebrate zoology and ecology
- root and rhizosphere ecology
- soil science
- soil biotechnology
- ecotoxicology
- nematology
- entomology
- plant pathology
- agronomy and sustainable agriculture
- nutrient cycling
- ecosystem modelling and food webs

The journal publishes original papers, short communications, viewpoints, letters to the editor, editorials, book reviews and announcements. Review articles on critical and emerging topics are especially welcomed.

### Submission of manuscripts

*Please Note :* As of January 2006, submission to this journal proceeds totally on-line. Use the following guidelines to prepare your article. Author Gateway page of this journal, will guide you stepwise through the creation and uploading of the various files. Once the uploading is done, our system automatically generates an electronic (PDF) proof, which is then used for reviewing. It is crucial that all graphical elements be uploaded in separate files, so that the PDF is suitable for reviewing. Authors can upload their article as a LaTex, Microsoft (MS) Word, WordPerfect, PostScript or Adobe Acrobat PDF document. All correspondence, including notification of the Editor's decision and requests for revisions, will be by e-mail.

Submission of an article implies that the work described has not been published previously (except in the form of an abstract or as part of a published lecture or academic thesis), that it is not under consideration for publication elsewhere, that its publication is approved by all authors and tacitly or explicitly by the responsible authorities where the work was carried out, and that, if accepted, will not be published elsewhere in the same form, in English or in any other language, without the written consent of the Publisher.

Upon acceptance of an article, authors will be asked to transfer copyright (for more information on copyright see http://authors.elsevier.com). This transfer will ensure the widest possible dissemination of information. A letter will be sent to the corresponding author confirming receipt of the manuscript. A form facilitating transfer of copyright will be provided. If excerpts from other copyrighted works are included, the author(s) must obtain written permission from the copyright owners and credit the source(s) in the article. Elsevier has preprinted forms for use by authors in these cases: contact ELSEVIER, Rights Department, P.O. Box 800, Oxford, OX5 1DX, UK; phone: (+44) 1865 843830, fax: (+44) 1865 853333, e-mail: permissions@elsevier.com

## Types of contribution

1. Original research papers (Regular Papers)
2. Review articles
3. Short Communications
4. Viewpoints
5. Letters to the Editor
6. Editorials
7. Book Reviews
8. Announcements

*Original research papers* should report the results of original research. The material should not have been previously published elsewhere, except in a preliminary form.

*Review articles* should cover a subject of active current interest. They may be submitted or invited.

A *Short Communication* is a concise, but complete, description of a limited investigation, which will not be included in a later paper. Short Communications should be as completely documented, both by reference to the literature and description of the experimental procedures employed, as a regular paper. They should not occupy more than 6 printed pages (about 12 manuscript pages, including figures, etc.).

The section *Viewpoints* offers comment or useful critique on material published in the journal or on soil ecological issues. Contributions to this section should not occupy more than 2 printed pages (about 4 manuscript pages).

*Books for review* may be sent to Professor J.P. Curry

Authors wishing to submit a *Letter to the Editor* or an *Editorial* should contact one of the Editors-in-Chief to discuss this.

## Enquiries

Authors can also keep a track on the progress of their accepted article, and set up e-mail alerts informing them of changes to their manuscript's status, by using the "Track a Paper" feature at http://www.elsevier.com. For privacy, information on each article is password-protected. The author should key in the "Our Reference" code (which is in the letter of acknowledgement sent by the publisher on receipt of the accepted article) and the name of the corresponding author. In case of problems or questions, authors may contact the Author Service Department, E-mail: authorsupport@elsevier.com.

Electronic Format Requirements for Accepted Articles **Electronic manuscripts have the advantage that there is no need for rekeying of text, thereby avoiding the possibility of introducing errors and resulting in reliable and fast delivery of proofs.**

For the initial submission of manuscripts for consideration, hardcopies are sufficient. Elsevier is now publishing all manuscripts using electronic production methods, and therefore *needs to receive the electronic files of the accepted version of your article.*

### General points

We accept most wordprocessing formats, but Word, WordPerfect or LaTeX is preferred. An electronic version of the text should be submitted together with the final hardcopy of the manuscript. The electronic version must match the hardcopy exactly. Always keep a backup copy of the electronic file for reference and safety. Label storage media with your name, journal title, and software used. Save your files using the default extension of the program used. No changes to the accepted version are permissible without the explicit approval of the Editor. Electronic files can be stored on diskette, ZIP-disk or CD (either MS-DOS or Macintosh).

***Wordprocessor documents***

It is important that the file be saved in the native format of the wordprocessor used. The text should be in single-column format. Keep the layout of the text as simple as possible. Most formatting codes will be removed and replaced on processing the article. In particular, do not use the wordprocessor's options to justify text or to hyphenate words. However, do use bold face, italics, subscripts, superscripts etc. Do not embed 'graphically designed' equations or tables, but prepare these using the wordprocessor's facility. When preparing tables, if you are using a table grid, use only one grid for each individual table and not a grid for each row. If no grid is used, use tabs, not spaces, to align columns. The electronic text should be prepared in a way very similar to that of conventional manuscripts (see also the Quickguide: http://www.elsevier.com). Do not import the figures into the text file but, instead, indicate their approximate locations directly in the electronic text and on the manuscript. See also the section on Preparation of electronic illustrations. To avoid unnecessary errors you are strongly advised to use the 'spellchecker' function of your wordprocessor.

Although Elsevier can process most wordprocessor file formats, should your electronic file prove to be unusable, the article will be typeset from the hardcopy printout.

## Preparation of manuscripts

1.  Manuscripts should be written in English. Authors whose native language is not English are strongly advised to have their manuscripts checked by an English-speaking colleague prior to submission.

    English language help service

    Upon request, Elsevier will direct authors to an agent who can check and improve the English of their paper (*before submission*). Please contact authorsupport@elsevier.com for further information.

2.  Submit the original and two copies of your manuscript. Enclose the original illustrations and two sets of photocopies (three prints of any photographs).The manuscript must be accompanied by a covering letter detailing what you are submitting (type of contribution, title, authors' names and affiliation, etc.). Please also indicate the author to whom we should address our correspondence in the case of multiple authors and include a contact address, telephone/fax numbers and e-mail address. Authors are requested to submit, with their manuscripts, the names and addresses of two potential referees.

3.  Manuscripts should be typewritten, typed on one side of the paper (with numbered lines), with wide margins and double spacing throughout, i.e. also for abstracts, footnotes and references. Every page of the manuscript, including the title page, references, tables, etc. should be numbered. However, in the text no reference should be made to page numbers; if

necessary, one may refer to sections. Underline words that should be in italics, and do not underline any other words. Avoid excessive usage of italics to emphasize part of the text.

4.  Manuscripts in general should be organized in the following order:

Title (should be clear, descriptive and not too long)

Name(s) of author(s)

Complete postal address(es) of affiliations

Full telephone, E-mail and Fax No. of the corresponding author

Present address(es) of author(s) if applicable

Complete correspondence address to which the proofs should be sent

Abstract

Keywords (indexing terms), normally 3-6 items

Introduction

Material studied, area descriptions, methods, techniques

Results

Discussion

Conclusion

Acknowledgements and any additional information concerning research grants, etc.

References

Tables

Figure captions

5.  In typing the manuscript, titles and subtitles should not be run within the text. They should be typed on a separate line, without indentation. Use lower-case lettertype.

6.  SI units should be used.

7.  If a special instruction to the copy editor or typesetter is written on the copy it should be encircled. The typesetter will then know that the enclosed matter is not to be set in type. When a typewritten character may have more than one meaning (e.g. the lower case letter l may be confused with the numeral 1), a note should be inserted in a circle in the margin to make the meaning clear to the typesetter. If Greek letters or uncommon symbols are used in the manuscript, they should be written very clearly, and if necessary a note such as "Greek lower-case chi" should be put in the margin and encircled.

8.  Elsevier reserves the privilege of returning to the author for revision accepted manuscripts and illustrations which are not in the proper form given in this guide.

## Abstracts

The abstract should be clear, descriptive and not longer than 400 words.

## Tables

1. Authors should take notice of the limitations set by the size and lay-out of the journal. Large tables should be avoided. Reversing columns and rows will often reduce the dimensions of a table.

2. If many data are to be presented, an attempt should be made to divide them over two or more tables.

3. Drawn tables, from which prints need to be made, should not be folded.

4. Tables should be numbered according to their sequence in the text. The text should include references to all tables.

5. Each table should be typewritten on a separate page of the manuscript. Tables should never be included in the text.

6. Each table should have a brief and self-explanatory title.

7. Column headings should be brief, but sufficiently explanatory. Standard abbreviations of units of measurement should be added between parentheses.

8. Vertical lines should not be used to separate columns. Leave some extra space between the columns instead.

9. Any explanation essential to the understanding of the table should be given as a footnote at the bottom of the table.

## Illustrations

1. All illustrations (line drawings and photographs) should be submitted separately, unmounted and not folded.

2. Illustrations should be numbered according to their sequence in the text. References should be made in the text to each illustration.

3. Each illustration should be identified on the reverse side (or - in the case of line drawings - on the lower front side) by its number and the name of the author. An indication on the top of the illustrations is required in photographs of profiles, thin sections, and other cases where doubt can arise.

4. Illustrations should be designed with the format of the page of the journal in mind. Illustrations should be of such a size as to allow a reduction of50%.

5. Lettering should be in Indian ink or by printed labels. Make sure that the size of the lettering is big enough to allow a reduction of 50% without becoming illegible. The lettering should be in English. Use the same kind of lettering throughout and follow the style of the journal.

6. If a scale is given, use bar scales on all illustrations instead of numerical scales that must be changed with reduction.

7. Each illustration should have a caption. The captions to all illustrations should be typed on a separate sheet of the manuscript.

8. Explanations should be given in the typewritten legend. Drawn text in the illustrations should be kept to a minimum.

9. Photographs are only acceptable if they have good contrast and intensity. Sharp and glossy copies are required. Reproductions of photographs already printed cannot be accepted.

### *Preparation of electronic illustrations*

Submitting your artwork in an electronic format helps us to produce your work to the best possible standards, ensuring accuracy, clarity and a high level of detail.

- Always supply high-quality printouts of your artwork, in case conversion of the electronic artwork is problematic.

- Make sure you use uniform lettering and sizing of your original artwork.

- Save text in illustrations as "graphics" or enclose the font.

- Only use the following fonts in your illustrations: Arial, Courier, Helvetica, Times, Symbol.

- Number the illustrations according to their sequence in the text.

- Use a logical naming convention for your artwork files, and supply a separate listing of the files and the software used.

- Provide all illustrations as separate files and as hardcopy printouts on separate sheets.

- Provide captions to illustrations separately.

- Produce images near to the desired size of the printed version.

Files can be stored on diskette, ZIP-disk or CD (either MS-DOS or acintosh).

A detailed guide on electronic artwork is available on our website: http://authors.elsevier.com/artwork

**You are urged to visit this site; some excerpts from the detailed information are given here.**

### *Formats*

Regardless of the application used, when your electronic artwork is finalised, please "save as" or convert the images to one of the following formats (Note the resolution requirements for line drawings, halftones, and line/halftone combinations given below.):

*EPS :* Vector drawings. Embed the font or save the text as "graphics".

*TIFF :* Colour or greyscale photographs (halftones): always use a minimum of 300 dpi.

*TIFF :* Bitmapped line drawings: use a minimum of 1000 dpi.

*IFF :* Combinations bitmapped line/half-tone (colour or greyscale): a minimum of 500 dpi is required.

DOC, XLS or PPT: If your electronic artwork is created in any of these Microsoft Office applications please supply "as is".

**Please do not :**

- Supply embedded graphics in your wordprocessor (spreadsheet, presentation) document;
- Supply files that are optimised for screen use (like GIF, BMP, PICT, WPG); the resolution is too low;
- Supply files that are too low in resolution;
- Submit graphics that are disproportionately large for the content.

### *Colour Reproduction*

Submit colour illustrations as original photographs, high-quality computer prints or transparencies, close to the size expected in publication, or as 35 mm slides. Polaroid colour prints are not suitable. **If, together with your accepted article, you submit usable colour figures then Elsevier will ensure, at no additional charge, that these figures will appear in colour on the web (e.g., ScienceDirect and other sites) regardless of whether or not these illustrations are reproduced in colour in the printed version.** For colour reproduction in print, you will receive information regarding the costs from Elsevier after receipt of your accepted article. For further information on the preparation of electronic artwork, please see http://authors.elsevier.com/artwork.

*Please note* **:** Because of technical complications which can arise by converting colour figures to 'grey scale' (for the printed version should you not opt for colour in print) please submit in addition usable black and white prints corresponding to all the colour illustrations.

## References

1. All publications cited in the text should be presented in a list of references following the text of the manuscript. The manuscript should be carefully checked to ensure that the spelling of author's names and dates are exactly the same in the text as in the reference list.

2. In the text refer to the author's name (without initial) and year of publication, followed - if necessary - by a short reference to appropriate pages. Examples: "Since Peterson (1988) has shown that..." "This is in agreement with results obtained later (Kramer, 1989, pp. 12-16)".

3. If reference is made in the text to a publication written by more than two authors, the name of the first author should be used followed by "et al.". This indication, however, should never be used in the list of references. In this list, names of first author and co-authors should be mentioned.

4.  References cited together in the text should be arranged chronologically. The list of references should be arranged alphabetically on authors' names, and chronologically per author. If an author's name in the list is also mentioned with co-authors the following order should be used: publications of the single author, arranged according to publication dates - publications of the same author with one co-author - publications of the author with more than one co-author. Publications by the same author(s) in the same year should be listed as 1974a, 1974b, etc.

5.  Use the following system for arranging your references:

    (a)  *For periodicals*

         Tietema, A., Riemer, L., Verstraten, J.M., van der Maas, M.P., van Wijk, A.J., van Voorthuyzen, I., 1992. Nitrogen cycling in acid forest soils subject to increased atmospheric nitrogen input. For. Ecol. Manage. 57, 29-44.

    (b)  *For edited symposia, special issues, etc. published in a periodical*

         Rice, K., 1992. Theory and conceptual issues. In: Gall, G.A.E., Staton, M. (Eds.), Integrating Conservation Biology and Agricultural Production. Agriculture, Ecosystems and Environment 42, 9-26.

    (c)  *For books*

         Gaugh, Jr., H.G., 1992. Statistical Analysis of Regional Yield Trials. Elsevier, Amsterdam, 278 pp.

    (d)  *For multi-author books*

         Baker, Jr., 1993. Insects. In: de Hertogh, A., Le Nard, M. (Eds.), The Physiology of Flower Bulbs. Elsevier, Amsterdam, pp. 101-153.

6.  Abbreviate the titles of periodicals mentioned in the list of references; according to the International *List of Periodical Title Word Abbreviations*.

7.  In the case of publications in any language other than English, the original title is to be retained. However, the titles of publications in non-Latin alphabets should be transliterated, and a notation such as "(in Russian)" or "(in Greek, with English abstract)" should be added.

8.  Work accepted for publication but not yet published should be referred to as "in press".

9.  References concerning unpublished data and "personal communications" should not be cited in the reference list but may be mentioned in the text.

## Formulae

1.  Formulae should be typewritten, if possible. Leave ample space around the formulae.

2.  Subscripts and superscripts should be clear.

3.  Greek letters and other non-Latin or handwritten symbols should be explained in the margin where they are first used. Take special care to

show clearly the difference between zero (0) and the letter O, and between one (1) and the letter l.

4. Give the meaning of all symbols immediately after the equation in which they are first used.

5. For simple fractions use the solidus (/) instead of a horizontal line.

6. Equations should be numbered serially at the right-hand side in parentheses. In general only equations explicitly referred to in the text need be numbered.

7. The use of fractional powers instead of root signs is recommended. Also powers of e are often more conveniently denoted by exp.

8. Levels of statistical significance which can be mentioned without further explanation are * $P<0.05$, ** $P<0.01$ and *** $P<0.001$.

9. In chemical formulae, valence of ions should be given as, e.g. $Ca^{2+}$ and not as $Ca^{++}$.

10. Isotope numbers should precede the symbols, e.g., $^{18}O$.

11. The repeated writing of chemical formulae in the text is to be avoided where reasonably possible; instead, the name of the compound should be given in full. Exceptions may be made in the case of a very long name occurring very frequently or in the case of a compound being described as the end product of a gravimetric determination (e.g., phosphate as $P_2O_5$).

## Footnotes

1. Footnotes should only be used if absolutely essential. In most cases it should be possible to incorporate the information in normal text.

2. If used, they should be numbered in the text, indicated by superscript numbers, and kept as short as possible.

## Nomenclature

1. Authors and editors are, by general agreement, obliged to accept the rules governing biological nomenclature, as laid down in the *International Code of Botanical Nomenclature*, the *International Code of Nomenclature of Bacteria*, and the *International Code of Zoological Nomenclature*.

2. All biotica (crops, plants, insects, birds, mammals, etc.) should be identified by their scientific names when the English term is first used, with the exception of common domestic animals.

3. All biocides and other organic compounds must be identified by their Geneva names when first used in the text. Active ingredients of all formulations should be likewise identified.

4. For chemical nomenclature, the conventions of the *International Union of Pure and Applied Chemistry* and the official recommendations of the *IUPAC-IUB Combined Commission on Biochemical Nomenclature* should be followed.

*Supplementary data*

Elsevier now accepts electronic supplementary material to support and enhance your scientific research. Supplementary files offer the author additional possibilities to publish supporting applications, movies, animation sequences, high-resolution images, background datasets, sound clips and more. Supplementary files supplied will be published online alongside the electronic version of your article in Elsevier web products, including ScienceDirect: http://www.sciencedirect.com. In order to ensure that your submitted material is directly usable, please ensure that data is provided in one of our recommended file formats. Authors should submit the material in electronic format together with the article and supply a concise and descriptive caption for each file. For more detailed instructions please visit http://www.elsevier.com.

## Proofs

When your manuscript is received at the Publisher it is considered to be in its final form. Proofs are not to be regarded as 'drafts'.

One set of page proofs in PDF format will be sent by e-mail to the corresponding author, to be checked for typesetting/editing. No changes in, or additions to, the accepted (and subsequently edited) manuscript will be allowed at this stage. Proofreading is solely your responsibility.

A form with queries from the copy editor may accompany your proofs. Please answer all queries and make any corrections or additions required.

The Publisher reserves the right to proceed with publication if corrections are not communicated. Return corrections within two working days of receipt of the proofs. Should there be no corrections, please confirm this.

Elsevier will do everything possible to get your article corrected and published as quickly and accurately as possible. In order to do this we need your help. When you receive the (PDF) proof of your article for correction, it is important to ensure that all of your corrections are sent back to us in one communication. Subsequent corrections will not be possible, so please ensure your first sending is complete. Note that this does not mean you have any less time to make your corrections, just that only one set of corrections will be accepted.

## Offprints

1.  The corresponding author, at no cost, will be provided with a PDF file of the article via e-mail or, alternatively, 25 free offprints of any paper supplied. 100 free offprints will be given to authors of a Review article. The PDF file is a watermarked version of the published article and includes a cover sheet with the journal cover image and a disclaimer outlining the terms and conditions of use.

*Applied Soil Ecology* **carries no page charges**

# CHAPTER - 23

# Arid Land Research and Management

## INSTRUCTIONS FOR AUTHORS

It is preferred that initial manuscripts are submitted electronically as an attachment to the Editor-in-Chief, Tibor Tóth, at **tibor@rissac.hu**. Please make sure that figures are included as separate files and not embedded within the text. The Editor will also accept manuscripts by regular mail at the following address: Research Institute for Soil Science and Agricultural Chemistry, Hungarian Academy of Sciences, (MTA TAKI-RISSAC), P.O. Box 35, Budapest, 1525, Hungary.

Finalized manuscripts should be submitted on a 9 cm (3½") disk, prepared in MS Word, and accompanied with a hard copy, including hard copies of figures.

Each manuscript must be accompanied by a statement that it has not been published elsewhere and that it has not been submitted simultaneously for publication elsewhere. Authors are responsible for obtaining permission to reproduce copyrighted material from other sources and are required to sign an agreement for the transfer of copyright to the publisher. All accepted manuscripts, artwork, and photographs become the property of the publisher.

### Further Instructions

Authors are strongly advised to consult **"Notes for Preparation of Successful Manuscripts: Instructions to Authors Elaborated"** and **"Short Guide to the Use of SI Units"**, both published in this journal, volume 15, issue 3, pages 275-284 (2001) before submitting manuscripts to the editorial office for consideration. The Editorial Board and the Publisher reserve the right to edit all manuscripts linguistically and stylistically.

### Manuscript

The manuscript should be typewritten, double-spaced, with margins of 2.5 cm. The preferred order of presentation is as follows: Title page (title, names of authors with their affiliations, acknowledgments, complete address of the corresponding author including e-mail address and/or a fax number, and a running title not exceeding 50 character spaces), Abstract, Keywords, Introduction, Materials and Methods, Results, Discussion, Conclusions, References, tables, figure legends, and copies of figures (illustrations). Do not underline anything (nor type in italics) unless the material is to be printed in italics, such as scientific binomial names of organisms and publication titles in References. Do not type any titles, headings or captions in capital letters. All headings for sections and subsections should be typed in lowercase letters and placed on separate lines.

The total number of double-spaced pages submitted, including the Abstract, References and tables should not exceed 18. Publication of longer manuscripts will be at the discretion of Editorial Board.

## Title

Title must be consistent with clarity and not exceed 15 words. Uninformative initial phrases such as "Effect of ..., Influence of ..., Examination of..., Studies on..." and similar wording must be avoided.

## Abstract

Abstract should not exceed 250 words. An abstract should provide the objective of research, followed by a few sentences describing methods. The main body should describe results and provide important numerical values. It ends with a brief conclusion statement. The Abstract is typed as a single paragraph and there are no references, tables nor abbreviations included.

## Keywords

Keywords are placed following the abstract. Authors must provide from three to ten alphabetized keywords or two-word key phrases that are not already listed in the title.

## Introduction

Introduction should be short and should describe matters directly pertinent to the subject of the paper. Avoid statements and discussion of generalized knowledge or history. Include a statement specifying why this research was done; include hypotheses addressed.

## Materials and Methods

All methods and techniques used must be noted, described and referenced in detail. List sources of organisms used. Describe the experimental design. Soil classification in FAO or USDA taxonomic nomenclature must be provided (include FAO or USDA reference). For field experiments the geographic location (longitude, latitude), elevation and pertinent climatic characteristics must be provided. Description of statistical methods used is required.

## Results

Write methods and results in the past tense. The mean values reported should be rounded off to three significant digits. Statistical evaluation of results must be included and documented. Indicate the number of replicates and sample sizes ("n" numbers). It is strongly suggested to describe the interpretation of your findings in a separate Discussion section and not to present a combined "Results and Discussion" section. The text must be written in a scientifically rigorous manner (just like the rest of the paper). Avoid subjective adjectives, adverbs and anthropocentric expressions.

## Conclusions

Describe your conclusions that the interpretation of your experimental results have indicated. This section is not a "summary" of the paper.

## References

Literature citations in the text should be by the author's surname and year. When references are by three or more authors list the first author's surname and "et al.": "Smith (1992) has shown... Previous work (Smith & Jones, 1995; Jackson, 1996a, b) has demonstrated... Jackson et al. (2000) have Shown..." The listing of surnames in References should be given in an alphabetical order with all authors and editors listed. Spell out everything in full. Full titles of publications and articles, and full pagination should be given. Follow the examples, note punctuation and placement of initials:

*Journal article :* Hamilton, M. A., D. T. Westermann, and D. W. James. 1993. Factors affecting zinc uptake in cropping systems. Soil Science Society of America Journal 57:1310-1315.

*Book :* Tate III, R. L., and D. A. Klein. 1985. Soil reclamation processes. Marcel Dekker. New York, USA.

*Chapter from a book and from proceedings of a meeting :* Bremner, J. M., and C. S. Mulvaney. 1982. Nitrogen - total, pp. 595 - 624, in A. L. Page, R. H. Miller, and D. R. Keeney, eds., Methods of soil analysis, part 2, 2nd edition. American Society of Agronomy, Madison, Wisconsin, USA. [Always cite individual chapters from an edited book, not just solely the book.]

## Report

Parker, J. C., and M. Th. van Genuchten. 1984. Determining transport parameters from laboratory and field tracer experiments. Virginia Agricultural Experiment Station Bulletin 84-3. Virginia Agricultural Experiment Station, Blacksburg, Virginia, USA. Supporting Online Material may be listed at the end of the alphabetized References.

## Tables

Tables should not be embedded in the text but each table should be included on a separate page (or files) following the Reference section. A descriptive title should appear above each table. All abbreviations should be described in the footnotes, indicated by superscript letters, beginning with superscript a in each table. The use of explanatory footnotes is encouraged. Tables and figures shall be understood independently without reference to the text.

Do not use vertical lines to separate columns. Delete all computer-generated horizontal lines. Horizontal lines are used to separate table headings only. Tables must be double-spaced.

## Figures (illustrations)

Number of figures should be limited to a minimum needed to clarify the text. Double documentation in tables and figures is not acceptable. Micrographs should have an internal magnification marker. Figures submitted with finalized manuscripts should be clean originals or digital files. Digital files are recommended for the highest quality reproduction and should follow these guidelines:

- 300 dpi or higher.
- sized to fit on a journal page.
- EPS, TIFF, or PSD format only.
- submitted as separate files, not imbedded in text files.

Color illustrations will be considered for publication; however, the author will be required to bear the full cost involved in their printing and publication. The charge for the first page with color is $900.00. The next three pages with color are $450.00 each. A custom quote will be provided for color art totaling more than 4 journal pages. Good-quality color prints or files should be provided in their final size. The publisher has the right to refuse publication of color prints deemed unacceptable.

## Legends to Figures (illustrations)

Each illustration should be provided with a legend sufficiently clear so that the meaning of data is understandable without reference to the text. A legend of symbols should be included in a graph. Legends should be typed, double-spaced on a separate page, placed after the tables.

## Units of Measurement and Abbreviations

Use of International System of Units (SI, Système international d'unités) is required. In the text, tables and figures use the minus index instead of a slash and conform to the SI abbreviations (see "The Short Guide to the Use of SI Units", this journal, vol. 15, issue 3, pp. 281-284, 2001). Use chemical designations instead of spelling out chemical names, e.g. $CO_2$ release, C and N cycling. Define each uncommon abbreviation the first time it is used, e.g. "modified universal buffer (MUB) was used."

The conventional designation of a binomial scientific name for an organism must be noted the first time it is introduced in the main text (except in the Title and Abstract). Fertilized values should be expressed in terms of elemental P, K, etc., not as oxides.

## Specific Suggestions

Type everything double-spaced, including title page, references, tables, titles for tables and legends to figures, footnotes. Always place zeros before decimal fractions: write 0.05, not .05. Do not use dashes in the tables: distinguish and specify ND (not determined), NA (not applicable), or 0 (not found). Do not use "levels" when you mean concentrations ("wt. per wt.") or amounts (total, or per area, per volume). Avoid the use of "rates" for concentrations and amounts used. Never use "ppm". Express concentrations in SI units. Use SI designations for units, e.g., Mg (not ton), mL (not ml), MPa (not bars), J (not cal.), kat (not U), cmolc kg-1 (not meq./100 g), etc. Do not use % values where the concentrations may be expressed in SI units, e.g., C, N, and water presence in soils. When using a word processor for typing the manuscript, do not use the "right margin justification."

## Notes

Notes, not exceeding 1500 words, are intended for the presentation of research reports that do not warrant full-length papers. Notes are peer-reviewed and are not considered as preliminary communications. A Note has an Abstract of not more than 100 words and Keywords. Section headings are not used in the body of Note. The number of tables and figures should not exceed three. The Reference section is similar to a full-length paper.

## Page Charges

Part of the publication cost is covered by a page charge of $ 40 per printed page. If these charges are paid, the corresponding author will receive 50 offprints of the article. The acceptance and publication of articles are not dependent to the payment of these charges.

# CHAPTER - 24

# Artha Vijnan

## NOTES TO CONTRIBUTORS

1. Two copies of the paper should be sent to the Managing Editor. The paper should be typed in double space on A4 size bond paper and should be accompanied by a statement giving the approximate number of words, which it contains. The paper should not preferably exceed a maximum of 7500 words.

2. When you are informed about the acceptance of your paper for publication, it will be helpful for expeditiously processing the paper further if you send to us a copy of the paper on a floppy, preferably in MS WORD 6.0 or WORD PERFECT 5.1. Please use a disk-mailer for sending the floppy.

3. The manuscript should include the title of the paper at the top of the first page followed by the name(s) of the author(s). The institutional affiliation(s) of the author(s) should be mentioned at the bottom of the first page.

4. The manuscript should be accompanied by an abstract of not more than 100 words.

5. Acknowledgements and related matters may be given before the references.

6. Footnotes should be numbered consecutively and should appear at the bottom of corresponding pages.

7. References should come after acknowledgements or at the end as the case may be.

8. The mathematical formulae and equations should be clearly written and numbered consecutively in the manuscript.

9. Graphs, charts, etc., should be neatly drawn and presented; preferably art work or laser print of the same should be sent.

10. Any manuscript which does not conform to the above instructions may be returned for necessary revisions before publication.

11. Contributors will receive one copy of the issue of Artha Vijnana in which the article appears and twenty off-prints of the paper.

## Notes to Publishers

For favour of review the publishers should send two copies of the book/report/ monograph two off-prints of the review will be sent to the publishers.

# CHAPTER - 25

# Asian Journal of Microbiology, Biotechnology and Environmntal Sciences

## INSTRUCTIONS FOR SUBMITTING A MANUSCRIPT

Please submit 2 copies (1+1) of your manuscript. The matter should by typed in double space on one side of the white paper. The authors can arrange their manuscripts in suitable headings. Short Communications can be submitted in running.

Figures should be drawn on tracing paper or white art sheet using black Indian ink. Only good Xerox copies of the figures can be submitted. The figures must be capable of taking a reduction by at least 4 times. All legends should be typed on separate sheet.

All references should be referred in the text by name and not by numbers. The references should be arranged in the end alphabetically as per authors surname in the following pattern. (Kindly mark the style for papers in journals, books and reports etc.)

1. Oirubi, C.C., b.Cowey and J.Y. Chiou. 1978. Leaf protein concentrateas a protein sournce in    diet for carp and rainbow trouts. Bull. Jap. Soc. Sc. Fish. 44 : 49-52.

2. Lyengar, L. 1998.  Biotechnology for wastewater treatment. P. 167-173 In: R.K. Trivedy (ed.) *Advances in Wastewater Treatment  Technologies* Vol. 1. Global Science Publications, Aligarh. India.

3. Macfadyen. A. 1970. Soil metabolism in relation to ecosystem energy flow and primary and secondary production In : Philipson J. (Ed.) *Methods of study in soil ecology.* IBP/UNESCO, Paris.

Authors will receive acknowledgment within 15 days and acceptence within 60 days (if accepted). Reprints shall reach authors within 4-5 months. All manuscripts and other editorial correspondence should be directed to : **Dr. SADHANA SHARMA, EXECUTIVE Editor, AJMBES C/O GLOBAL SCIENCE PUBLICATIONS 23, Maharshi Dayanand Nagar, Surendra Nagar, ALIGARH 202 001 (U.P). Ph : (0571) 404271.**

# CHAPTER - 26

# Australian Journal of Plant Physiology

## NOTICE TO AUTHORS

### Assignment of Copyright

When submitting your manuscript, please enclose a completed Assignment Form. These are also available from the Editorial Office or the form printed in the journal may be photocopied.

### Journal Policy and Notes

We consider papers in any of the main areas of plant physiology, including biochemistry, biophysics, cell biology, genetics, plant-environment interactions, plant-microbe interactions and molecular biology.

Types of papers published include research papers, either experimental or theoretical, critical review articles, viewpoints, research notes, hypotheses and comments. **Review articles** should indicate fruitful areas of further research. **Viewpoints** and **Hypotheses** provide a means for authors to explore contemporary issues with more freedom than in the conventional research paper. **Research notes** will be short papers typically not greater than four published pages, reporting experimental data that may not justify a full paper. For example, a note may be (i) the reporting of a DNA sequence and/or full amino acid sequence of a protein, with data pertaining to its importance to plant function; or (ii) the reporting of a technical advance that provides a new approach for the analysis of plant function. **Comments** on published papers should be confined to the substance of the paper; the authors of the paper referred to will be offered the opportunity to respond.

If you are interested in preparing a review article or viewpoint, please discuss the matter with an appropriate member of the Advisory Committee or with the Managing Editor.

All papers are peer-reviewed. Submission of a paper is taken to mean that the results reported have not been published and are not being considered for publication elsewhere. The journal assumes that all authors of a multi-authored paper agree to its submission. The journal will use its best endeavors to ensure that work published is that of the named authors except where acknowledged and, through its reviewing procedures, that any published results and conclusions are consistent with the primary data. The journal takes no responsibility for fraud or inaccuracy on the part of its contributors.

In the interests of rapid reviewing, we encourage referees to send in their reports by fax or email; we may, therefore, be unable to return all the original submitted manuscripts in the event of the paper being rejected. Original illustrations will be returned after publication if this is requested at the time of submission.

There are no page charges. Normally, no free reprints are provided; an order form for reprints will be supplied with the page proofs.

## Guidelines for the Preparation of Manuscripts

### General Presentation

Present the work concisely and clearly in English. The Introduction should not exceed what is necessary to indicate the reason for the work and its essential background. Sufficient experimental detail should be given to enable the work to be repeated. The Discussion should explain the significance of the results.

Supplementary material may be lodged as an Accessory Publication with the Managing Editor, provided that it is submitted with the manuscript for inspection by the referees. Such material will be made available on request and a note to this effect should be included in the paper. Where practicable and appropriate Accessory publications will also be made available via the **AJPP** web site.

Much useful advice on preparing manuscripts will be found in F.B. Salisbury (1996) 'Units, symbols and terminology for plant physiology'. (Oxford University Press: New York) [ISBN 0 19 509445 X]

### Manuscripts

Submit three clearly legible copies. Type them with double-spacing throughout, making the left-hand margin at least 3 cm wide, on good quality paper. Please use only Times New Roman type 1 font.

Word for Windows is preferred but most packages (e.g. other versions of Microsoft Word or WordPerfect) are acceptable. If you have none of these, please send an RTF (Rich Text Format) file. We can usually receive files by email although they sometimes are corrupted. Please send figures as separate files.

Place tables, figures and captions to figures after the text, and number all pages of the manuscript consecutively. Please ensure that figure captions appear on the same page as the figure. Number all illustrations in a common sequence. Refer to each figure and table in the text. Submit one complete set of lettered and mounted photographs line-work with each copy of the manuscript. Take particular care that photographs accompanying each copy of the manuscript show all relevant features.

When the manuscript has been reviewed and subsequently revised, the author will be requested to submit a computer disk bearing the final version of the manuscript, as well as a hard copy of the same manuscript.

Please refer to recent issues of AJPP, or to the free samples on this site, to note details of headings, table, illustrations, style and layout. Observance of these and the following details will shorten the time between submission and publication. Poor preparation and unnecessary length are impediments to rapid publication.

When you submit a manuscript please include on the first page the telephone number, facsimile number and email address of the corresponding author, as well as your postal address; we may need to contact you urgently.

## Title

Make it concise and informative; it should contain the main keywords. Extra keywords may also be suggested. If you include a botanical name in the title, omit the authority, but mention it in the Abstract and in the 'Materials and methods'. Please also supply an abridged title, for use as a running head, that does not exceed 8 words in length.

If the paper is one of a numbered series give a footnote at the bottom of the first page, referring to the previous part. If a part not yet published needs to be consulted for a proper understanding of the submitted paper, then send us two copies of that manuscript for use by referees.

## Authors and Addresses

Please include first name, initial and surnames for all authors, and a current mailing address for each. Provide a fax number if the corresponding author does not have an email address.

## Abstract

State the scope of the work and the principal findings in fewer than 200 words. The abstract should be complete enough for use by abstracting services. Avoid acronyms and references. Scientific names of plants should be accompanied by their authority.

## References

In the text, cite references chronologically; do not number them. Make sure that all references in the text are listed at the end of the paper and vice versa. At the end of the paper, list references in alphabetical order.

Use 'and' to link the names of two coauthors in the text, and use 'et al.' where there are more than two. Take special care in checking the accuracy of the references; we take no editorial responsibility for them. Do not include papers that have not been accepted for publication; cite them either as 'unpublished data' or 'pers. Comm.' if they must be included at all. Authors must provide written proof of acceptance for any papers cited as 'in press'.

Give titles of books and names of journals in full. Include the title of the paper in all journal references, and provide first and last page numbers for all entries.

Refer to the latest issues of AJPP for the style used in citing references to books and other literature. Examples of common references can be found at Referencing Style.

## Use of Referencing Software

To obtain the style file for AJPP, please go to the following websites:

If using Reference Manager or ProCite software, visit *http:/www.isiresearchsoft.com*

If using EndNote software, visit *http:/www.Crandon.com.au.*

## Units

Use the SI system where appropriate, especially for exact measurement of physical quantities. However, non-SI units such as day and year are acceptable. Give measurements of radiation as irradiance or photon flux density, or both, and specify the waveband of the radiation. Photon flux density units should be used in papers concerned with the quantum efficiency of plant photo-processes. Do not use luminous flux density units (e.g. lux). Use the negative index system. For example g m$^{-2}$, kg ha$^{-1}$, m m$^{-2}$ s$^{-1}$.

## Statistical Evaluation of Results

Describe the experimental design and analysis in sufficient detail to allow them to be evaluated, and support this description with references if necessary. State the number of individuals; mean values and measures of variability. State clearly whether you have included the standard deviation or the standard error of the mean.

## Mathematical Formulae

Correctly align and adequately space all symbols. If you must hand-write symbols, insert them carefully and make identifying marks in the margin in pencil. Avoid two-line mathematical expressions in the running text. Display each long formula on a separate line with at least two lines of space above and below it.

## Enzyme Nomenclature

Names should conform to Recommendations (1992) of the Nomenclature Committee of the International Union of Biochemistry and Molecular Biology on the Nomenclature and Classification of Enzymes, as published in 'Enzyme Nomenclature 1992' (Academic Press: San Diego) or as available on the Enzyme Nomenclature Database. Where an enzyme is referred to only in the course of discussion, or is obtained from commercial sources and is used solely as a reagent, it is enough to use the recommended name without the identifying EC number. For an enzyme that is not the main subject of the paper, the recommended name should be used but the EC number should be given at the first mention in the body of the paper. Where an enzyme is the main subject of a paper, the recommended name should be used but the EC number, source, and either the systematic name or reaction equation given at its first mention in the text. If you wish to use a name other than the recommended name, at the first mention of the alternative name identify it by giving the recommended name and its EC number. Inform the Managing Editor of your reasons for wishing to use the alternative name.

## Sequences

DNA sequences and primers should be specified in the following manner, using the ' (prime) sign in preference to a '(single quotation mark):

5'-GGGATGAGGCACAATCCC-3'

## Chemical Nomenclature

Follow the recommendations of the IUPAC-IUB Commission on Biochemical Nomenclature when naming compounds such as amino acids, carbohydrates, lipids, steroids, vitamins, etc. Refer to other biologically active compounds, such as metabolic inhibitors, plant growth regulators, buffers, etc. (in accordance with IUPAC Chemical Nomenclature homepage), once and then by their most widely accepted common name. You may use trade names or letter abbreviations of the chemical where there is no common name.

Describe solutions of common acids and bases in terms of normality (N), e.g. I N KOH, and those of salts in terms of molarity (M), making sure you use the 'Small caps' font option for 'N' and 'M' Define% as (w/w), (w/v), or (v/v). Represent ions as follows: $Na^+$, $Mn^{3+}$, $Br^-$, $PO_4^{3-}$.

Isotopically labeled compounds should be indicated by writing the chemical formulae, for example: $^{14}CO_2$, $^2H_2O$. For other molecules, place the isotopic symbol in square brackets attached to the name or the formula without a hyphen or space, for example: $[^{14}C]$ glucose, $[^{32}P]$ATP.

### *Tables*

Refer to every table in the text. Number each with an Arabic numeral and supply a heading. If using Microsoft word, please useTable Formatting (i.e. use table cells). Otherwise please use tabs (not spaces or hard returns) when setting up columns.

Avoid long titles by incorporating an explanatory note in the headnote, which should be started on a new line from the title of the table. Include in the headnote, where applicable, the levels of probability attached to statistics in the body of the table, and any other information relevant to the table as a whole.

Avoid footnotes where possible; use them only to refer to specific data points in the body of the table. Use A and B etc. for footnotes; use *, **, *** only to define probability levels.

Insert horizontal rules above and below the column headings and across the bottom of the table; do not use vertical rules. The first letter only of headings of rows and columns should be capitalized. Include the symbols for the units of measurement in parentheses below the column heading. Use standard SI prefixes with units in the column headings to avoid an excessive number of digits in the body of the table.

### *Figures*

Refer to every figure or illustration in the text. Prepare line drawings, photographs and lettering according to the instructions in 'Submission of Electronic Files' below. Unsatisfactory figures will be returned for correction. The Managing Editor may be consulted for further guidance. Please prepare your illustrations electronically (both line diagrams and photographs).

## Submission of Electronic Files

### *Text*

Word for Windows (preferred) and most word-processing packages (e.g. Microsoft Word, Word Perfect) are acceptable. If you have none of these, please send a RTF (Rich Text Format) file, remembering that certain formatting (notably special characters, superscripts and subscripts) may drop out.

An author stylesheet together with instructions on its use are available from the journal web site at www.publish.csiro.au/journals/ajpp/nta.html.

## Figures

### *Line Diagrams*

When preparing your figures, labeling must be in Helvetica or Arial, Symbol and Zapf Dingbat (type 1) fonts and conform to journal style. The thickness of all lines on line diagrams *must* be no less than 1 pt. Where possible, avoid the use of hatching or program-generated fills in bar graphs; when using shading, limit use to 30%, 60%, and 100% black. Remove all colour enhancements from drawings. Save diagrams at a resolution of at least 600 dpi.

Please supply electronic files of all figures in your manuscript, where possible.

Adobe Illustrator is the preferred program. Computer-generated figures prepared using either a draw or chart/graph program are also acceptable only if saved as follows:

- Excel charts saved as Excel 97;
- Power Point files saved as native PowerPoint files;
- CorelDraw files saves as native CorelDraw files and as adobe illustrator EPS (Encapsulated post script) files;
- SigmaPlot Version 5 for Windows should be pasted into a Microsoft Word Document and saved as Word 97;
- For all other programs, please also save as EPS, PostScript or high resolution PDF files (some programs can save usable PICT files, but please forward a sample first);
- Figures embedded in a Microsoft Word document are only acceptable if you are unable to save as above.

Avoid using 3D surface area charts where possible because print quality is often poor. Avoid assigning light colours such as yellow and pale blue in charts because these colours print poorly when converted to greyscale images.

Please provide one good set of hard copies, which can be scanned if there are any problems using the electronic files.

Refer to the Notice to Authors and the free sample papers, both of which are available from the *AJPP* web site (www.publish.csiro.au/jounals/ajpp).

### *Photographs*

Photographic scans can also be submitted; Adobe Photoshop is the preferred program. If figures are created in a paint program, line art should be saved at 600 dpi in TIFF or EPS format; greyscale and colour images (CMYK format) should be saved at 300 dpi in TIFF or EPS format. All scans should be at the intended print size (85 mm wide for one column and up to a page width of 175 mm). Electronic photographic work can be submitted on PC-ZIP or MAC-ZIP discs.

When submitting your paper for publication, please send only paper copies of the figures or illustrations. The electronic files of both the figures and text should be sent after your paper has been refereed and any necessary changes have been carried out.

## Proofs and Reprints

Proofs should be checked and returned by fax and airmail to the Editors as soon as possible after receipt. Reprint order forms and prices are enclosed with the proofs and should be returned to the Managing Editor with the proofs. Free reprints are not normally provided. A table of reprint prices is available.

## Summary and Checklist for Preparation of Manuscripts

Submit manuscripts in hard copy in triplicate, making sure a good set of figures is included with each copy of the manuscript. When the manuscript is returned to the authors with the referees' reports for revision, the editor will make notes on presentation, layout and general matters concerning style, and these should be incorporated into the revised paper. A disk will be requested only once the paper is accepted in principle. Please ensure that the version on the disk is identical to the hard copy. Please name your file with the surname of the first author and the 'PP' file number assigned at submission. Please give the name of the file and the word-processing package used.

1.  Type the manuscript double-spaced throughout, including references, figure caption and tables.

2.  Type the title and all headings with only the first initial in upper case.

3.  Main headings (Introduction, Materials and methods, Results, Discussion, Acknowledgments, References) are set in bold roman type, left justified. Minor headings are set in light italic type, aligned at the left.

4.  Use the following conventions: 'from...to'. 'between...and',' range x-y'; use L for litre (hence mL, etc.); use single not double quotation marks; abbreviate the units hour(s) as h, minute(s) as min, second(s) as s, e.g.4 h, 5 min, 3 s.

5.  Check that all references mentioned in the text are in the References, and vice versa.

6.   List references in the text in chronological order, separated by semi-colons. List references in the References list in alphabetical order. In the text, do not use a comma between the author's name and the date. Italicise a, b, c etc. where several references are the same year.

7.  In the References, use italic type for the journal name and use bold type for the journal volume number.

8.  Give full journal and book titles in the References list.

9.  Spell out numbers lower than 10 unless accompanied by a unit, e.g. 2 mm, 15 mm, two plants, 15 plants, but 2 out of 15 plants, DO NOT leave a space between a numeral and the unit %, $\%_0$ or $^0$c.

10.  Prepare figures with symbols and letters appropriate for the size reduction intended and the figure caption on the same page.

11.  Type the title of each table as a separate paragraph from the actual table. Put explanatory matter referring to the table as a whole in the headnote, which should be in a separate paragraph from the title.

12.  Indicate approximate positions of figures and tables in the manuscript margin.

13.  Suggest a running head for the paper of not more than 8 words.

14.  Return the requested number of revised manuscripts; also return the original manuscript annotated by the editor.

## Correspondence

When you submit a manuscript, please supply us with the telephone number, facsimile number, email address and postal address of the corresponding author; we may need to contact you urgently.

## Address for Submissions

*Australian Journal of Plant Physiology*
**CSIRO** PUBLISHING
PO Box 1139 (150 Oxford St)
Collingwood VIC 3066
Australia

Fax +61 3 9662 7611
Telephone +61 3 9662 7625

# CHAPTER - 27

# Bhagirath

## INFORMATION TO CONTRIBUTORS

The authors and authorities contributing articles and other material for publication in 'BHAGIRATH' are requested to observe the following guidelines which will enable prompt attention and publication of material found suitable for the journal:

1.  Contribution to 'BHAGIRATH' should normally be exclusive to the journal. Material already published, except news of interest, will not be accepted.

2.  Articles should be concise, popular in presentation and contain fresh material or present a new approach to available information, existing problem, etc. relating to River Valley Projects, resource development and allied subjects.

3.  The matter, free from typographical and other errors, should not normally exceed ten pages of double-spaced typed matter, should carry helpful sub-heads, and generally conform to the pattern of articles in the Journal.

4.  Text of the articles should be sent in duplicate.

5.  Text should be free from specialized technical terms. Where unavoidable, they should be unobtrusively explained.

6.  Data should be in metric units, according to Indian Standard Conversions.

7.  Articles should be illustrated, as far as possible with relevant photographs artist's sketches, maps, diagrams, etc. in a manner intelligible to the educated layman.

8.  Pictures, maps and sketches should be in black and white only. Photographs should be on contrasty/glossy paper. The tracings should be in Indian ink on tracing cloth or paper. The drawings/maps should be (i) without colour, (ii) no rail/road lines, (iii) without scale, and (iv) larger letterings.

9.  Sketches, diagrams and maps should not exceed double the type-area of the 'BHAGIRATH' page. The letterings should be large enough to be legible after reduction to half the 'BHAGIRATH' page or less. They should be free from all information, lines and letterings, which are not necessary for subject matter of the article. Crowding as well as large blank areas should be avoided. A balanced layout should be aimed at.

10. Honorarium will be paid to contributors as per prescribed rules.

All contributions and editorial communications should be addressed to :

Editor, Bhagirath

Central Water Commission,

Ministry of Water Resources,

425 (N), Sewa Bhavan, R.K.Puram

New Delhi –110066

# CHAPTER - 28

# Biology and Fertility of Soils

## ONLINE MANUSCRIPT SUBMISSION, REVIEW AND TRACKING SYSTEM FOR THE JOURNAL

We trust that you will find this Online Manuscript Submission, Review and Tracking System very user friendly. To make your start even easier, please find below a few instructions:

### New Authors

Please click the 'Register' button from the menu above and enter the requested information. Upon successful registration you will be sent an e-mail with instructions to verify your registration.

### Note

- When you have received an e-mail from us with an assigned user ID and password, DO NOT REGISTER AGAIN. Just log in to the system as 'Author'.

### Authors

Please click the 'Login' button from the menu above and log in to the system as 'Author'. Then submit your manuscript and track its progress through the system. A wide range of submission file formats is supported, including: Word, RTF, TIFF, GIF, JPEG, EPS, Excel and Powerpoint. PDF is not an acceptable file format.

### Note

- Please upload your manuscript only ONCE on to the system. After uploading your manuscript, it will be automatically formatted as a PDF file, and you will be sent an e-mail requesting that you approve your submission. Please return to the main menu and APPROVE your submission accordingly.

### Returning Authors

Please use the provided username and password and log in as 'Author' to track your manuscript or to submit a NEW manuscript. (*Do not register again as you will then be unable to track your manuscript*).

## Reviewers

Please click the 'Login' button from the menu above and log in to the system as 'Reviewer'. You may view and/or download manuscripts assigned to you for review, submit your comments for the editors and the authors, and track the progress of your manuscripts through the system.

## Note

- Please click the 'Accept' or 'Decline' button as soon as possible after receipt of the e-mail asking you to review a manuscript.

## To change your username and password

Log in to the system and select 'Update My Information' from the menu above. At the top of the Update My Information screen, click the 'Change Password' button and follow the directions.

## Forget your password?

If you have forgotten your password, click the 'Login' button and click 'Forget Your Password?' at the bottom of the Login screen and follow the directions.

# CHAPTER - 29

# Breeding Science

## INSTRUCTIONS FOR AUTHORS

This Journal publishes original contributions related to breeding. Four types of manuscripts may be accepted: Reviews, Research papers, Research communications and Notes. Since the objective of the Research communications is to bring new findings quickly to the attention of the research community, these may be published somewhat more rapidly than Research papers. Notes will cover new breeding materials or techniques. If a Research paper is longer than 6 printed pages, the author will be charged for excess pages. For a Research communication, full page charges should be paid. If a Note is longer than 4 printed pages, the author will be charged for excess pages. Reviews summarize recent progress or historical events related to research areas in the field of plant breeding. All the Review articles are solicited. Page charge is not required for Review articles. Page charge is 9,000 yen per page. Four sheets with double-space text (26 lines)correspond to one printed page. At least one author must be a member (except for groups and supporting members) of the Japanese Society of Breeding. All the manuscripts should be written in English.

## I.  General Instructions

1.  Please check the most recent issues of Breeding Science for current format and style.

2.  All sections of the manuscript should be typed double-spaced on one side only of A4 (21 × 29.5 cm) paper with margins of at least 3 cm all round. A word at the end of a line should not be divided.

3.  Manuscripts should be arranged in the following order : Title Page, Summary, Introduction, Materials and  Methods, Results, Discussion, Acknowledgment (s), Literature Cited, Figure Legends, Tables and Figures. All pages must be consecutively numbered except for figures.

4.  Standard international units (SIU) should be used. Latin names are written in italics. The name of the author of a scientific name and words such as var. are written in Roman characters. The first time an abbreviation is used, it is in parenthesis after the non-abbreviated name.

## II. Title Page (page 1)

At the top, write each author's last name and a running title (about 10 words). Next write title (begin each word with capital letters except for prepositions, articles and conjunctions. Do not divide into a main title and a subtitle), author(s) full name, name and address of the institution where the work was performed and postal number. Superscript number after an author's name should be used to indicate a different address. Numbers of figures and tables should be indicated on the title page. The title page should include telephone and fax numbers and E-mail address of the corresponding author. Name of more than two Review Editors requested by authors should be written in the title page, too.

## III. Summary Page (page 2)

Summary is a single concise paragraph (less than 350 words). The author should describe the objective of the study, materials, methods and results in the summary without using references. Summary should include key words characterizing the scope of the paper. Up to seven Key Words should be provided following the summary.

## IV.  Text (page 3, etc.)

Main headings (Introduction, Materials and Methods, Results, Discussion and Acknowledgment(s) are not numbered but written in Gothic characters (begin with capital letters). Second-level headings should be written in italic characters (capitalize first letter of first word). All figures and tables  must be mentioned in the text. The author should cite references by author's last name and year of publication. When there are two authors, write the last name of each author, but when there are more than two, write *"et al "*after the first author's last name.

## V.  Literature Cited

Reference are limited to those cited and should contain complete titles and inclusive page numbers. References should be listed alphabetically, by the author(s)' last name, according to the following rules and examples. When there are more than two references by the same author, they should be listed by the year. "In press"  citations must be accepted for publication and the name of the journal or publisher must be included in Literature Cited. It is the authors' responsibility to obtain permission from colleagues to include their work as a personal communication.

Khalafalla, M.M. and K. Hattori (2000) Differential *in vitro* direct shoot regeneration responses in embryo axis and shoot tip explants of faba bean. Breed. Sci. 50: 117-122.

Kurata,N. and T. Sasaki (1997) Molecular analysis of the rice genome. *In* "DNA Markers: Protocols, Applications, and Overviews" Caetano-Anolles, G. and P.M. Gresshoff (eds), Wiley-Liss, New York. p. 271-282.

Bos. and P. Caligari (1995) Selection methods in plant breeding. Chapman & Hall, London. 347 p.

## VI. Figure Legends

Figure legends should be numbered consecutively with Arabic numerals and typed continuously. Each should begin with a short title for the figure. All symbols and abbreviations used in the figures should be explained.

## VII. Tables

Each table must start on a separate sheet. Tables should be numbered with Arabic numerals. The first words of the titles and each entry in each column should be capitalized. Vertical lines should not be used. When the author wishes to make a comment as a footnote, he/she should use a superscript number in parentheses where the comment as directed. The symbols* and # should not be used.

## VIII. Figures

Figures are black and white half-tones (photographs), drawings or graphs. Authors should consult the Editorial Office about color figures. Figures with marks, letters etc. should be drawn on appropriate size so that they can be printed in the Journal on a scale of 1 : 2. Authors should prepare figures using professional standards. Failure to provide good-quality copies of figures may be grounds for returning a manuscript without review. The title and footnotes should not be written on the original figures but on a copy of the figure. Write the number, the title and author(s) name in the corner of the figure. Figures should be received flat.

## IX. Notes

Notes do not include summary. Key Words should be written on the Title page. Text should be written without starting items such as Introduction, Materials and Methods, Results, Discussion, etc. The rest of the Note should be written according to the above-mentioned instructions.

## X. Submission of Manuscripts

Three clear copies and original manuscript with figures should be submitted to

Editorial Office of the Breeding Science,

C/o Dr. Masumi Katsuta

National Institute of Crop Science

2-1-18 Kannondai, Tsukuba, Ibaraki 305-8518, Japan

## XI. Disk Submission of Manuscripts

Breeding Science requests authors to submit the revised version of an accepted manuscript on 3.5 inches floppy disk prepared with any one of following word processing programs; Word, Ichitoro, Word Perfect and LaTex. Text file is also acceptable. The disk must contain all final revisions and must be accompanied by a hard copy printout. Tabular material should be included on the disk as a Text file or an Excel file.

## XII.  Review Procedure and Publications

A manuscript submitted to the Editorial Office will be sent to the Review Editor requested by authors after checking the style. All manuscripts will be reviewed by at least two referees with expertise in the research area. The Review Editor determines acceptance of a manuscript on the basis of reviewers' comments. Final acceptance of a manuscript is contingent upon strict compliance with Journal format and style. As a rule, the order of publication corresponds to the date of receipt. Manuscripts will not be returned except the photographs and figures upon the author's request.

## XIII.  Reprints

100 copies will be supplied free of charge. Additional reprints will be available at cost. Charge will be indicated at the proofreading stage

## XIV. Copyright

The copyright on any published materials in Breeding Science is owned by the Japanese Society of Breeding. Anyone who intends to use these materials must obtain the authorization from the Japanese Society of Breeding.

# Chapter - 30

# Calcutta Statistical Association Bulletin

## INSTRUCTIONS TO AUTHORS

Authors are requested to submit THREE typescripts of their articles (typed on one side of paper with double spacing and wide margins) to the Editor, Calcutta Statistical Association Bulletin, C/o Department of Statistics, University of Calcutta, 35, Ballygunge Circular Road, Kolkata – 700 019. Each communication should contain an Abstract at the beginning. Keywords and Phrases and AMS Subject Classification should be given.

Communication to Application of Statistics Section should ideally cover not more than fifteen printed pages inclusive of representative data sets. At the time of submission the authors should specifically mention that the communication is meant for the section 'Applications of Statistics'.

References should be given according to the following style :

Billingsley, P. (1968) : *Convergence of Probability Measures.* Wiley, New York.

Blackman, J. (1955)  : On the approximation of a distribution function by an empirical distribution. *Ann. Math. Statist.* **27** 642-669.

Abbreviation of names of journals should be those accepted internationally (see Mathematic Subject Classification 2000 available at the Editorial Office of Mathematical Reviews or at http://www.ams.org.msc).

*Only articles submitted exclusively to the Bulletin will be considered for publication.*

To avoid uncertainties, authors are advised to send copies of articles per registered mail.

Authors are requested to be particularly careful in the preparation of their scripts, as galley proofs of papers accepted for publication are not normally sent to authors. Typescripts of papers found unacceptable, are not returned to overseas addresses. Authors are encouraged to submit papers written in latex.

Two copies of the issue containing the articles are provided free of charge to the communicating author.

# CHAPTER - 31

# Clay Research

## INSTRUCTIONS FOR CONTRIBUTORS

CLAY RESEARCH is the official publication of THE CLAY MINERALS SOCIETY OF INDIA and is published twice a year, in June and December. The journal undertakes to publish articles of interest to the international community of clay scientists, and will cover the subject areas of mineralogy, geology and geochemistry, crystallography, physical and colloid chemistry, physics, ceramics, civil and petroleum engineering and soil science. The journal is reviewed in *Chemical Abstracts. Mineralogical Abstracts, and Soils and Fertilizers.*

Papers (in English) should be submitted to the Publication Office, the Clay Minerals Society of India. Division of Soil Resource Studies, National Bureau of Soil Survey and Land Use Planning, Amaravathi Road, Nagpur –440 010. At least one of the authors should be member of THE CLAY MINERALS SOCIETY OF INDIA. Submission is an undertaking that the manuscript has not been published or submitted for publication elsewhere.

Manuscripts should not exceed sixteen typed (double spaced pages including tables and illustrations. The **original and two copies of text and illustrations should be submitted.**

Form Manuscripts should be typewritten, double spaced on white paper, with wide margins. Intending contributors should consult a recent issue of CLAY RESEARCH for the standard format and style. The manuscript should have the sections ABSTRACT, introductory portion (united), MATERIALS AND METHODS, RESULTS and DISCUSSION and REFERENCES.

**Title** page should contain manuscript title, full name(s) of author(s), address (es) of the institution(s) of the author(s), a short running title not exceeding 60 characters including spaces, footnotes if any to the title, and complete mailing address of the person to whom communications should be sent.

**Abstract** should be a condensation of the ideas and results of the paper. It should not exceed 250 words. Do not make reference to the literature in the abstract.

**Tables** should have the simplest possible column headings. Type each table on a separate page; indicate, location in the text by marking in the margin of text page.

**Figures** should be self-illustrative, drawn with black Indian ink on tracing paper or white Board. The lettering should be large enough to permit size reduction to one journal page column width (about 7.0 cm) without sacrificing legibility. The original tracing should be submitted. The size of the drawing should not exceed 24 × 17 cm. Give the numbered legend on a separate sheet, not on the figure itself. Data available in the tables should not be duplicated in the form of illustrations. Indicate the location of the figure in the text by marking in the margin of the page.

**Photographs** should be in the form of glossy prints with strong contrast. In photomicrographs, the scale in micron or other suitable unit should be drawn on the print. Give the numbered legend on a separate sheet. Indicate the location of the photograph in the text by making in the margin of the text page.

References should be cited in the text by the name(s) of author(s) if two or less, and year of publication. If there are more than two authors, give the name of the first author followed by 'et al' and year. Full references giving author(s) and initial(s), year, title of paper, (journal, volume, number if paged separately), first and last pages should be listed alphabetically at the end of the paper. Journal title should be abbreviated in accordance with the world list of scientific periodicals and its sequences. Examples are

Grim, R.E., Bray, R.H. and Bradley, W.F. 1937. The mica in argillaceous sediments. *Am. Miner.* 22:813-829.

Brindley, G.W. 1961. Chlorite minerals. In (G.Brown, Ed.) The X-ray Identification and Crystal Structures of Clay Minerals, Mineralogical Society, London.pp. 242-296.

Theng, B.K.G. 1974. *The Chemistry of Clay Organic Reactions,* Adam. Hilger, London, 343, pp.

**Review** Every manuscript submitted to CLAY RESEARCH is independently reviewed by one or more referees. Acceptance or rejection of a manuscript is the responsibility of the Editor.

**Reprints** No free reprints are supplied to authors. Order for priced reprints should be sent when required by the Editor.

# CHAPTER - 32

# Communicator

## JOURNAL OF INDIAN INSTITUTE OF MASS COMMUNICATIONS

**Manuscript Information**

Authors should send four (4) copies of the manuscript to Abhilasha Kumari, Editor, Communicator, Indian Institute of Mass Communication, Aruna Asaf Ali Marg, New Delhi 110 067, India (Tel.: 26168122 Fax: 91-11-26107462); E-mail: abhilashakumari@hotmail.com. Submissions should conform to the stylistic guidelines of the Manual of the American Psychological Association (Fourth edition). Mailing address and brief biographical  notes on all authors should be provided on a separate page, in addition to a 100-word abstract and a list of key words for indexing. Successful submissions generally run no longer than 30 typed double-spaced pages. The entire manuscript should have one and a half-inch margin on all sides. Tables should be clear and understandable independent of the text. Figures should be sent camera, ready, clear and legible when reduced. References should be carefully edited to ensure consistency with APA guidelines. Authors are responsible for gaining permission for including any copy righted material that needs permission, including quotations of more than 300 words. The manuscript has to be sent on a computer disk along with the hard copy. Submission implies commitment to publish in the journal. Authors submitting manuscripts to the journal, should see that the manuscript have not been published elsewhere in substantially similar form or with substantially similar content. All manuscripts are subjected to review.

**Subscription**

Indian Institute of Mass Communication publishes *Communicator* twice a year in January and June. The annual subscription for the journal is Rs. 80/-. The cost of an individual copy is Rs. 40/-. Subscription outside India is $ 30/- annually and $ 12/- for an individual copy. All cheques and drafts should be drawn in favour of Indian Institute of Mass Communication and addressed to Abhilasha Kumari, Editor, *Communicator*, Indian Institute of Mass Communication, Aruna Asf Ali Marg, New Delhi 110 067, India (Tel;: 26168122,26160987;Fax:91-11-26107462); E-mail: abhilashakumari@hotmail.com Similarly, all inquiries about back issues can be made to the same address.

## Permission Request, Reprints and Photocopies



## Advertising

Current rates and specifications may be obtained by writing to Abhilasha Kumari, Editor, *Communicator,* Indian Institute of Mass Communication, Aruna Asf Ali Marg, New Delhi 110 067, India (Tel;: 26168122,26160987;Fax:91-11-26107462); E-mail: abhilashakumari@hotmail.com

## Claims

Claims for undelivered copies must be made not later than six months following the month of publication. The publisher will supply missing copies when losses have been sustained in transit and when reserve stocks permit.

## Postmaster

Six weeks advance notice must be given when notifying a change of address. Please send an old address label along with the new address to ensure proper identification. Send address change to Abhilasha Kumari, Editor, *Communicator,* Indian Institute of Mass Communication, Aruna Asf Ali Marg, New Delhi 110 067, India (Tel;: 26168122,26160987;Fax:91-11-26107462); E-mail: abhilashakumari@hotmail.com.

# CHAPTER - 33

# Crop Protection

## GUIDE FOR AUTHORS

The Editors of *Crop Protection* especially welcome papers describing aninter disciplinary approach showing how different control strategies can be integrated into practical pest management programmes. *Crop Protection* particularly emphasizes the practical aspects of control in the field, and includes work which may lead in the near future to more effective control. The journal does not duplicate the many existing excellent biological science journals, which deal mainly with the more fundamental aspects of plant pathology, applied zoology and weed science. *Crop Protection* covers all practical aspects of pest, disease and weed control, including the following topics:

- Control of diseases of crop plants caused by microorganisms
- Control of animal pests of world crops
- Control of weeds
- Assessment of pest and disease damage
- Epidemiology of pests and diseases in relation to control
- Development of pesticides
- Biorational pesticides
- Effects of plant growth regulators
- Pesticide application methods
- Genetic engineering applications
- Environmental effects of pesticides
- Importance and control of postharvest crop losses
- Economic considerations
- Biological control
- Agronomic control methods
- Integrated control
- Pest management
- Interrelationships between different control strategies

- Abiotic damage
- Food safety

The editors of *Crop Protection* invite workers concerned with pest, disease and weed control to submit suitable contributions on any topic falling within the aims and scope of the journal.

Please follow these instructions carefully to ensure that the review and publication of your paper is as swift and efficient as possible. These notes may be copied freely.

## Types of contribution

Contributions falling into the following categories will be considered for publication:

- Original high-quality research papers (preferably no more than 20 double line spaced manuscript pages, including tables and illustrations)

- Short communications. These should not exceed 8-10 double line spaced manuscript pages excluding references and legends. Submissions should include a short Abstract not exceeding 10% of the length of the communication and which summarizes briefly the main findings of the work to be reported. The bulk of the text may be in a continuous form but generally will follow the usual format that does not require numbered sections such as Introduction, Materials and methods, Results, and Discussion. However, a Cover page, Abstract and a list of Keywords are required at the beginning of the communication and Acknowledgements and References at the end. These components are to be prepared in the same format as used for full-length research papers. Occasionally authors may use sub-titles of their own choice to highlight sections of the text.

- *Review articles :*   *Crop Protection* also publishes, book reviews, conference reports and a calendar of forthcoming events. Please contact one of the Principal Editors.

## Submission of manuscripts

*Please Note* : As of January 2006 submission to this journal proceeds totally on-line. Use the following guidelines to prepare your article. Via the homepage of this journal (http://www.elsevier.com/locate/cropro), you will be guided stepwise through the creation and uploading of the various files. Once the uploading is done, our system automatically generates an electronic (PDF) proof, which is then used for reviewing. It is crucial that all graphical elements be uploaded in separate files, so that the PDF is suitable for reviewing. Authors can upload their article as a LaTex, Microsoft (MS) Word, WordPerfect, PostScript or Adobe Acrobat PDF document. All correspondence, including notification of the Editor's decision and requests for revisions, will be by e-mail.

Dr. D.L. Cole (Plant pathology, nematology)

Prof.W.D. Hutchison (Biological Control and Integrated Management of Arthropod Pests

Prof. G.A. Matthews (Pesticide applications, entomology, integrated pest management, tropical pest management)

Dr. R.G. Turner (Weed science, European horticulture and general agronomy

Papers in agricultural economics and vertebrate control should be sent to one of the above Editors.

*Repeat experiments* Manuscripts that report original research should not be submitted unless experiments have been repeated **at least twice** or, in the case of field experiments, relate to two seasons. In most cases, three or more replications will be necessary for appropriate statistical analysis. In exceptional circumstances, studies that do not meet these criteria may be acceptable, but the relevant Principal Editor should be consulted prior to submission.

Submission of an article implies that the work described has not been published previously (except in the form of an abstract or as part of a published lecture or academic thesis), that it is not under consideration for publication elsewhere, that its publication is approved by all authors and tacitly or explicitly by the responsible authorities where the work was carried out, and that, if accepted, it will not be published elsewhere in the same form, in English or in any other language, without the written consent of the Publisher.

Contributions are normally received with the understanding that they comprise original, unpublished material and are not being submitted for publication elsewhere. Translated material, which has not been published in English, will also be considered. **All submissions should be accompanied by a written declaration, signed by all authors, that the paper has not been submitted for consideration elsewhere.** Authors are solely responsible for the factual accuracy of their papers. The receipt of manuscripts will be acknowledged.

## Author enquiries

For enquiries relating to the submission of articles (including electronic submission where available) please visit this journal's homepage at http://www.elsevier.com/locate/cropro. You can track accepted articles at http://www.elsevier.com/trackarticle and set up e-mail alerts to inform you of when an article's status has changed, as well as copyright information, frequently asked questions and more.

For privacy, information on each article is password-protected. The author should key in the "Our Reference" code (which is in the letter of acknowledgement sent by the publisher on receipt of the accepted article) and the name of the corresponding author.

In case of problems or questions, authors may contact the Author Service Department, E-mail: authors@elsevier.co.uk; Fax: +44 (0) 1865 843905.

## Review process

All contributions are reviewed by two or more referees to ensure both accuracy and relevance, and revisions to the script may thus be required. On acceptance, contributions are subject to editorial amendment to suit house style. When a manuscript is returned for revision prior to final acceptance, the revised version must be submitted as soon as possible after the author's receipt of the referee's reports. Revised manuscripts returned after four months will be considered as new submissions subject to full re-review.

## Electronic Format Requirements for Accepted Articles

Electronic manuscripts have the advantage that there is no need for rekeying of text, thereby avoiding the possibility of introducing errors and resulting in reliable and fast delivery of proofs.

### *General points*

We accept most wordprocessing formats, but Word, WordPerfect or LaTeX is preferred. An electronic version of the text should be submitted together with the final hardcopy of the manuscript. The electronic version must match the hardcopy exactly. Always keep a backup copy of the electronic file for reference and safety. Label storage media with your name, journal title, and software used. Save your files using the default without the explicit approval of the Editor. Electronic files can be stored on 3½ inch diskette, ZIP-disk or CD (either MS-DOS or Macintosh).

### *Wordprocessor documents*

It is important that the file be saved in the native format of the wordprocessor used. The text should be in single-column format. Keep the layout of the text as simple as possible. Most formatting codes will be removed and replaced on processing the article. In particular, do not use the wordprocessor's options to justify text or to hyphenate words. However, do use bold face, italics, subscripts, superscripts etc. Do not embed 'graphically designed' equations or tables, but prepare these using the wordprocessor's facility. When preparing tables, if you are using a table grid, use only one grid for each individual table and not a grid for each row. If no grid is used, use tabs, not spaces, to align columns. The electronic text should be prepared in a way very similar to that of conventional manuscripts ( see also the Guide to Publishing with Elsevier: http://www.elsevier.com/wps/find/authorshome.authors/howtosubmitpaper). Do not import the figures into the text file but, instead, indicate their approximate locations directly in the electronic text and on the manuscript. To avoid unnecessary errors you are strongly advised to use the 'spellchecker' function of your wordprocessor.

## Preparation of manuscripts

1.  Manuscripts should be written in English. Authors whose native language is not English are strongly advised to have their manuscripts checked by an English-speaking colleague prior to submission.

    *Authors in Japan please note:* Upon request, Elsevier Japan will provide authors with a list of people who can check and improve the English of their paper (*before submission*). Please contact our Tokyo office: Elsevier Japan, 4F Higashi-Azabu,1-chome Bldg, 1-9-15, Higashi-Azabu, Minato-ku, Tokyo 106-0044; Japan; Tel. (+81) 3-5561-5032; Fax: (+81) 3-5561-5045; E-mail: jp.info@elsevier.com.

2. The manuscript must be accompanied by a covering letter detailing what you are submitting (type of contribution, title, authors' names and affiliation, etc.). Please also indicate the author to whom we should address our correspondence in the case of multiple authors and include a contact address, telephone/fax numbers and e-mail address. Authors are requested to submit, with their manuscripts, the names and addresses of four potential referees.

3. Manuscripts should be prepared with numbered lines, with wide margins and double spacing throughout, i.e. also for abstracts, footnotes and references.**Every page of the manuscript, including the title page, references, tables, etc. should be numbered. Authors are requested to submit, with their manuscripts, the names and contact details of four potential referees.** However, in the text no reference should be made to page numbers; if necessary, one may refer to sections. Underline words that should be in italics, and do not underline any other words. Avoid excessive usage of italics to emphasize part of the text. Manuscripts not conforming to these standards may be returned to the authors for correction.

4. Manuscripts in general should be organized in the following order:
Title (should be clear, descriptive and not too long)
Name(s) of author(s)
Complete postal address(es) of affiliations\
Present address(es) of author(s) if applicable
Abstract
Keywords (indexing terms), normally 3-6 items
Introduction
Material and methods
Results
Discussion
Conclusion (If needed)
Acknowledgements and any additional information concerning research grants, etc.
References
Tables
Figure captions
Figures

5. In typing the manuscript, titles and subtitles should not be run within the text. They should be typed on a separate line, without indentation. Use lower-case letter type.

6. SI units should be used.

7. If a special instruction to the copy editor or typesetter is written on the copy it should be encircled. The typesetter will then know that the enclosed matter is not to be set in type. When a typewritten character may have more than one meaning (e.g. the lower case letter l may be confused with the numeral 1), a note should be inserted in a circle in the margin to make the meaning clear to the typesetter. If Greek letters or uncommon symbols are used in the manuscript, they should be written very clearly, and if necessary a note such as "Greek lower-case chi" should be put in the margin and encircled.

8. Elsevier reserves the privilege of returning to the author for revision accepted manuscripts and illustrations that are not in the proper form given in this guide.

## Abstracts

The abstract should be clear, descriptive and not longer than 400 words.

## Tables

1. Authors should take notice of the limitations set by the size and layout of the journal. Large tables should be avoided. Reversing columns and rows will often reduce the dimensions of a table.

2. If many data are to be presented, an attempt should be made to divide them over two or more tables.

3. Tables drawn by hand, from which prints need to be made, should not be folded.

4. Tables should be numbered according to their sequence in the text. The text should include references to all tables.

5. Each table should be typewritten on a separate page of the manuscript. Tables should never be included in the text.

6. Each table should have a brief and self-explanatory title.

7. Column headings should be brief, but sufficiently explanatory. Standard abbreviations of units of measurement should be added between parentheses.

8. Vertical lines should not be used to separate columns. Leave some extra space between the columns instead.

9. Any explanation essential to the understanding of the table should be given as a numbered or lettered footnote at the bottom of the table.

## Illustrations

1.  All illustrations (line drawings and photographs) should be submitted separately and unmounted.

2.  Illustrations should be numbered according to their sequence in the text. References should be made in the text to each illustration.

3.  Each illustration should be identified by its number and the name of the author. An indication of the top of the illustrations is required in photographs of profiles, thin sections, and other cases where doubt can arise.

4.  Illustrations should be designed with the format of the page of the journal in mind. Illustrations should be of such a size as to allow a reduction of 50%.

5.  Lettering should be in Indian ink or by printed labels. Make sure that the size of the lettering is big enough to allow a reduction of 50% without becoming illegible. The lettering should be in English. Use the same kind of lettering throughout and follow the style of the journal.

6.  If a scale should be given, use bar scales on all illustrations instead of numerical scales that must be changed with reduction.

7.  Each illustration should have a caption. The captions to all illustrations should be typed on a separate sheet of the manuscript.

8.  Explanations should be given in the typewritten legend. Drawn text in the illustrations should be kept to a minimum.

9.  Photographs are only acceptable if they have good contrast and intensity. Sharp and glossy copies are required. Reproductions of photographs already printed cannot be accepted.

### *Preparation of electronic illustrations*

Submitting your artwork in an electronic format helps us to produce your work to the best possible standards, ensuring accuracy, clarity and a high level of detail.

-   Always supply high-quality printouts of your artwork, in case conversion of the  electronic artwork is problematic.

-   Make sure you use uniform lettering and sizing of your original art work.

-   Save text in illustrations as "graphs" or enclose the font.

-   Only use the following fonts in your illustrations : Arial, Courier, Helvetica, Time Symbol.

-   Number the illustrations according to their sequence in the text.

-   Use a logical naming convention for your artwork files, and supply a separate listing of the files and the software used.

-   Provide all illustrations as separate files and as hardcopy printouts on separate sheets.

- Provide captions to illustrations separately.
- Produce images near to the desired size of the printed version.

Files can be stored on diskette, ZIP-disk or CD (either MS-DOS or Macintosh).

A detailed guide on electronic artwork is available on our website: http://authors.elsevier.com/artwork

You are urged to visit this site; some excerpts from the detailed information are given here.

### *Formats*

Regardless of the application used, when your electronic artwork is finalised, please "save as" or convert the images to one of the following formats (Note the resolution requirements for line drawings, halftones, and line/halftone combinations given below.):

*EPS :* Vector drawings. Embed the font or save the text as "graphics".

*TIFF :* Colour or greyscale photographs (halftones): always use a minimum of 300 dpi.

*TIFF :* Bitmapped line drawings: use a minimum of 1000 dpi.

*TIFF :* Combinations bitmapped line/half-tone (colour or greyscale): a minimum of 500 dpi is required. DOC, XLS or PPT: If your electronic artwork is created in any of these Microsoft Office applications please supply "as is".

### Please do not

- Supply embedded graphics in your wordprocessor (spreadsheet, presentation) document;
- Supply files that are optimised for screen use (like GIF, BMP, PICT, WPG); the resolution is too low;
- Supply files that are too low in resolution;
- Submit graphics that are disproportionately large for the content.

### *Colour Reproduction*

Submit colour illustrations as original photographs, high-quality computer prints or transparencies, close to the size expected in publication, or as 35 mm slides. Polaroid colour prints are not suitable. If, together with your accepted article, you submit usable colour figures then Elsevier will ensure, at no additional charge, that these figures will appear in colour on the web (e.g., ScienceDirect and other sites) regardless of whether or not these illustrations are reproduced in colour in the printed version. For colour reproduction in print, you will receive information regarding the costs from Elsevier after receipt of your accepted article. For further information on the preparation of electronic artwork, please see http://authors.elsevier.com/artwork.

***Please note :*** Because of technical complications which can arise by converting colour figures to 'grey scale' (for the printed version should you not opt for colour in print) please submit in addition usable black and white prints corresponding to all the colour illustrations.

## References

1. All publications cited in the text should be presented in a list of references following the text of the manuscript. The manuscript should be carefully checked to ensure that the spelling of author's names and dates are exactly the same in the text as in the reference list.

2. In the text refer to the author's name (without initial) and year of publication, followed - if necessary - by a short reference to appropriate pages. Examples: "Since Peterson (1988) has shown that..." "This is in agreement with results obtained later (Kramer, 1989, pp. 12-16)". (Give page numbers for books only.)

3. If reference is made in the text to a publication written by more than two authors the name of the first author should be used followed by "et al.". This indication, however, should never be used in the list of references. In this list names of first author and co-authors should be mentioned.

4. References cited together in the text should be arranged chronologically. The list of references should be arranged alphabetically on authors' names, and chronologically per author. If an author's name in the list is also mentioned with co-authors the following order should be used: publications of the single author, arranged according to publication dates - publications of the same author with one co-author - publications of the author with more than one co-author. Publications by the same author(s) in the same year should be listed as 1994a, 1994b, etc.

5. Use the following system for arranging your references:

    (a)  *For periodicals*
       Hettiaratchi, D.R.P., 1993. The development of a powered low draught tine cultivator. Soil Tillage Res. 28, 159-177.

    (b)  *For edited symposia, special issues, etc., published in a periodical*
       Benites, J.R., Ofori, C.S., 1993. Crop production through conservation-effective tillage in the tropics. In: Lal, R. (Ed.), Soil Tillage for Agricultural Sustainability. Proceedings of the 12th Conference of ISTRO, 8-12 July 1991, Ibadan, Nigeria. Soil Tillage Res. 27, 9-33.

    (c)  *For books*
       Russell, E.W., 1973. Soil Conditions and Plant Growth. 10th ed. Langmans, London.

   (d)    *For multi-author books*

Kuipers, H., Koolen, A.J., 1989. Interface between implements, tillage and soil structure. In: Larson, W.E., Blake, G.R., Allmaras, R.R., Voorhees, W.B., Gupta, S.C. (Eds.), Mechanics and Related Processes in Structured Agricultural Soils, NATO ASI Series E, Vol. 172. Kluwer Academic Publishers, Dordrecht, Netherlands, pp. 105-120

6. Abbreviate the titles of periodicals mentioned in the list of references according to the International *List of Periodical Title Word Abbreviations*.(http://www.issn.org/lstwa.html).

7. In the case of publications in any language other than English, the original title is to be retained. However, the titles of publications in non-Roman alphabets should be transliterated, and a notation such as "(in Russian)" or "(in Greek, with English abstract)" should be added.

8. Work accepted for publication but not yet published should be referred to as "in press".

9. References concerning unpublished data and "personal communications" should not be cited in the reference list but may be mentioned in the text.

## Formulae

1. Formulae should be typewritten, if possible. Leave ample space around the formulae.

2. Subscripts and superscripts should be clear.

3. Greek letters and other non-Roman or handwritten symbols should be explained in the margin where they are first used. Take special care to show clearly the difference between zero (0) and the letter O, and between one (1) and the letter l.

4. Give the meaning of all symbols immediately after the equation in which they are first used.

5. For simple fractions use the solidus (/) instead of a horizontal line.

6. Equations should be numbered serially at the right-hand side in parentheses. in general only equations explicitly referred to in the text need be numbered.

7. The use of fractional powers instead of root signs is recommended. Also powers of e are often more conveniently denoted by exp.

8. Levels of statistical significance which can be mentioned without further explanation are * $P<0.05$, ** $P<0.01$ and *** $P<0.001$.

9. In chemical formulae, valence of ions should be given, e.g. as $Ca^{2+}$ and $CO_3^{2}$ rather than as $Ca^{++}$ and $CO_3{}^{-}$.

10. Isotope numbers should precede the symbols, e.g. $^{18}O$

11. The repeated writing of chemical formulae in the text is to be avoided where reasonably possible; instead, the name of the compound should be

given in full. Exceptions may be made in the case of a very long name occurring very frequently or in the case of a compound being described as the end product of a gravimetric determination (e.g phosphate as $P_2O_5$).

11. Footnotes should only be used if absolutely essential. In most cases it should be possible to incorporate the information in normal text.

12. If used, they should be numbered in the text, indicated by superscript numbers, and kept as short as possible.

## Nomenclature

1. Authors and Editor(s) are, by general agreement, obliged to accept the rules governing biological nomenclature, as laid down in the *International Code of Botanical Nomenclature*, the *International Code of Nomenclature of Bacteria*, and the *International Code of Zoological Nomenclature.*

2. All biotica (crops, plants, insects, birds, mammals, etc.) should be identified by their scientific names when the English term is first used, with the exception of common domestic animals.

3. All biocides and other organic compounds must be identified by their Geneva names when first used in the text. Active ingredients of all formulations should be likewise identified.

4. For chemical nomenclature, the conventions of the *International Union of Pure and Applied Chemistry* and the official recommendations of the *IUPAC-IUB Combined Commission on Biochemical Nomenclature* should be followed where the compound is novel. For compounds more than two years old please use the approved name as given in the *Pesticide Manual*

*Application of pesticides* Full details must be given of techniques used to apply pesticides (e.g. type of equipment, type of nozzle, pressure, volume of spray, etc.) and of the amount of active ingredient applied per unit area.

### *Supplementary data*

Elsevier now accepts electronic supplementary material to support and enhance your scientific research. Supplementary files offer the author additional possibilities to publish supporting applications, movies, animation sequences, high-resolution images, background datasets, sound clips and more. Supplementary files supplied will be published online alongside the electronic version of your article in Elsevier web products, including ScienceDirect: http://www.sciencedirect.com. In order to ensure that your submitted material is directly usable, please ensure that data is provided in one of our recommended file formats. Authors should submit the material in electronic format together with the article and supply a concise and descriptive caption for each file. For more detailed instructions please visit our artwork instruction pages at http://www.elsevier.com/artworkinstructions.

## Copyright

Upon acceptance of an article, authors will be asked to transfer copyright (for more information on copyright see http://authors.elsevier.com). This transfer will ensure the widest possible dissemination of information. A letter will be sent to the corresponding will be provided. If excerpts from other copyrighted works are included, the author(s) must obtain written permission from the copyright owners and credit the source(s) in the article. Elsevier has preprinted forms for use by authors in these cases: contact ELSEVIER, Rights Department, P.O. Box 800, Oxford, OX5 1DX, UK; phone: (+44) 1865 843830, fax: (+44) 1865 853333, e-mail: permissions@elsevier.com

## Proofs

When your manuscript is received at the Publisher it is considered to be in its final form. Proofs are not to be regarded as 'drafts'.

One set of page proofs in PDF format will be sent by e-mail to the corresponding author, to be checked for typesetting/editing. No changes in, or additions to, the accepted (and subsequently edited) manuscript will be allowed at this stage. Proofreading is solely your responsibility.

A form with queries from the copy editor may accompany your proofs. Please answer all queries and make any corrections or additions required.

The Publisher reserves the right to proceed with publication if corrections are not communicated. Return corrections within two working days of receipt of the proofs. Should there be no corrections, please confirm this.

Elsevier will do everything possible to get your article corrected and published as quickly and accurately as possible. In order to do this we need your help. When you receive the (PDF) proof of your article for correction, it is important to ensure that all of your corrections are sent back to us in one communication. Subsequent corrections will not be possible, so please ensure your first sending is complete. Note that this does not mean you have any less time to make your corrections, just that only one set of corrections will be accepted.

## Offprints

1. Twenty five offprints will be supplied free of charge.
2. One hundred free offprints will be supplied to the first author of a review article.
3. Additional offprints can be ordered on an offprints order form, which is included with the proofs.
4. UNESCO coupons are acceptable in payment of extra offprints.

# CHAPTER - 34

# Crop Science

## INSTRUCTIONS TO AUTHORS

### General Requirements

Full papers must be either reports of original research, critical reviews, or interpretivearticles. The journal also publishes crop registration papers, short communications, book reviews, and letters to the editor. Submissions to *Crop Science* must not be previously published in or simultaneously submitted to any other scientific or technical journal. For the policy regarding publishing in nontechnical outlets, see *Publications Handbook and Style Manual* (ASA–CSSA–SSSA, 2004).

### Scope

*Crop Science* is the normal channel for publication of papers in plant genetics; breeding; cytology; metabolism; physiology; ecology; turfgrass; weed science; crop quality, production, and utilization; genomics, molecular genetics, and biotechnology; and plant genetic resources. Articles reporting experimentation or research in field crops or reviews or interpretation of such research will be accepted for review as papers. For research involving controlled environments, see www.asa-cssa-sssa.org/publications/pdf/CESGuide.pdf for guidelines. Short articles concerned with experimental techniques, apparatus, or observation of unique phenomena will be accepted for review as short communications. Letters to the editor are welcomed and are published subject to review and approval of the editor. When letters concern previous articles, the authors will be invited to reply; letter and reply are published together.

### Submission Procedures and Preparation Full-Length Manuscripts and Reviews

All manuscripts should be submitted through the online submission tool (http://mc.manuscriptcentral.com/crop). Detailed instructions can be found at this site, along with instructions related to logging on to the *Crop Science* Manuscript Central system.

### Creating the manuscript files.

Submit the main text document in a common word processing file. LaTeX or other typeset formats are not allowed. Manuscript Central will convert your original files into PDF format; please check this PDF "proof" before submitting. Check any Greek characters and figures carefully. If you have a character conversion, fix your word processing file by embedding fonts (in Word, go to Tools/Options/"Save" tab, and check "Embed Truetype Fonts"). If there is an error in the PDF you cannot fix, mention it in the cover letter so the editors and eventually Headquarters will be aware of the problem.

On the first page, give the title, a byline with the names of all authors, an author–paper documentation footnote, a list of all nonstandard abbreviations used in the paper (standard abbreviations available on p. 22 of the style manual, www.asa-cssasssa. org/publications/style/Chapt02W.pdf), and any other necessary footnotes. An abstract is required and is normally the second manuscript page. After the title page and abstract, the usual order of sections is an untitled introduction (which includes the literature review), Materials and Methods, Results, Discussion, Conclusions (optional), Acknowledgments (optional), and References, followed by any figure captions and the tables. Results and Discussion may be combined and conclusions can be given at the close of the Discussion section. Start each section (including figure captions and tables) on a new page and number all pages.

## Figures

Submit figures in high-resolution, individual files (one figure per file). All panels of one figure need to be in the same single file and on the same page if possible. Check your figures in the PDF proof generated by Manuscript Central, as the figures in the PDF may be used for publication. TIFF or EPS files are best for resolution (don't insert these files into a word processing document because this will reduce resolution). Width of figures should approximate desired print size, i.e., 3 ¼ inches for a one column figure, 7 inches for a two column figure.

*PowerPoint files should be avoided.* Photographs and drawings for graphs and charts should be prepared with good contrast of dark and light. Give careful attention to the width of lines and size and clarity of type and symbols. Variables (e.g., $r, x, y$) should be italicized. A figure caption should be brief, but sufficiently detailed to tell its own story. Specify the crop or soil involved, the major variables presented, and the place and year. Identify curves or symbols in a legend within the figure itself, not in the caption. Define abbreviations in the caption. Define symbols used in the caption or in the legend. Be sure to indicate the scale for micrographs, either in the illustration or the caption.

## Tables

Prepare tables with the tables feature in your word processor; do not use tabs, spaces, or graphics boxes. Each datum needs to be contained in an individual cell. Number tables consecutively. Table heads should be brief but complete and self-contained. Define all variables and spell out all abbreviations. Tables should be placed at the end of the main text document. The *, **, and *** are always used in this order to show statistical significance at the 0.05, 0.01, and 0.001

probability levels, respectively, and cannot be used for other notes. Significance at other levels is designated by a supplemental note. Lack of significance is usually indicated by NS.

For table footnotes, use the following symbols in this order: †, ‡, §, ¶, #, ††, ‡‡,... Cite these symbols just as you would read a table—from left to right and from top to bottom, and reading across all spanner and subheadings for one column before moving on to the next. An exponential expression (e.g., × 10-3) in the units line is often necessary to keep the length of data values reasonably short. This ambiguous expression must be referenced with an explanatory note.

## Title and byline

A title gives the reader a clear idea of what the article is about; it should be brief and informative. Use common names for crops and avoid abbreviations. The usual limit for titles is 10 to 12 words (not counting "and," "of," and similar conjunctions and prepositions). Titles in a numbered series of articles may be longer. Below the title, list the names of all authors. Place an asterisk after the name of the corresponding author (i.e., the person from whom reprints are to be requested).

## Author–paper documentation

The author–paper documentation is a single paragraph. The first sentence lists the authors (without professional titles) and their complete, current addresses. If a paper has only one author, or if all authors are from the same department and institution, omit the names (i.e., give the address only).

*Publisher :* AGRONOMY; Journal: CROPSCI:Crop Science; Copyright: 2007 Volume: 47; Issue: all; Manuscript: instructions to authors; DOI: ; PII: <txtPII> TOC Head: Will notify...; Section Head: ; Article Type: OTHER Page 3 of 4 The second sentence lists institutional sponsors, with the institutional article number of similar contribution acknowledgment. Add such an acknowledgment if an author has moved and using the current address leaves no other mention of the involvement of the former institution.

Other information such as granting, funding, or dissertation status may follow.

End the author–paper documentation paragraph with these two statements: "Received ___________. *Corresponding author (e-mail)." The date received will be filled in by an editor.

## Abbreviations

Prepare a list in alphabetical order of abbreviations used in your article. Do not include SI units, chemical abbreviations, or most common abbreviations such as those listed in the style manual.

## Footnotes

Footnotes are discouraged in text, but may be used when needed (typically for a product disclaimer). Number any footnotes consecutively.

## Abstract

Abstracts are a single self-contained paragraph of no more than 1500 characters – including word spaces – for papers or 750 characters for Notes. Abstracts should contain the rationale, objectives, methods, results, and their meaning or scope of application. Be specific. Identify the crops or organisms involved, the soil type, chemicals, and other details that are pertinent to the results. Do not cite references.

## Nomenclature and identification of materials

Give the complete binomial and authorities at first mention (in Abstract or text) of plants, pathogens, and insects or pests.

## Units of measure

The SI system (Système International de Unités) is required in *Crop Science*. Other units may be indicated in parentheses after the SI unit if this helps understanding or is needed for replication of the work.

## References

The author–year notation system is required; do not use numbered notation. In the list, arrange references alphabetically by author. All single-author entries precede multiple-author entries for the same first author. Use chronological order only within entries with identical authorship (alphabetizing by title for same-author, same-year entries). Add a lowercase letter a, b, c, etc. to the year to identify same-year entries for text citation. Do this also for any multiple-author entries that would otherwise result in identical citations in the text.

## Cover submissions

If you have any images which highlight your paper, you may submit them along with your paper in the Manuscript Central system. Please be sure to label as "image" (not "figure"). Cover images need to be at least 300 dpi at actual size; further electronic image specifics can be found at www.asa-cssa-sssa.org/ publications/pdf/guiddigitalimage.pdf. Otherwise, a slide or glossy print provides high resolution and can be submitted directly to the Headquarters office on acceptance of your manuscript (attn: Managing Editor, *Crop Science*, 677 S. Segoe Road, Madison, WI 53711–1086). Be sure to label it with the manuscript number and title and provide a descriptive caption which will aid in the selection process.

## Supplemental materials

If you wish to include supplemental materials, you need to include these files with your submission. Please label the file as "Supplemental File" when you upload.

## Revisions

All revisions to the manuscript during the review process will be made by the author only, and revisions will be given the same manuscript number, with an R number on the end (e.g., CROP-2006–04–0017-ORA.R1). Each revision has the opportunity for another round of review—the manuscript status "awaiting reviewer selection" is automatic and does not indicate a resubmission.

## Publication Charges and Length of Manuscript

Full-length manuscripts accepted for publication in *Crop Science* are assessed a publication charge of $450 for members and $700 for nonmembers for the first seven pages; a charge of $100 per page after seven pages is also assessed. Authors are charged $10 per illustration, $10 per table, and $1000 per page of color.

## Accepted Manuscripts

When your manuscript is accepted, you will receive notification from your technical or associate editor, and the accepted files (word processing, PDF, and figure files in any format) will automatically be sent to Headquarters. You will hear from Headquarters on receipt of your files. A hard copy is no longer required; the figure files submitted will be used for press. The only item you will need to send to Headquarters on acceptance is the completed Publish and Republish form (www.asa-cssa-sssa.org/publications/pdf/permissiontoprint.pdf).

## Useful References

ASA–CSSA–SSSA. 2004. Publications handbook and style manual. 3rd ed. [Online]. Available at www.asa-cssa-sssa.org/publications/style/[verified 2 Feb. 2006]. ASA, CSSA, and SSSA, Madison, WI. Crop Science Society of America, Terminology Committee. 1992. Glossary of crop science terms. Available at www.crops.org/cropgloss/[verified 2 Feb. 2006]. CSSA, Madison, WI.

USDA-ARS-National Genetic Resources Program. 2005. Germplasm Resources Information

Network (GRIN) database. Available at www.ars-grin.gov/cgi-bin/npgs/html/taxgenform.pl [verified 2 Feb. 2006]. National Germplasm Resources Laboratory, Beltsville, MD. USDA-NRCS Soil Survey Division. 2005. USDA-NRCS Official Soil Series Descriptions Enter By Name. Available at http://ortho.ftw.nrcs.usda.gov/cgi-bin/osd/osdname.cgi [verified 2 Feb. 2006].

# CHAPTER - 35

# Economic and Political Weekly

## NOTES TO CONTRIBUTORS

Here are some guidelines of authors who wish to make submission to the journal.

**Special Articles**

*EPW* welcomes original research papers in any of the social sciences.

- Articles must be no more than 8,000 words, including notes, references and tables. Longer articles will not be processed.

- Contributions should be sent in a hard copy format accompanied; by a floppy/CD VERSION. A soft copy can also be sent by email. Hard and soft copy versions of articles are essential for processing.

- Special articles should be accompanied by an abstract of a maximum of 150-200 words.

- Papers should not have been simultaneously submitted for publication to another journal or newspaper. If the paper has appeared earlier in a different version, we would appreciate a copy of this along with the submitted paper.

- Graphs and charts prepared in MS Office (Word/Excel) or equivalent software are preferable to material prepared in jpeg or other formats.

- Every effort is taken to complete early processing of the papers we receive. Since we receive more than 35 articles every week and adequate time has to be provided for internal reading and external refereeing. It can take up to four months for a final decision on whether the paper is accepted for publication.

- Articles accepted for publication can take up to six to eight months from the date of acceptance to appear in the *EPW*. Every effort will, however, be made to ensure early publication. Papers with immediate relevance for policy would be considered for early publication. Please note that this is a matter of editorial judgment.

## Commentaries

**EPW** invites short contributions to the 'Commentary' section on topical social, economic and political developments. These should ideally be between 1,000 and 2,500 words and exclusive to the EPW.

Short contributions may be sent by email.

## Book Reviews

*EPW* sends out books for review. It does not normally accept unsolicited reviews. However, all reviews that are received are read with interest and where a book has not been sent out for review, the unsolicited review is on occasion considered for publication.

## Letters

Readers of *EPW* are encouraged to send comments and suggestions (300-400 words) on published articles to the letters column. All letters should have the writer's full name and postal address.

## Discussion

*EPW* encourage researchers to comment on Special Articles. Submissions should be 1,000 to 2,000 words.

## General Guidelines

- **Writers are requested to provide full details for correspondence: postal address, day –time phone numbers and email address.**

  (The email address of writers in the Special Article, Commentary and Discussion sections will be published at the end of the article).

- Authors are requested to prepare their soft copy versions in text formats. PDF versions are not accepted by the *EPW*. Authors are encouraged to use UK English spellings (Writers using MS Word or similar software could change the appropriate settings in the Language menu of the application).

- Contributors are requested to send articles that are complete in all respects, including references, as this facilitates quicker processing.

- When there are major developments in the field of study after the first submission, authors can send a revised version. *EPW* **requests writers not to send revised versions based on stylistic changes/additions, deletions of references, minor changes, etc, as this poses challenges in processing.**

- All submissions will be acknowledged immediately on receipt with a reference number. Quoting the reference number in inquiries will help.

- *EPW* posts all published articles on its web site and may reproduce them on CDs.

**Address for communication:**

*Economic and Political Weekly,*
Hitkari House,
284 Shahid Bhagatsingh Road,
Mumbai 400 001, India.
Email : edit@epw.org.in,
Epw.Mumbai@gmail.com

# CHAPTER - 36

# European Journal of Agronomy

## The Official Journal of the European Society for Agronomy

## GUIDE FOR AUTHORS

*The European Journal of Agronomy*, the official journal of the European Society for Agronomy, publishes original research papers reporting experimental and theoretical contributions to crop science in the following fields:

- crop physiology
- crop production and management
- agroclimatology and modelling
- plant soil relationships
- crop quality and post-harvest physiology
- farming and cropping systems
- agroecosystems and the environment

In determining the suitability of submitted articles for publication, particular scrutiny will be placed on the degree of novelty and significance of the research and the extent to which it adds to existing knowledge in Agronomy. Confirmatory research and results from routine cultivar or Agronomy trials will not normally be considered for publication. Review articles are normally written on invitation from the Editor-in-Chief. Authors intending to prepare review papers for the Journal are advised to consult the Editor-in-Chief before writing their reviews.

## Types of contribution

1. *Original research papers (regular papers).* Original research papers should report the results of original research. The material should not have been previously published elsewhere, except in a preliminary form.

2. *Review articles.* Review articles should cover subjects falling within the scope of the journal which are of active current interest. They are normally written upon invitation by the Editor-in-Chief. Intending authors should first consult the Editor-in-Chief.

## Online Submission of manuscripts

Submission of an article implies that the work described has not been published previously (except in the form of an abstract or as part of a published lecture or academic thesis), that it is not under consideration for publication elsewhere, that its publication is approved by all authors and tacitly or explicitly by the responsible authorities where the work was carried out, and that, if accepted, it will not be published elsewhere in the same form, in English or in any other language, without the written consent of the Publisher.

Upon acceptance of an article, authors will be asked to transfer copyright (for more information on copyright see http://authors.elsevier.com. This transfer will ensure the widest possible dissemination of information. A letter will be sent to the corresponding author confirming receipt of the manuscript. A form facilitating transfer of copyright will be provided.

If excerpts from other copyrighted works are included, the author(s) must obtain written permission from the copyright owners and credit the source(s) in the article. Elsevier has preprinted forms for use by authors in these cases: contact Rights Department, P.O. Box 800, Oxford, OX5 1DX, UK; phone: (+44) 1865 843830, fax: (+44) 1865 853333, e-mail: permissions@elsevier.com

Papers for consideration should be submitted to:Elsevier Editorial System

Submission to this journal proceeds totally on-line. Use the following guidelines to prepare your article. Via the homepage of this journal (http://ees.elsevier.com/euragr/ you will be guided stepwise through the creation and uploading of the various files.

The system automatically converts source files to a single Adobe Acrobat PDF version of the article, which is used in the peer-review process. Please note that even though manuscript source files are converted to PDF at submission for the review process, these source files are needed for further processing after acceptance. All correspondence, including notification of the Editor's decision and requests for revision, takes place by e-mail and via the author's homepage, removing the need for a hard-copy paper trail.

### Electronic format requirements for accepted articles

We accept most wordprocessing formats, but Word, WordPerfect or LaTeX is preferred. Always keep a backup copy of the electronic file for reference and safety. Save your files using the default extension of the program used.

### *Wordprocessor documents*

It is important that the file be saved in the native format of the wordprocessor used. The text should be in single-column format. Keep the layout of the text as simple as possible. Most formatting codes will be removed and replaced on processing the article. In particular, do not use the wordprocessor's options to justify text or to hyphenate words. However, do use bold face, italics, subscripts, superscripts etc. Do not embed 'graphically designed' equations or tables, but prepare these using the wordprocessor's facility. When preparing tables, if you are using a table grid, use only one grid for each individual table and not a grid for each row. If no grid is used, use tabs, not spaces, to align columns. The electronic text should be prepared in a way very similar to that of conventional manuscripts (see also the Guide to Publishing with Elsevier: http://www.elsevier.com/wps/find/authorshome.authors/howtosubmitpaper). Do not import the figures into the text file but, instead, indicate their approximate locations directly in the electronic text and on the manuscript. See also the section on Preparation of electronic illustrations.

To avoid unnecessary errors you are strongly advised to use the 'spellchecker' function of your wordprocessor

## Preparation of manuscripts

1. It is essential to give a fax number and e-mail address when submitting a manuscript. Articles must be written in good English. Authors whose native language is not English are strongly advised to have their manuscripts checked by an English-speaking colleague prior to submission.

   Language Polishing. Authors who require information about language editing and copyediting services pre- and post-submission please visit http://www.elsevier.com/wps/find/authorshome.authors/languagepolishing or contact authorsupport@elsevier.com for more information. Please note Elsevier neither endorses nor takes responsibility for any products, goods or services offered by outside vendors through our services or in any advertising. For more information please refer to our Terms & Conditions http://www.elsevier.com/wps/find/termsconditions.cws_home/termsconditions.

2. Manuscripts should be prepared with numbered lines, with wide margins and double spacing throughout, i.e. also for abstracts, footnotes and references. **Every page of the manuscript, including the title page, references, tables, etc. should be numbered.** However, in the text no reference should be made to page numbers; if necessary, one may refer to sections. Underline words that should be in italics, and do not underline any other words. Avoid excessive usage of italics to emphasize part of the text.

3. Manuscripts in general should be organized in the following order:
   - Title (should be clear, descriptive and not too long)
   - Name(s) of author(s)
   - Complete postal address(es) of affiliations
   - Full telephone and Fax no. of the corresponding author and one or more other authors (if any)
   - Present address(es) of author(s) if applicable
   - Abstract
   - Key words (indexing terms), normally 3–6 items

- Introduction
- Material studied, area descriptions, methods, techniques
- Results
- Discussion
- Conclusion
- Acknowledgements and any additional information concerning research grants, etc.
- References
- Tables
- Figure captions

4. In typing the manuscript, titles and subtitles should not be run within the text. They should be typed on a separate line, without indentation. Use lower-case lettertype.

5. SI units should be used.

6. Units and their abbreviations should be those approved by ISO (International Standard 1000:92

7. SI units and recommendations for the use of their multiples and of certain other units). Abbreviate units of measure only when used with numerals.

8. If a special instruction to the copy editor or typesetter is written on the copy it should be encircled. The typesetter will then know that the enclosed matter is not to be set in type. When a typewritten character may have more than one meaning (e.g. the lower case letter l may be confused with the numeral 1), a note should be inserted in a circle in the margin to make the meaning dear to the typesetter. If Greek letters or uncommon symbols are used in the manuscript, they should be written very clearly, and if necessary a note such as *"Greek lower-case chi"* should be put in the margin and encircled.

9. Elsevier reserves the privilege of returning to the author for revision accepted manuscripts and illustrations which are not in the proper form given in this guide.

## Abstracts

The abstract should be clear, descriptive and not longer than 400 words.

## Tables

1. Authors should take notice of the limitations set by the size and lay-out of the journal. Large tables should be avoided. Reversing columns and rows will often reduce the dimensions of a table.

2. If many data are to be presented, an attempt should be made to divide them over two or more tables.

3. Drawn tables, from which prints need to be made, should not be folded.

4.  Tables should be numbered according to their sequence in the text. The text should include references to all tables.

5.  Each table should be typewritten on a separate page of the manuscript. Tables should never be included in the text.

6.  Each table should have a brief and self-explanatory title.

7.  Column headings should be brief, but sufficiently explanatory. Standard abbreviations of units of measurement should be added between parentheses.

8.  Vertical lines should not be used to separate columns. Leave some extra space between the columns instead.

9.  Any explanation essential to the understanding of the table should be given as a footnote at the bottom of the table.

## Preparation of electronic illustrations

Submitting your artwork in an electronic format helps us to produce your work to the best possible standards, ensuring accuracy, clarity and a high level of detail.

### *General points*

- Make sure you use uniform lettering and sizing of your original artwork.
- Save text in illustrations as "graphics" or enclose the font.
- Only use the following fonts in your illustrations: Arial, Courier, Helvetica, Times, Symbol.
- Number the illustrations according to their sequence in the text.
- Use a logical naming convention for your artwork files, and supply a separate listing of the files and the software used.
- Provide all illustrations as separate files and as hardcopy printouts on separate sheets.
- Provide captions to illustrations separately.
- Produce images near to the desired size of the printed version.

A detailed guide on electronic artwork is available on our website: http://authors.elsevier.com/artwork

You are urged to visit this site; some excerpts from the detailed information are given here.

## Formats

Regardless of the application used, when your electronic artwork is finalised, please "save as" or convert the images to one of the following formats (Note the resolution requirements for line drawings, halftones, and line/halftone combinations given below.):

*EPS :* Vector drawings. Embed the font or save the text as "graphics".

*TIFF :* Colour or greyscale photographs (halftones): always use a minimum of 300 dpi.

*TIFF :* Bitmapped line drawings: use a minimum of 1000 dpi.

*TIFF:* Combinations bitmapped line/half-tone (colour or greyscale): a minimum of 500 dpi is required.

***DOC, XLS or PPT :*** If your electronic artwork is created in any of these Microsoft Office applications please supply "as is".

**Please do not**

- Supply embedded graphics in your wordprocessor (spreadsheet, presentation) document;
- Supply files that are optimised for screen use(like GIF,BMP,PIC,WPG) the resolution is too low ;
- Supply files that are too low in resolution;
- Submit graphics that are disproportionately large for the content. Colour illustrations

Please make sure that artwork files are in an acceptable format (TIFF, EPS or MS Office files) and with the correct resolution. If, together with your accepted article, you submit usable colour figures then Elsevier will ensure, at no additional charge, that these figures will appear in colour on the Web (e.g., ScienceDirect and other sites) regardless of whether or not these illustrations are reproduced in colour in the printed version. For colour reproduction in print, you will receive information regarding the costs from Elsevier after receipt of your accepted article. Please indicate your preference for colour in print or on the Web only. For further information on the preparation of electronic artwork, please see http://www.elsevier.com/artworkinstructions.

*Please note :* Because of technical complications which can arise by converting colour figures to 'grey scale' (for the printed version should you not opt for colour in print) please submit in addition usable black and white prints corresponding to all the colour illustrations.

## Supplementary files

Preparation of supplementary data. Elsevier accepts supplementary material to support and enhance your scientific research. Supplementary files offer the author additional possibilities to publish supporting applications, movies, animation sequences, high-resolution images, background datasets, sound clips and more. Supplementary files supplied will be published online alongside the electronic version of your article in Elsevier Web products, including ScienceDirect: http://www.sciencedirect.com. In order to ensure that your submitted material is directly usable, please ensure that data is provided in one of our recommended file formats. Authors should submit the material in electronic format together with the article and supply a concise and descriptive caption for each file. For more detailed instructions please visit our artwork instruction pages a http://www.elsevier.com/artworkinstructions.

## References

1. All publications cited in the text should be presented in a list of references following the text of the manuscript. The manuscript should be carefully checked to ensure that the spelling of authors names and dates are exactly the same in the text as in the reference list.

2. In the text refer to the author's name (without initial) and year of publication, followed if necessary by a short reference to appropriate pages. Examples: "Since Peterson (1988) has shown that..." "This is in agreement with results obtained later (Kramer, 1989, pp. 12-16)".

3. If reference is made in the text to a publication written by more than two authors the name of the first author should be used followed by "et al.". This indication, however, should never be used in the list of references. In this list names of first author and co-authors should be mentioned.

4. References cited together in the text should be arranged chronologically. The list of references should be arranged alphabetically on authors' names, and chronologically per author. If an author's name in the list is also mentioned with co-authors the following order should be used: publications of the single author, arranged according to publication dates - publications of the same author with one co-author - publications of the author with more than one co-author. Publications by the same author(s) in the same year should be listed as 1974a, 1974b, etc.

5. Use the following system for arranging your references:
    (a) *For periodicals*
        Muniez-Jolain, N.G., Ney, B., Duthion, C. 1993. Sequential development of flowers and seeds on the main stem of an intermediate soybean. Crop Sci. 33, 768-771.

    (b) *For edited symposia, special issues, etc., published in a periodical*
        Rice, K., 1992. Theory and conceptual issues. In: Gall, G.A.E., Staton, M. (Eds.), Integrating Conservation Biology and Agricultural Production. Agriculture, Ecosystems and Environment 42, 9-26.

    (c) *For books*
        Gaugh, Jr., H.G., 1992. Statistical Analysis of Regional Yield Trials. Elsevier, Amsterdam.

    (d) *For multi-author books*
        Snobas, B.A., Wilkins, D.E., Hadjichristodoulou, A., Haddad, N.I., 1988. Stand establishment in pulse crops. In: Summerfield, R.J. (Ed.), World Crops: Cool Season Food Legumes. Kluwer Academic Publishers, Dordrecht, pp 257-259.

6. In the case of publications in any language other than English, the original title is to be retained. However, the titles of publications in non-Latin alphabets should be transliterated, and a notation such as "(in Russian)" or "(in Greek, with English abstract)" should be added.

7. Work accepted for publication but not yet published should be referred to as "in press".

8. References concerning unpublished data and "personal communications" should not be cited in the reference list but may be mentioned in the text.

## Formulae

1. Leave ample space around the formulae.

2. Subscripts and superscripts should be clear.

3. Take special care to show clearly the difference between zero (0) and the letter O, and between one (1) and the letter l.

4. Give the meaning of all symbols immediately after the equation in which they are first used.

5. For simple fractions use the solidus (/) instead of a horizontal line.

6. Equations should be numbered serially at the right-hand side in parentheses. In general only equations explicitly referred to in the text need be numbered.

7. The use of fractional powers instead of root signs is recommended. Also powers of $e$ are often more conveniently denoted by exp.

8. Levels of statistical significance which can be mentioned without further explanation are $*P< 0.05$, $**P< 0.01$ and $***P< 0.001$.

9. In chemical formulae, valence of ions should be given as, e.g. $Ca^{2+}$ not as $Ca^{++}$.

10. Isotope numbers should precede the symbols, e.g. $^{18}O$.

11. The repeated writing of chemical formulae in the text is to be avoided where reasonably possible; instead, the name of the compound should be given in full. Exceptions maybe made in the case of a very long name occurring very frequently or in the case of a compound being described as the end product of a gravimetric determination (e.g. phosphate as $P_2O_5$).

## Footnotes

1. Footnotes should only be used if absolutely essential. In most cases it should be possible to incorporate the information in normal text.

2. If used, they should be numbered in the text, indicated by superscript numbers, and kept as short as possible.

## Nomenclature

1. Authors and editors are, by general agreement, obliged to accept the rules governing biological nomenclature, as laid down in the *International Code of Botanical Nomenclature,* and the *International Code of Nomenclature of Bacteria.*

2. All biotica (crops, plants, insects etc.) should be identified by their scientific names when the English term is first used, with the exception of common domestic crops.

3. All biocides and other organic compounds must be identified by their Geneva names when first used in the text. Active ingredients of all formulations should be likewise identified.

4. For chemical nomenclature, the conventions of the *International Union of Pure and Applied Chemistry* and the official recommendations of the *IUPAC-IUB Combined Commission on Biochemical Nomenclature* should be followed.

## Copyright

1. An author, when quoting from someone else's work or when considering reproducing an illustration or table from a book or journal article, should make sure that copyright is not being infringed.

2. Although in general an author may quote from other published works, permission from the holder of the copyright should be obtained if substantial extracts are taken or tables, plates, or other illustrations are reproduced. If the copyright-holder is not the author of the quoted or reproduced material, it is recommended that the permission of the author should also be sought.

3. Material in unpublished letters and manuscripts is also protected and must not be published unless permission has been obtained.

4. A suitable acknowledgement of any borrowed material must always be made.

## Proofs

One set of page proofs in PDF format will be sent by e-mail to the corresponding author (if we do not have an e-mail address then paper proofs will be sent by post). Elsevier now sends PDF proofs which can be annotated; for this you will need to download Adobe Reader version 7 available free from http://www.adobe.com/products/acrobat/readstep2.html. Instructions on how to annotate PDF files will accompany the proofs. The exact system requirements are given at the Adobe site: http://www.adobe.com/products/acrobat/acrrsystemreqs.html#70win. If you do not wish to use the PDF annotations function, you may list the corrections (including replies to the Query Form) and return to Elsevier in an e-mail. Please list your corrections quoting line number.

If, for any reason, this is not possible, then mark the corrections and any other comments (including replies to the Query Form) on a printout of your proof and return by fax, or scan the pages and e-mail, or by post. Please use this proof only for checking the typesetting, editing, completeness and correctness of the text, tables and figures. Significant changes to the article as accepted for publication will only be considered at this stage with permission from the Editor. We will do everything possible to get your article published quickly and accurately. Therefore, it is important to ensure that all of your corrections are sent back to us in one communication: please check carefully before replying, as inclusion of any subsequent corrections cannot be guaranteed. Proofreading is solely your responsibility. Note that Elsevier may proceed with the publication of your article if no response is received.

## Offprints

The corresponding author, at no cost, will be provided with a PDF file of the article via e-mail or, alternatively, 25 free paper offprints. The PDF file is a watermarked version of the published article and includes a cover sheet with the journal cover image and a disclaimer outlining the terms and conditions of use.

One hundred free offprints will be supplied to the first author of a review article.

Additional offprints can be ordered on an offprints order form, which is included with the proofs, UNESCO coupons are acceptable in payment of extra offprints.

## Author Services

Authors can track accepted articles at http://www.elsevier.com/trackarticle and set up e-mail alerts to inform you of when an article's status has changed, as well as copyright information, frequently asked questions and more.

Contact details for questions arising after acceptance of an article, especially those relating to proofs, are provided after registration of an article for publication.

**The European Journal of Agronomy carries no page charges**

# CHAPTER - 37

# European Journal of Soil Science

## INTRUCTIONS TO AUTHORS

(More extensive Guidance to Authors, a style file, a sample article and other useful electronic files are at *www.blackwellpublishng.com/ejs*)

The Journal exists to report the latest significant results of research in all aspects of soil Science and its applications, to describe new techniques, to provide up-to-date, authoritative and critical reviews, and to publish advanced educational papers. The Editor welcomes comments on papers in the Journal, which may be published as Letters to the Editor together with any replies from the authors, Contributions are invited from all countries; the preferred language of publication is English, but papers in French are considered. **Submissions must be accompanied by a covering letter stating that the paper has not been published previously elsewhere nor is being considered by another journal, in any language, and that all authors have seen and agreed to the version submitted.** There are not page charges excepts for colour (see below).

## Submissions

**The Journal encourages electronic submission of papers as Word files: please refer to the Journal website for more details.** Papers should be no longer than about 6500 words plus figures, tables and references. Authors will be asked to reduce papers that are significantly longer than this. The Journal does not publish 'Short Communications' as such,  but brief papers of about 2000 words or less are welcome; they should conform to the scientific , stylistic and format requirements of longer papers. Papers should be double-spaced and in single column format throughout, with ample margins. Number the pages, including those listing references, figure captions and tables, in a single sequence and use continuous line numbers, Follow the style of presentation used in recent issued of the Journal*. Please use a Times or similar typeface in 12 pt font. Provide a succinct title containing the keywords, and a short running head title of no more than 50 characters including spaces. At the beginning of the paper, provide an informative summary of no more than 250 words, outlining the scope and giving only the principal findings of the work. Papers in French must have an English summary in addition to a resume.

The Journal accepts supplementary material. For further information see the website.

## Units,Symbols And Spelling

Use SI units throughout, printing them in upright type (e.g. mmol kg). Symbols for measured or calculated variables, e.g. $K$ for hydraulic conductivity, should be typed in italics or underlined. English spelling should follow the *Concise Oxford Dictionary.*

## Citations In The Text

Do not cite unnecessarily and cite works only if you have read them. For each reference to the literature give the author(s) and they year; for example. Chertkov (2001). McGarry & Yule (2002), (Ghassemi *et al.* 1977; Siemens, 2003), with the parentheses placed according to context. Note that for a work by two authors cite both and for a work by more than two cite only the first followed by *et al.* Where several references appear together they are arranged chronologically and then alphabetically. Two or more papers by the same author(s) in the same year are differentiated by adding 'a', 'b' etc. after the year. Beware of transferring references obtained from searches of electronic databases; they are often truncated.

## Reference List

Give full details in the references at the end of the paper. For periodicals give all authors' names, year of publication, title of article and full name of periodical, together with volume number and full page members. For articles or chapters in books give the book title, editor(s), page numbers, publisher and place of publication . Do not cite unpublished documents other than theses. Arrange the references in alphabetical order by author; where the same author appears more than once, single-author papers come before two-author papers, which come before multi-author papers. Within each class arrange papers in chronological order.

The following examples are for guidance.

Chertkov, V.Y. 2002. Modelling cracking stages of saturated soils as they dray and shrink.*European Journal of Soil Science*, 53, 105-118.

Ghassemi, F.,Jakeman, A.J. & Nix, H.A. 1995. *Salinisation of Land and Water Resources*. CAB International, Wallingford.

McGarry, D.& Yule, D.F.2002.Shrinkage. In: *Encyclopedia of Soil Science* (ed. R.Lal), pp. 1197-1200. Marcel Dekker, New York

Siemens, J. 2003. *Controls of carbon, nitrogen, and phosphorus fluxes in vadose zone and groundwater of protected watersheds in Minister (Germany). Doctoral dissertation,* Technische Universitat, Berlin.

## Tables

Prepare tables separately from the text, and list each table separately with its heading and with double spacing if there is room. Design your tables so that they will fit either in a single column of the printed Journal or across two columns. Avoid large tables that are likely to occupy more than one page. Provide all columns with headings, with the first letter of each heading capitalized. Consult a recent issue of the Journal or the sample article* for examples of layout and format. Footnotes of tables should be referred to by superscripts letters ([a,b,] etc.).

## Figures

(See also Electronic artwork below) Please provide good quality images of figures. Figures will be reduced to approximately half their original size; they should be prepared for this reduction and also take into account the page size of the Journal. Final size illustrations can be single column (83 mm), double column (175 mm), or intermediate (115 mm), Lettering should be sans serif and of a size that will be 2-3 mm in height when reduced. Lines should be thick enough to be at least 0.33 mm in width when reduced, and symbols should be 2-3 mm across. Preferred symbols on graphs are Î.n.?.g.?.?. List figure captions separately. Note that in the full-text online version of the Journal figure captions may be truncated in abbreviated links to the full-screen version. Therefore, the first 100 characters of any caption should inform the reader of key aspects of the figure.

Photographs. Black and white photographs will be accepted when they illustrate some essential some point.

Color. Please see the form on the Journal website. Authors will be charged some or all of the cost. Colour in the online version of a paper is free of charge.

## Electronic Copies

All papers are typeset from an electronic version, details of which will be confirmed when your paper is finally accepted.

## Electronic Artwork

Please save vector graphics (e.g. line artwork) in Encapsulated Postscript Format (ESP), and bitmap files (e.g. half-tones) in Tagged Image File Format (TIFF). We cannot accept PowerPoint files. Detailed information on our digital illustration standards is available at *www.blackwellpublishing.com/authors/digill.asp*.

## Proofs and Offprints

Page proofs are supplied to authors for the purpose of correcting printing errors; all other corrections must be kept to an absolute minimum, as alterations at this stage are expensive and time-consuming. Authors should specify on the script the e-mail address to which proofs should be sent. An e-mail alert containing a link to a website from which the proof should be downloaded as a PDF (portable document format) file will be sent to this address. Authors will be provided with electronic offprint of their paper. When authors receive proofs they will also receive order forms for purchasing paper offprint, which should be returned to the offprint printer*at the same time as corrected proofs are returned to the Editor.

## Disposal of Original Materal

Blackwell Publishing will dispose of all hardcopy or electronic material submitted two months after publication. If you require the return of any material submitted, please inform the Editorial Office or Production Editor as soon as possible if you have not yet done so.

## Early Online Publication

The Journal is covered by Blackwell Publishing's *OnlineEarly* service. *OnlineEarly* articles are published online on Blackwell Synergy (*www.blackwell-synergy.com*) in advance of their publication in a printed issue. They are complete and final: they have been fully reviewed, revised and edited for publication, and the authors' final corrections incorporated. Because they are in final form, no changes can be made after online publication. The nature of *OnlineEarly* articles means that they do not yet have volume, issue. After print publication, and DOI remains valid and can continue to be found at *www.doi.org/faq.htm*l.

## Exclusive Licence Form

*Authors are required to sign an Exclusive License Form (ELF) which is a condition of publication. Signature of the ELF is a condition of publication and papers will not be passed to the publisher for production unless a signed form has been received. ELF does not affect ownership of copyright in the material. (Government employees need to complete the Author Warranty sections, although copyright in such cases does not need to be assigned.) After submission authors will retain the right to publish their paper in various medium/circumstances. You must send the signed original of the ELF when you send your paper for the first time.

*Consult the files available at *www.blackwellpublishing.com/ejs.*

*COS Printers Pte Ltd, 9 Kian Teck Crescent, Singapore 628875.

# CHAPTER - 38

# Everyman's Science

## GUIDELINES FOR SUBMISSION OF MANUSCRIPTS

1. Everymans's Science intends to propagate the latest message of science in all its varied branches to its readers and through them, to every one interested in Science or Engineering or Technology. Research articles usually meant for publication in periodicals devoted to particular branches of Science & Technology and addressed to specialised sections of the readers, are not appropriate for Everyman's Science. Instead, popular or easily intelligible expositions of new or recent developments in different branches of Science & Technology are welcome.

2. Manuscripts should be typewritten on one side of the paper with double spacing. Articles should be written generally in non-technical language and should not ordinarily *exceed 2000* words. Articles must be understandable by the average enthusiastic readers with some modest scientific background but outside the field. It should not be a review article in a specialised area. Without being too technical, it must also reflect state of the art situation in the field. *A summary* in 50 words should be submitted along with the paper highlighting the importance of the work. *Two copies* of the manuscript complete in all respects should be submitted. The title should be written in capital letters and name(s) of the author(s) should be given along with the Department, Institution, City and Country of each author.

3. *Illustration and Tables :* The size of illustration should be such as to permit reduction to about one-third. Legends and captions should be typed on a separate sheet of paper. Photographs should be on glossy paper with strong contrast in black and white. Typed tables should be in separate pages and provided with titles and their serial numbers. The exact position for the placement of the tables should be  marked in the script. Authors are specially requested to reduce the number of tables, illustrations and diagrams to a minimum ( maximum of 3).

4. *References :* References to be given on a selective basic, (maximum of 10 ) and the order of placement should be numerically with (a) name(s) of the author(s) (surname last), (b) name of journal in abbreviated form

according to the 'World list of Scientific Periodicals' and in italics, (c) volume number (in bold) (d) page number and (e) year of publication.

For citations of books the author's name should be followed by the (a) title of the book, (b) year of publication or edition or both, (c) page number, (d) name of publishers, and (e) place of publication.

5.  The Indian Science Congress Association and the Editors of Everyman's Science  assume no responsibility for statements and opinions advanced by the contributors to the journal.

## Reprints

The communicating author will receive one copy of the journal and 10 reprints free of cost.

All manuscripts and correspondences should be addressed to the Hony, Editor, Everyman's Science, The Indian Science Congress Association 14, *Dr. Biresh Guha Street, kolkata-700 017.* Email : iscacal@vsnl.net., Fax : 91-33-2240-2551.

# CHAPTER - 39

# Experimental Agriculture

## INSTRUCTION TO CONTRIBUTORS

### Submission of Manuscripts

Contributions for consideration for publication, accompanied by a covering letter or email, should be submitted to the following address:

Prof: M.K.V.Carr, Editor, Experimental Agriculture

Crop and Water Management Systems   (International) Ltd.,

Pear Tree Cottage, Frog Lane, Ilmington, Shipston-on-STOUR,

Warwickshire CV36 4LG, UK.

Email: *mikecarr.rtcs@freeUK.com*

Please consult the Instruction to Authors published in the first issue of each Volume and also available at *http://uk.cambridge.org/journals/eag/eagifc.htm.* Potential contributors are asked to give careful attention to these instructions. This will greatly assist the editors and thus speed the processing of their contributions.

### Editorial Policy

With a focus on the tropics and sub-tropical regions of the world, Experimental Agriculture publishes the results of original research on field, plantation and herbage crops grown for food or feed, or for industrial purposes, and on farming systems, including livestock and people. It reports experimental work designed to explain how crops respond to the environment in biological and physical terms, and on the social and economic issues that may influence the uptake of the results of research by policy makers and farmers. The journal also publishes accounts and critical discussions of new quantitative and qualitative methods in agricultural research, and of contemporary issues arising in countries where agricultural production needs to develop rapidly. There is a regular book review section and occasional, often invited, reviews of research. Each paper is critically reviewed by referees, one of whom, where appropriate, may be asked to comment specifically on biometric aspects. On their advice the Editor accepts or rejects the paper, or returns the typescript to the author(s) for revision.

The minimum standards for a paper to be considered by the Editor and the referees are set out below:

- The whole of the typescript, including the summary, table and figure caption and references, must be double spaced.

- The title page, all headings and the reference must conform to the style of Experimental Agriculture.

- Each table and figure must be on a separate sheet.

- Master copies of the original artwork together with photocopies must be provided for each figure.

- Figures should be in a sans serif typeface.

- The preferred position of each table and figure must be marked in the text of the typescript.

- The statistical treatment of experimental data must conform to the instructions given in Riley, J.(2001). Presentation of statistical analyses. Experimental Agriculture 37: 115-123.

# CHAPTER - 40

# Farming Systems

## INSTUCTIONS TO AUTHORS

It is the policy of the Editorial Board to publish papers based on original research in any branch of Agricultural Science related to farming systems. The Editorial Board has the right to accept or reject a paper and the Board does not shoulder any responsibility for the opinion(s) expressed in the paper by the author(s). Once a paper is accepted for publication, it should not be published elsewhere either in the same or abridged form or in any other language, without the written permission of the editor.

Manuscripts should be typed double space, on bond paper (size 8 ½ × 11 inch) on one side of the page only with at least 1 ½ inch margin on the left and ½ inch margin on the right, and the original and the first copy of the typed paper should be sent. The language should be simple and care should be taken to check up the spellings, punctuations, etc. All tables should be serially numbered, should not be too lengthy and should have a heading stating concisely the contents. Each table should be typed on separate sheet and not with the running matter.

The paper in general be divided into the following parts in the same sequence (i) Abstract (ii) Introduction or Introductory Paragraph (iii) Material and Methods (iv) Results and Discussion and (v) References. Acknowledgement if any may be included in the last paragraph of the text before the' References' part.

The abstract not exceeding 150 words should comprise a brief and factual summary of the contents and conclusions of the paper and refer to any new information it may contain indicating the relevance of such information.

'Introduction' should be brief and relevant to the subject matter covered by the research work. While brief review of the literature should be incorporated in the 'Introduction'. 'Material and Methods' should be precise and whenever the methods of other authors are followed, it would suffice if reference to their paper is made, instead of repeating the procedure. The 'Results' should govern the presentation of experimental data only and each of the experiments should be properly titled. Common names of plant species, micro-organisms, insects etc., should be supported with authentic latest Latin names, which should be underlined in the typescript. When such names are first mentioned, the full generic name and the specific name with the authority should be mentioned and in subsequent references, the generic name be abbreviated and the authority deleted. The 'Discussion' should be brief, but should properly interpret all experimental findings.

The illustrations and text-figures should be clear and capable of reproduction in print. They should be drawn neatly on white drawing paper of Bristol boar in Indian ink with serial numbers in clear, uniform, large enough letters to enable suitable reduction. Original Photo prints duly indicating the magnification should be on glazed paper with proper contrast and at least of half plate size pasted on thick mount. Photographs should be incorporated only when essential. Legends to figures and/or photo prints should be typed on separate sheet.

'References' should be typed double space along with the body of the paper but starting on a fresh page. These should be listed in the alphabetical and chronological order. In the text the references should be cited as Rao (1956) OR (Rao, 1956). Chessin and Scott (1955) or (Chessin and Scott, 1955). Porter et al. (1960) or (Porter et al. 1960) when there are more than two authors. While listing the 'References' the following examples should be followed:

Mathur, R.S., 1956, *U.P. Agriculture Anim. Husb., Uttar Pradesh,* 6: 12-114.

Jackson, M.L., 1967, *Soil Chemical Analysis,* Prentice Hall of India Pvt., Ltd., New Delhi, pp. 498.

Channe Gowda, H., 1971, *Master's Thesis,* Univ. Agric. Sci., Bangalore.

The names of journals, should be abbreviated as given in the fourth edition (1964-65) of the 'World List of Scientific Periodicals.' If an article has not been seen in the original, the fact should be indicated against such references. The papers cited in the text should only be included in the 'References'.

The conventional symbols and abbreviations may be used in the manuscript in addition to the following: degree Celsius $^0$mC; hour/s, hr; hectare, ha; minute/s, min; second/s, sec; molar, m; milliliter, ml (avoid using cc); milligram, mg; micro, m; microgram, mcg or mg; 100%, 100 percent (in running matter); Kilogram/s, kg; etc.

Abbreviations should be used sparingly and could be done with if of advantage to the reader. Short words for units of measurements such as hectare, acre, day, inch, mile, tonne, week, etc., should ordinarily be spelled out. Do not use abbreviations in the title or abstract of the paper.

Ordinarily the length of the article should not exceed 10 type written pages, including the Tables and Figures.

The senior author shall be entitled for the supply of 10 reprints of the papers, free of cost.

Additional reprints can be obtained on payment of actual cost. The request for reprints should be sent immediately after the communication regarding acceptance is received. The contribution of papers is open to the subscribers only.

All correspondence relating to the Journal and papers for publication should be addressed to the Director, Institute for Studies on Agriculture and Rural Development, Belgaum Road, Dharwad – 580 008.

# CHAPTER - 41

# Filed Crops Research

## Manuscript Submission

Authors are encouraged to submit their manuscripts to the journal electronically, by using the EISubmit submission tool at *htt://authors.Elsevier.com/journal/fcr.* After registration, authors will be asked to upload their article and associated artwork. The submission tool will generate a PDF file to be used for the reviewing process.

Please follow the instructions for preparation of manuscripts and the Guide to Online submission at the above web address.

- *USA mailing notice :* Field Crops Research (0378-4290) is published monthly with 3 additional issues by Elsevier B.V.(P.O.Box 211, 1000 AE Amsterdam, The Netherlands). Annual subscription price in the USA US$2216 (valid in North, Central and South America), including air speed delivery. Application to mail at periodical postage rate paid at Jamaica, NY 11431.

  *USA POSTMASTERS :* Send address changes to Field Crops Research, Publications Expediting, Inc.,200 Meacham Ave, Elmont, NY 11003.

  *AIRFREIGHT AND MAILING* in the USA by Publications Expediting Inc., 200 Meacham Avenue, Elmont, NY 11003.

# Chapter - 42

# Financing Agriculture

## GUIDELINES FOR AUTHORS

Editorial Focus and Readership

Financing Agriculture is an in house journal of Agricultural Finance Corporation Ltd. It covers all aspects of agriculture: policy and planning; development and management of agriculture and allied activities; rural socio-economic development; environmental considerations; including the role of agriculture and agro-based activities as also institutional credit in maintaining a sustainable base for agricultural production at the micro and macro levels.

Articles submitted should be of interest to the broadest possible cross-section of *Financing Agriculture's* readership, which represents the following groups.

- Government officials, particularly in Central and State services and the administrative structures responsible for overall agricultural and rural development, as well as staff of banks, financial institutions dealing with agriculture and allied activities as also rural development.
- Agricultural colleges, institutes and universities, including their research scholars.
- Individual professionals.

## Submission of Articles

*Financing Agriculture* prefers to Publish original articles written specifically for the journal, as opposed to papers prepared for presentation or publication elsewhere. The journal occasionally accepts previously published articles if they are judged to be of particular value.

*Financing Agriculture* reserves the right to edit all copy as deemed appropriate for length and the overall style of the journal. The Editors will attempt to maintain the style and point of view of the author(s).

## Copy Specifications

Articles should be between 2000 and 4000 words in length. However, shorter or longer articles may be considered or specifically requested in certain cases.

Text should be submitted in double spaced typewritten format.

Please provide full name of authors(s), title and contact address (including pincode, telephone, fax and e-mail if available) of the author(s) and including a brief abstract or summary of the article.

## Illustrative Material

Authors are encouraged to provide supporting illustrative material with manuscripts. Tables, graphs, maps and drawings should be separate from the body of the text, with the point of reference indicated in the text. Redundant, long tables should be avoided. Authors should provide precise data for possible re-elaboration.

Photographs are printed in black and white; however, originals may be submitted either as black and white or colour prints. Original material will be returned on request.

## Style

Articles should be written in plain, concise language and in a style that is accessible and interesting to agriculture professionals in general, and not only to specialists in the topic concerned. Jargon should be avoided and technical terms that may be unfamiliar to readers should be defined the first time they appear.  Footnotes should be avoided as far as possible.

All measurements should be given in the metric system. Numerical data should be given in hundreds, thousands, lakhs and crores as the case may be.

## Abbreviations

Abbreviations and acronyms should be defined the first time they are used; for example the National Bank for Agriculture and Rural Development (NABARD)

## Bibliography

Articles should be accompanied by appropriate bibliographies.  The name of the author(s) and date of publication should be indicated at appropriate points in the text with the full reference given in a separate list at the end of the article.

Authors may please note that :

- A token honorarium of Rs. 500/- (Rupees five hundred only) will be paid for an article accepted for publication.

- The author will receive five complimentary copies of the issue.

- Decision of the Board of Editors about acceptance of the article is final.

Articles should be sent to:

The Editor,

Agricultural Finance Corporation Ltd.

Dhanraj Mahal, 1st Floor, Chhatrapati Shivaji Maharaj Marg,

Mumbai – 400 001, Tel : 022 – 2202 8924, Fax : 022 – 2202 8966

E-mail : *afc@vsnl.com*, E-mail : *piplapure@afcmails.com*

# CHAPTER - 43

# Forage Research

## Instructions to Authors

FORAGE RESERARCH publishes research articles of original value in the field of plant breeding and genetics, agronomy, biochemistry, nutrition, plant protection, physiology and production and processing technology of forage crops and also covers other related disciplines of sufficient interest to a scientist concerned with aspects of forage production and its utilization. It also includes review articles and short communications. Authors(s) must be a member of the Indian Society of Forage Research except in case of invited articles. A volume consists of four numbers.

Manuscript, neatly typed in double space, on one side of the bond paper ( 8½ " × 11") with reasonable margin and thoroughly revised, should be sent in duplicate. All correspondence regarding publication should be addressed to the Editor, Forage Research, CCS Haryana Agricultural University, Hisar – 125 004. It will be presumed that the author(s) have obtained the official approval, wherever necessary, and the papers are understood to be offered to FORAGE RESEARCH exclusively. The responsibility for statements, whether of fact or opinion, would rest entirely with the writers thereof. Papers should not exceed the limit of six printed pages otherwise authors would be required to meet the additional charges.

Manuscripts must be in English and subdivision of articles into Summary (not more than 200 words), Introduction, Materials and Methods, Results, Acknowledgements and References is recommended. Results and Discussion, preferably be given in a single section. Authors' full name(s), and complete address should be included on the title page and the name of the author to whom the proofs are to be sent should be given. A short title not exceeding 30 letters should also be provided on the title page.

Tables should be typed with double spacing on separate pages and should be provided with headings. Places at which tables are to be inserted should be indicated in the text.

Figures should be in a form suitable for reproduction drawn in Indian ink on drawing, butter paper or tracing paper with lettering, etc. and the reduction should be clearly indicated. Photographs should be properly arranged and large glossy prints of good quality not exceeding 1/3 of the text should be provided. The additional plates will be published at author's cost. These should be clear and relevant to the subject. Legends for figures should be typed on a separate page. Tables and illustrations should not reproduce the same data.

Two copies of revised manuscripts after revision/corrections as suggested by the referee should be sent also with the original copy accompanied by a soft copy on CD PM5/MS word.

References should include the author's name (family name if first author precedes initials, initials precede family names of all co-authors), year of publication, title of publication (abbreviated in accordance with the fourth edition of the World List of Scientific Periodicals), volume number and first and last page numbers. References to books should, in addition, include the editor's name, the edition number, where appropriate, the publisher's name and place.

### Example
Yadav, R.,S.K. Puhuja, and J.V Singh, 2005 : *Forage Res.*, 30 : 20-245.

### Book Reference
Ronney, W.L., and C.W.Smith, 2000 ; In S*orghum : Origin, History, Technology and Production,* and R.A. Frederiksen (eds.). Wiley, New York. Pp. 329-347.

# CHAPTER - 44

# Genetics

## Instructions For Contributors

GENETICS publishes contributions that present the results of original research in genetics and related scientific disciplines. Although GENETICS is an official publication of the Genetics Society of America, contributors are not required to be members of the Society: publication in the journal is open to members and nonmembers alike. These instructions describe how to correctly format a manuscript for submission to GENETICS.

## Submiting to Genetics

*Do not submit manuscript directly to a member of the Editorial Board.* You must upload your manuscript online at submit genetics. Org. Please ensure that your manuscript conforms to the proper format, described below. When you upload your manuscript, please provide the names of at least two associate editors. If not, we will assign one according to subject matter expertise. If you have trouble submitting your paper electronically, please contact the Editorial Office at

*Genetics-gsa@andrew.cmu.edu*

(412) 268.1812 (phone)

(412) 268.1813 (fax)

or write

Managing Editor

GENETICS

Mellon Institute, Box I

4400 Fifth Avenue

Pittsburgh, Pennsylvania 51213-2683

## Preparation of Research Articles

Manuscripts must be :

- written in English with American spelling and correct grammar and punctuation

- in double-spaced, 12-point type throughout, including the Literature Cited section, appendices, tables, and legends
- printed on a single side of letter-sized paper with 1-inch margins(double-sided manuscripts may be refused)
- held together with clips, not staples
- marked with consecutive page numbers, beginning with the cover page.

*Do not* number lines in the margins. Start each element of the manuscript on a new page:

1. The **title page** must contain :
   - a concise and informative title that includes the organism under study
   - the authors' names
   - the authors' institutional affiliations, including department; institution; and city, state or province, country if outside the United States, and postal code. Do not include districts or street addresses.

Indicate different affiliations with the symbols *, , ,§, **, , , §§.

Indicate an author's present address with a numbered footnote.

Indicate a dedicatory footnote (if desired) with boldface type.

List sequence accession numbers in an unnumbered footnote on page 1 using the following wording:

Sequence data from this article have been deposited with the EMBL/GenBank Data Libraries under accession nos. XXXXXX-XXXXX.

A paper otherwise found acceptable will not be published until accession numbers are provided. See the section below, **Sequences**, for more information on obtaining accession numbers.

2. **Page 2** must contain :
   - a short running head of about 35 characters, including spaces
   - up to five keywords or phrases
   - one corresponding author's name, mailing address including street name and number, phone and fax numbers, and e-mail address.

The corresponding author is the person responsible for checking the page proofs, ordering offprints, and arranging for the payment of page and author alteration charges. (Only one author name may appear in the correspondence footnote.)

3. The **Abstract** must be a single paragraph that :
   - does not exceed 200 words
   - does not contain references

4.  The **text** must be as succinct as possible. Present and discuss results just once, never in both the Results and Discussion sections. Discuss a possible explanation only if the results presented make a difference by allowing a resolution or posing a conundrum or paradox; ideas should not be discussed solely for the sake of completeness.

Manuscripts should conform to the style in recent issues of GENETICS, with particular  attention to genetic symbols, in-text references, and the Literature Cited section. Differentiate  between letters and numbers where they might be easily misrepresented, as with the typed letter 1 and the number 1 or the letter O and the number 0.

*Text citations :*   In reference citations with two authors, include both names. In citations with three or more authors, name the first author and us "*et al.*" Cite only articles that are published or in press. When citing personal communications or unpublished results, list all names including initials; do not use  "*et al.*"

*Numbers :* In the text, write out numbers nine or less except as part of a date, a fraction or decimal, a percentage, or a unit of measurement. Use Arabic numbers for those larger than nine, except as the first word of sentence; however, try to avoid starting a sentence with such a number.

*Abbreviations :* Use abbreviations of the customary units of measurement only when they are preceded by a number: "3 min" but "several minutes." Write "percent" as one word, except when used with a number: "several percent" but "75%." To indicate temperature in centigrade, use ° (for example,37°); include a letter after the degree symbol only when some other scale is intended (for example, 45° K).

*Binomial names :*   Italicize names of organisms only when the species is indicated: Neurospora, but *Neurospora crassa* or N. *crassa*.

Italicize the first three letters of the names of restriction sites, as in *Hind*III. Write the names of strains in roman except when incorporating specific genotypic designations.

*Genotypes :* Italicize genotype names and symbols, including all components of alleles, but not "+" indicating wild type and not when the name of a gene is the same as the name of an enzyme. Carefully distinguish between genotype and phenotype in both writing and symbolism.

*Headings :* Text should be divided into the following sections: Abstract, Introduction, Materials and Methods, Results, and Discussion.

Use only the following levels of headings:

- Level 1: Centered, all capital letters

**Example**

**Discussion**
Use this level only for major sections of text, such as Materials and Methods and the Discussion.

- Level 2 : Freestanding flush-left boldface

**Example**

**Background and Analysis**

Use this level only to group two or more closely related level 3 headings in long articles.

- Level 3 : Paragraph-initiating boldface, followed by a colon

**Example**

*Text :* Manuscripts must be as succinct as…..

This is the most frequently used subheading.

- Level 4 : Paragraph-initiating italic

**Example**

*Binomial names :* Italicize names of organisms…..

Only level 4 headings may be numbered, but *only* when the numbers must be cited in the text.

5. The Acknowledgments (optional) must be a single paragraph that :
   - immediately follows the discussion
   - includes all references to grant support and institutional publication codes
   - includes the full names of all granting agencies (example: National Institutes of Health).

6. The literature Cited section lists only articles that are published or in press.
   - Citations should follow the format of a recent issue of GENETICS; cite parenthetical references in text chronologically.
   - Insert a space between an author's initials.
   - Order citations alphabetically by first author.
   - For multiple citations with the same first author, first list single-author entries by year using 1996a, 1996b (etc) as needed. Then list two-author entries alphabetically by second author. Finally, list entries by three or more authors(cited in the text as "First *et al.*1996") only by year and without regard to number of authors or alphabetical rank of authors beyond the first.
   - For articles with more than five authors, list the first five names and then "*et al.*"

*Sample journal article citation*

BRIDGES, C. B., and E. G. ANDERSON, 1925 Crossing over in the X chromosomes of triploid females of *Drosophila melanogaster.* Genetics 10: 418-441.

(Note spaces between authors' initials and after the boldface colon)

***Sample book citation***

STURTEVANT, A. H., and G. W. BEADLE, 1939 *An Introduction to Genetics*. W. B. Saunders, Philadelphia.

***Sample chapter-in-book citation***

BEADLE, G. W., 1957 The role of the nucleus in heredity, pp.3-22 in *The Chemical Basis of Heredity*, edited by W. D. McELORY and B. GLASS. Johns Hopkins Press, Baltimore.

7.  Appendices should be reserved for lager bodies of data or arcane mathematical derivations that would disrupt the main text. Do not use appendices for briefer items requiring, for example, fewer than two typed pages.

8.  The tables must each start on a new pages.
    - Give each a concise title.
    - Double space all parts
    - Number each consecutively with Arabic numerals. Do not number consecutive tables 1A, 1B, etc., although interior parts of a table can be labeled A,B, etc., if necessary for easy reference in the text (but avoid this whenever possible).
    - Define the boxhead and the bottom of the table with horizontal rules. Use shorter horizontal rule within the boxhead to indicate unambiguously which subheadings are subordinate to a higher-level heading. Use no vertical or diagonal rules. Use no horizontal rules between the boxhead and the closing rule. Give each column a title in the boxhead. Except in unusual circumstances, each boxhead entry should refer to material beneath it and not to material to the right.
    - A table legend, if any, should precede footnotes.
    - Indicate footnotes, typed directly below the table, with lowercase, superscript italic letters. Use *,**, and *** to indicate conventional levels of statistical significance, explained below the table.
    - Do not include shading, color type, line drawings, graphics or other illustrations within tables. Instead, prepare a separate, numbered figure to accompany the table.
    - Table files must be Word or Excel documents and not in Picture Format.

9.  The figure legends must be with a brief title leading into text. All conventional symbols used to indicate figure data points are available for typesetting; unconventional symbols should not be used.

10. The figures must be numbered consecutively using Arabic numerals. Provide reproduction-quality figures as well as figures on disk.
    - Write the name of the author(s) and the figure number on the back corner of each figure.

- Indicate "up" with an arrow on the back of each figure 1A, 1B, etc. Parts of a figure can be labeled A,B,etc., if necessary for easy reference in the text.
- If a figure is submitted in unattached parts, include a sketch of how these should be arranged in the printed version.
- Halftones should be high-contrast, is particularly important for chromatographs, such as gel separations.
- Make the size and proportion of each group of figures suitable for reproduction at, or reduction to, 3 ½ or 7 inches wide.
- Color photographs should be avoided if possible.
- Drawings, graphs, mating-type charts, complex chemical formulas, and other sketches should  be treated as numbered figures.
- Be sure to distinguish between similar characters, such as genotypes, use italic rather than underlined roman characters.
- Do not use bold fact labels.
- Be sure that the axes of graphs are exactly perpendicular and that all lines are of equal density.
- The size of numbers and letters in a figure should be suitable for reduction to no greater than 10 points and no less than 6 points. Use similar size lettering throughout a figure and in similar figures.
- Some figures summarize large data sets of interest to other investigators. Examples include genetic maps based on DNA polymorphisms. The data sets supporting such figures should be placed on the journal's website as supplemental data so that future researchers can build upon these studies.

11. Sequences may appear in text or in figures.
    - DNA must be sequenced on both strands.
    - New nucleotide data must be submitted and deposited in the DDBJ/ EMBL/Gen Bank databases and an accession number obtained before the paper can be accepted for publication. Submission into any one of the three collaborating databanks ensures data entry in all. Sequence data may be submitted on the World Wide Web:

    DDBJ via SAKURA at *http://sakura.ddbj.ing.ac.jp/*
    EMBL via WEBIN at  *http://www.ebi.ac.uk/embl/Submission/webin.html*
    GenBank via BankIt at *http://www.nebi.nlm.gov/BankIt/*or vai Sequin *at http:/ /www.ncbi.nlm.nih.gov/Sequin/*

    - DNA,RNA, or protein sequences corresponding to equal or greater than 50 nucleotides must be entered into an appropriate data bank and the accession number must be provided before an article is published. If a proof is found to lack an accession number, the article will be held back from publication until the number is provided.

- Long sequences, such as those requiring more than two pages to reproduce, will not be published unless the reviewers and the Associate Editor agree that publication is necessary. An author wishing to include a long sequence not recommended for publication must agree to pay increased (unsubsidized) page charges for the extra pages.

- Complete DNA sequences will no longer be published in GENETICS. Relevant comparisons will be published if an accession number for the sequence in question is supplied.

- Authors must indicate that an unpublished sequence will be made available on disk or as hard copy for anyone requesting it.

## Submission of Other Articles

***Resubmission of rejected manuscripts*** is permitted a single time, unless repeated resubmission is encouraged by the Associate Editor. When submitting a previously rejected paper to a new Associate Editor, authors must fully disclose the paper's history with the journal.

***Perspectives*** contributions are encouraged, but because these articles are scheduled many months in advance and must adhere to stringent copy deadlines, a Perspectives articles may not be uploaded electronically. Authors receive free offprints and page charges are waived. The editors are

James F. Crow, Genetics Department

William F. Dove, McArdle Laboratory

University of Wisconsin

Madison, Wisconsin 53706

Genetics Education articles provide a scholarly forum for persons who have created new teaching instruments in their roles as educators. Authors are invited to submit articles that focus on effective ways for students at all levels to learn principles of genetics and to appreciate the implications of research in genetics. Genetics Education articles may be uploaded electronically. The editor is

Patricia J. Pukkila

Department of Biology

University of North Carolina

Chapel Hill, North Carolina 27599-3280

***Notes*** are intended for the presentation of brief observations that do not warrant full-length articles. Submit Notes as you would a full-length article (including electronically); they will receive the same review. They are not considered preliminary communications but should complete the study.

- Each Note must have an abstract of 50 or fewer words.

- Do not use section headings in the body of the Note; report introduction, results, and discussion in a single section. Level 3 headings are permissible.

- The text should be kept to a  minimum and if possible should not exceed 1000 words; the number of figures and tables should also be kept to a minimum.

- Materials and methods should be described in the figure legends or table footnotes.

- Place acknowledgments at the end without a heading.

- The Literature Cited section is identical to that of full-length articles.

***Letters to the Editor***  dealing with research and theory in basic genetics or with social issues of particular interest to geneticists are welcomed. Letters to the Editor should be uploaded electronically. Constructive comments on the subjects of articles from recent issues of GENETICS are appropriate. Figures, complex tables, and complex mathematical formulas should be avoided. The Editorial Board will choose to publish those letters it considers most pertinent to the interests of the readers.

## Editorial Review and Acceptance

All articles will be examined by one or more reviewers selected for their competence in the subject matter of the article. Acceptance will be based on scientific merit, clarity of expression, objectivity of writing, and suitability for GENETICS. An article may be accepted in its original form or subject to revision by the author. The reviewers' comments will be provided to the author. Reviewers will not be identified except at their own request.

Manuscripts will not be returned to authors unless requested by the corresponding author at the time of submission.

Each published article will show the date that the original manuscripts was received and the date that the final version was accepted at the GENETICS Editorial Office.

Label all *computer disks* with the first author's last name, platform (PS or Mac), and the software used.

1. The manuscript disk must contain that text of the article, but not the figures. PDFs are not sufficient but should accompany a mathematics file.

2. The disk for mathematics manuscripts should contain a Microsoft Word file if possible (to prevent any delay in publication) or all associated LaTeX files if that software was used (including the Literature Cited section) and a PDF if any math-specific software (e.g., TeX, Mathematica) was used. Rich text format (RTF) files are not acceptable.

3. Electronic images must be provided on a second disk, separate from the text. Publication-quality original artwork (hard copy) must be submitted

with the disk and must be identical to the files (inconsistencies may result in delays or additional charges). Art files must conform to the following minimum resolution specifications:

- grayscale and color images: 300 dpi
- combination grayscale and color images: 500-900 dip
- line art (Bitmap) images : 900-1200 dpi.

## Suggestions for Preventing the Return of a Manuscript

While manuscripts with perfect format are unprecedented and many errors can easily be corrected by the copy editor, the following suggestions are likely to prevent a manuscript from being returned to the author for correction after it has passed scientific review but before it can be sent to the printer:

- Use double spacing *everywhere* to provide space for the copyeditor to write typesetting instructions. This is particularly important in the literature Cited section and the table. "One-and half" spacing is inadequate.

- Print on one side of the paper only; do not submit double-sided manuscripts.

- Type long tables on several pages. Use wide paper or several standard-size sheets joined together for wide tables.

- Distinguish between phenotype and genotype for the reader not well versed in the subject. *Italicize* genotypes in the text, tables, legend, and figures. Because figures are reproduced photographically, genotypes in figures must be presented in italics. The genotype consists of gene names and allele designations; chromosome numbers or names need not be in italic type. Strain names should be presented in Roman unless the strain designation is the same as the genotype.

## Providing Unique Research Materials

Because the discoveries of science require continual verification, and progress in science depends so strongly upon prior investigations, it is crucial that key research materials developed by one scientist be made readily available to others. By publishing in GENETICS, authors describing unique research materials agree to provide them at reasonable cost to colleague who request them. Examples of research materials are strains, gene clones, antibodies (including cell lines producing monoclonal antibodies), and computer programs. A colleague is any active investigator, whether or not in training. The donor may require the recipient to agree neither to use such materials for commercial purposes nor to transfer them to a third party without the consent of the donor.

## Proofs

- The corresponding author will receive page proofs, the offprint order form, and proofreading instructions electronically as a PDF.

- Proofs must be printed (single, not double sided), corrected, and mailed to the printer within 2 days after receipt using a trackable express service.
- The article number must accompany all correspondence and appear on page 1 of proof.
- The entire article should be proofread carefully, with extra attention to equations and symbols.
- The article may be delayed to a later issue if the corrected proofs are not returned promptly.
- All corrections must be sent together, simultaneously; additional corrections may cause extra charges and delay in publication.
- The original manuscript or figures can be returned to the author after publication, but such a request must be made before the manuscript has appeared in print.
- "Notes added in proof" require editorial approval. A copy should be sent to the Managing Editor at the same time that proofs are returned to the printer to avoid postponing the article to a later issue.

## Charges to the Author

- Page charges are $65 per printed page. While no article will be refused because of an author's inability to page charges, authors citing grant support are assumed able to pay them. Authors lacking the requisite funds should explain their situation when submitting articles.
- Members of the Genetics Society of America will pay $100 for each color figure in an article, while nonmembers will be charged $400 for the first color figure and $250 for each subsequent color figure in an article.
- Authors are expected to pay all the extra costs of printing and handling an article that are made necessary by author alterations-changes in galley proof not due to printer's error-or by changes made necessary because the printer could not interpret the author's intention owing to poorly prepared copy, incorrect file formatting, extra rounds of corrections, or inadequately organized tables. Such charges will not be waived.
- Offprint order blanks will be sent to the author with the proofs. A price schedule is included on the order blank; it is also available from the GENETICS Editorial Office.
- All charges will be billed with one invoice soon after the article has been printed.

## Correspondence Regarding manuscripts

Manuscripts must be uploaded electronically at *http://submit.genetics.org* or submitted to the Editorial Office:

GENETICS Editorial Office
Mellon Institute, Box I
4400 fifth Avenue
Pittsburgh, AP 51213-2683
Telephone : (412) 268-1812
Fax :  (412) 268-1813
E-mail : *genetics-gsa@andrew.cmu.edu*

Correspondence regarding manuscripts may be sent directly to an appropriate member of the Editorial Board or to the GENETICS Editorial Office.

# CHAPTER - 45

# Genome

## Instructions to Authors

Genome publishes, in English or French, scientific reports on the mechanisms of genetics, cytogenetics, and evolution. The reports are expected to be based on the most advanced techniques that are appropriate for the experiments. Traditionally, Genome publishes a significant number of articles concerned with the basic aspects of heredity and evolution in fungi, plants, and animals of societal importance. Reports that solely address quality improvement are referred to more suitable journals. Associate editors in Australia, Canada, the Netherlands, Germany, the United Kingdom, and the United States receive and handle manuscripts directly and thereby promote rapid processing of papers. Authors are requested to suggest suitable referees and to submit their names, addresses, phone number, fax number, and e-mail address with manuscript. Manuscripts may be sent to the Editor or to an appropriate member of the Editorial board. When submitting a manuscript to an Associate Editor, a copy of the covering letter, the title page, and the abstract should be sent to the Editor.

## Types of papers

**Articles** report original results or concepts. Notes are brief scientific reports of a preliminary nature consisting of a title, abstract, text, and references. The materials and methods are included in the text or figure legends. The length should not exceed 4 printed pages or 12 typed manuscript pages. **Techniques** is a section of the Journal devoted to short reports on significant new techniques in cytogenetics. **Comments** offer amplification, alternate explanations, or corrections of original papers, preferably through the presentation of new evidence. **Reviews** are published by invitation. All contributions are evaluated by referees. Suggestions for review topics and authors should be addressed to:

Dr. R.S. Singh

Associate Editor, Genome

McMaster University

1280 Main Street West

Hamilton, ON L8S 4K1, Canada

E-mail: sing@mcmaster.ca

## Page charges

There are no page charges for publication in Genome.

## To submit by mail
## New manuscripts

Authors should submit **the original copy and two duplicates of their manuscript (including tables and figures)** to the Editor (address follows) or to a member of the Editorial Board (see front matter in each issue of the Journal):

Peter Moens, Editor, Genome, Department of Biology, York University, 4700 Keele St., North York, ON M3J 1P3, Canada (416-736-5358; fax: 416-736-5731; e-mail: genime@yorku.ca).

## Revised manuscripts

The revised manuscript (original and one copy on paper and on disk or by e-mail, indicating the work processing software used) and the original illustrations plus one duplicate set (photocopy is acceptable) must be sent to the Editor together with a covering letter outlining the precise disposition of all comments and criticisms.

## Accepted manuscripts

Author are requested to provide the **final accepted manuscript only, both in hard-copy format and  in electronic format (on  disk or by e-mail)**.  If providing files on disk, text files and figure files should be submitted on separate disks. All disks must be labeled clearly with the authors' names, software used, version number and platform should be provided in a word-processing format (any form of WordPerfect; Microsoft Word, or TeX). TeX ,macros for preparing papers for submissions are available at ftp://ftp.tex.ac.uk/tex-archive/macros/latex/contrib/nrc/, ftp://ftp.dante.de/tex-archive/macros/latex/contrib/nrc/, and ftp://ctan.tug.org/tex-archive/macros/latex/contrib/nrc/.  For **figures**, see the section "Preparation of Electronic Graphic Files."

## Other information regarding submission

### Cover letter

The corresponding author should send a cover letter with the submission, signed by all authors, that

(i)    states the type of paper being submitted (e.g. article, note, review, etc.),

(ii)    includes the full name and complete contact information (including e-mail address) for each co-author,

(iii)    warrants that the manuscript represents original work and that it is not being considered for publication, in whole or in part, in another journal, book, conference proceedings, or government publication with a substantial circulation (see Ethics section, Duplicate and prior publication),

(iv)  warrants that all previously published work cited in the manuscript has been fully acknowledged (see Publication process section, Permission to reproduce copyright material),

(v)  warrants that the manuscript is one of a kind, or part of a study or thesis form which other manuscripts may be generated,

(vi)  warrants that all of the authors have contributed substantially to the manuscript and approved the final submission,

(vii)  explains any real or perceived conflicts of interest (see Ethics section, Conflict of interest and disclosure), and

(viii)  lists the names, addresses, telephone and fax numbers, and e-mail addresses of four to six persons who are qualified to act as referees.

## Copyright forms

The submission package must include copyright release forms signed by all authors (see Publication process section, Copyright transfer).

## Preprints

To facilitate the review process, the author(s) must also provide two preprints of any relevant papers that have been submitted, are in press, or have been recently published. This is especially important if such papers are refereed to in the manuscript. To facilitate the review process, the author(s) must also provide copies of related manuscripts not already published, publications containing significant overlap with the submission, as well as a written explanation for the overlap (see Ethics section, Duplicate and prior publication)

## For databases

Authors of manuscripts reporting nucleic Acid sequences must submit the relevant data to the GenBank, EMBL, or DDBJ databases, whichever is most convenient. Date can be submitted by e-mail or on disk. Details regarding submission can be obtained form the relevant databases. Electronic mail addresses are as follows:

GenBank : gb@ncbi.nlm.nih.gov

EMBL : datasubs@embl-heidelberg.de

DDBJ : ddbjsub@ddbj.nig.ac.jp

National Center for Biotechnology Information

Bldg. 38A, Room 8N-803

Bethesda, MD 20894, U.S.A.

The accession number of the sequence must be provided in the manuscript.

Authors are encouraged to register their new bimolecular interactions, complexes, and pathways in the Bimolecular Interaction Network Database

(BIND). The BIND database archives bimolecular interactions including protein, DNA, RNA, ligand, or molecular complexes, pathways. BIND can be accessed at *http://www.blueprint.org/*

## Editorial process

### Receipt of manuscripts

Receipt of each manuscript is acknowledged by e-mail to the corresponding author within three working days. The manuscript is read and examined for conformity to these Instructions to Authors by the technical editor. Failure to meet the criteria outlined may result in return of the manuscript for correction before evaluation.

### Correspondence policy

Authors, Institutional Director, and Editorial Managers should note that it is the strict policy of Genome to correspond only with the authors through the designated corresponding author of a paper. The Editor regards a submitted manuscript as a confidential document and seeks to ensure that the authors retain control of the reports obtained during the evaluation process.

### Peer review/evaluation

**Peer review** – The Editor assigns management of the peer review process to an Associate Editor responsible for the subject area of the paper. However, the Editor will return un reviewed those manuscripts that do not fall within the Journal's scope or character and those that exceed the Journal's guidelines for prior publication. Papers submitted for inclusion in Journal supplements are treated with the same rigor of review as articles in regular issues.

The Associate Editor selects a minimum of two reviewers selected for their knowledge of, and their experience in, the subject treated in the manuscript. Reviewers are invited, in confidence, to recommend on the suitability of the submission and provide comments for the authors and the Associate Editor. The Associate Editor retains full responsibility, however, for all decisions regarding the manuscript. Authors are invited to suggest reviewers who are competent to examine their manuscript, but the Associate Editor is not limited to such suggestions. Reviewers are informed that they have received privileged documents for assessment of scientific merit and are expected to provide reasonable arguments to support their evaluations. Identities of reviewers will not be released to authors without the written consent of the reviewer. The review process is expected to be completed within eight weeks, but conflicting recommendations and other unpredictable events may cause some delay.

### Recommendations for acceptance, revision, and rejection

Associate Editors and reviewers are asked to make one of four recommendations: accept, accept after minor revision, accept after major revision, do not accept. Reviewers may also advise that a paper is more suitable to a specialist or local journal. Except where remarks are professionally inappropriate, all reviewers' comments are sent to authors.

The decision to accept a paper is made primarily on scientific content. However, authors should recognize that unclear writing and (or) data presentation often contribute to refusal of manuscripts. The decision to ask for revisions is made in light of the reviewers' comments and recommendations, and after evaluation by the Associate Editor. Authors are allowed 28 days to undertake revisions. Revised manuscripts that do not meet this deadline will be treated as new submissions and may be subject to further review. Papers requiring a new experimental work or major rewriting will be rejected, and the authors will be encouraged to submit a new manuscript when the required amendments have been completed. Authors should attempt to meet all the objections raised by reviewers, especially where clarification is sought. Editorial items must be completed as directed.

The final decision on acceptance or rejection is made by the Editor on the advice of the Associate Editor. This decision, together with any relevant reasons, will be communicated by letter from the Editor to the corresponding author. One copy of the original submission is retained by the Editor. In the case of papers that are not acceptable or are withdrawn, this manuscript and a copy of all reviews and correspondence are retained, for reference (in case of resubmission), for one year after the date of submission.

**Publication process**

The Editorial Office checks all accepted manuscripts of conformation to the Instructions to Authors and to ensure that all necessary paperwork is present. Any areas that are identified as problematic will be addressed by the Editorial Office in consultation with the corresponding author. Once the Editorial Office has resolved any problems with the manuscript and the original signed Assignment of Copyright forms have been received from all authors, the manuscript is forwarded to NCR Research Press in Ottawa for publication. The papers are prepared for publication by a professional copy editor responsible for ensuring that the final printed work is consistent in form and style.

Once the paper has been accepted, all correspondence should be with NRC Research Press, National Research Council of Canada, Ottawa, ONKIA 0R6, .Canada (fax: 613-925-7656; e-mail; pubs@nrc-cnrc.gc.ca; URL: http://pubs.nrc-crnc.gc.ca). NRC Research Press may make editorial changes as required, but will not make substantive changes in the content of a paper without consultation with the author and the Editors.

**Galley proofs**

A galley proof, illustration proofs, the copy-edited manuscript, and a reprint order form are sent to the corresponding author. **Galley proofs must be checked very carefully, as they will not be proofread by NCR Research Press,** and must be returned within 48 hrs of receipt. The proof stage is not the time to make extensive corrections, additions, or deletions, and the cost of changes introduced at the proof stage and deemed to be excessive will be charged to the author. Questions concerning galley proofs should be addressed to Sarah Currie (613-991-5399; fax;613-952-7656; e-mail: sarah.currie@nrc-cner.gc.ca).

## Reprints
If reprints are desired, the reprint order form must be filled out completely and returned with payment (cheque, credit card number, purchase order number, or journal voucher) together with the corrected proofs and manuscript. Orders submitted after the Journal has been printed are subject to considerably higher prices. **The journal does not provide free reprints, and reprints are not mailed until a purchase order number or payment is received**

## Permission to reproduce copyright material
Whenever a manuscript contains material (tables, figures, charts, etc.) that has been previously published and, hence, is protected by copyright, it is the obligation of the author to secure written permission from the holder of the copyright to reproduce the material for both the print and electronic formats. These letters must accompany the submitted manuscript.

## Copyright transfer
All authors are required to complete a copyright transfer form assigning all rights to NRC. Copyright transfer forms are available from the Editor, in the first issue of the volume, or on the Web site of journal (*http://pubs.nrc-cnrc.gc.ca/ rp/rptemp/gen_copyright_e.pdf*).

## Permission to reprint material published in NRC journals
Requests for permission to reproduce or republish the paper, in whole or in part, should be sent to NRC Research Press, Judy Gorman (tel.: 613-993-0151; e-mail; judy.Gorman@nrc-cnrc.gc.ca).

## Ethics
The ethical standards expected of authors, referees, and editors are described in the NRC Research Press Publication Policy (on the Journal Web site at *http:// pubs.nrc-cnrc.gc.ca/cgi-bin/rp/rp2_cust_e?pubpolicy*, OR available upon request).

## Duplicate and prior publication
The Editorial Board considers a paper not eligible for publication in most of the content of the paper that (i) is under consideration for publication or is published in a journal, or book chapter; (ii) is under consideration for publication or is published in conference proceedings or a government publication, with a substantial circulation (distributed to 100 or more individuals over a wide area). Authors may place a draft of a submitted article on their Web site or their organization's server, provided that the draft is not amended once accepted for publication. We encourage authors to insert hyperlinks from preprints to the final published version on the NRC Research Press Web site (*http://pubs.nrc-cnrc.gc.ca*). Abstracts or extended abstracts related to conferences do not

constitute prior publication. Extended abstracts are usually under 2000 words and do not include presentation of detailed tables and graphics of the results of the study.

## Assurance of authorship

In the cover letter, the corresponding author must affirm that all of the authors have read and approved the manuscript.

## Conflict of Interest and disclosure

The Editor recongnizes that authors and peer reviewers may have real or perceived conflicts of interest arising from intellectual, personal or financial circumstances of their research. Submitted manuscripts should include full disclosure of funding sources for the research and the letter of transmission should include an explanation of any real or perceived conflicts of interest that may arise during the peer review process. Failure to disclose such conflicts may lead to refusal of a submitted manuscript.

## Experiment involving humans or animals

All authors, regardless of their country of origin, who describe experiments on animals are required to give assurance in the Materials and methods that the animals were cared for in accordance with the Guide to the Care and Use of Experimental Animals (Vol.1, 2$^{nd}$ ed., 1993, and Vol.2, 1984, available from the Canadian Council on Animal Care, Constitution Square, Tower 2, Suite 315 Albert Street, Ottawa, No K1R 1B1, Canada, or on their Web site at *www.ccas.ca* or the Guide for the Care and Use of Laboratory Animals (1996, published by National Academy Press 2101 Constitution Ave NW, Washington, DC 20055, U.S.A.) and that their use of animals was reviewed and approved by the appropriate animal care review committee at the institution(s) where the experiments were carried out.

Authors who describe experiments on humans are required to provide assurance in the manuscript that appropriate standards for human experimentation have been followed, that the experiment has been reviewed and approved by their institution's ethics review committee, and that the subjects have given informed consent prior to participating in the study.

## Photo manipulation

Authors should be aware that the Journal considers digital images to be data. Hence, digital images submitted should contain the same data as the original image captured. Any manipulation using graphical software should be identified in the methods, including both the name of the software and the techniques used to enhance or change the graphic in any way. Such a disclaimer ensures that the methods are repeatable and ensures the scientific integrity of the work. The removal of artifacts or any (nonintegral) data held in the image is discouraged.

## The manuscript

The manuscript should be typewritten, double-spaced, on paper 8.5 × 11 in. (or ISO A4). Typing should be on one side of the page only. Each page should be numbered, beginning with the title page. For material that is to be set in italics, use an italic font; do not underline. Use capital letters only when the letters or words should appear in capitals.

All manuscripts should contain a title page (p. 1), an abstract (p. 2) followed by Instruction (p. 3), Materials and methods. Results, Discussion, and Acknowledgments sections, plus references, tables, figure captions, and appendices, in that order. (See descriptions of each part of the manuscript, below.) Tables and captions for illustrations should be on separate pages.

## Title

Both titles and abstracts provide information for contemporary alerting and information retrieval services, and should therefore be informative but brief.

## Abstracts

An abstract is required for every contribution and should contain accurate descriptive works that will draw the reader to the content. This is particularly important because contemporary alerting services and search engines will search this text. It should not be more than 200 words and should present the paper content accurately and supplement, not duplicate, the title in this respect. Authors able to submit abstracts in both English and French are encouraged to do so. They will be translated. References should not be cited in the abstract unless they are essential, in which case full bibliographic information must be provided.

## Text

The text should be written and arranged to ensure that the observations reported may be reproduced and (or) evaluated by readers. Sources of biological materials, experimental methods, should be described. Sources of commercially available laboratory or field equipment and fine chemicals should be indicated in parentheses; list the company name, city, and country.

## Introduction

Limit the introduction largely to the scope, purpose, and rationale of the study. Restrict the literature review and other background information to that needed in defining the problem or setting the work in perspective. An introduction generally need not exceed 375-500 words.

## Material and methods

The degree of reproducibility of experiments should be indicated either in general statements in Materials and methods and Results or, preferably, as statistical or numerical data cited in tabular or graphic form.

The experimental, or computational, material must be sufficiently detailed to permit reproduction of the work, but must be concise and avoid lengthy descriptions of known procedures; the latter should be specified by appropriate references.

Identify figures that have been digitally enhanced or modified, and provide the software and techniques used.

## Results

Limit the results to answers to the questions posed in the purpose of the work and condense them as comprehensively as possible. Material supplementary to the text can be archived in the report literature or a recognized data depository and referenced in the text (see Supplementary material section)

## Discussion or conclusion

Limit the Discussion to giving the main contributions of the study and interpreting particular findings, comparing them with those of other workers. Emphasis should be maintained in synthesis and interpretation exposition of broadly applicable generalizations and its principles. If the Discussion is brief and straightforward, it can be combined with the Results section.

## References

Each reference must be cited in the text using the surnames of the authors and the year, for example, (Walpole 1985) or Green and Brown (1990). If there are three or more authors, the citation should give the name of first author followed by et al. (e.g. Green et al. 1991). If references occur that are not uniquely identified by the authors' names and year, use a, b, c, etc. after the year, for example, Green 1983a, 1983b; Green and Brown 1988a, 1988b, for the text citation and in the reference list.

### Unpublished reports, private communications, and In press references

References to unpublished reports, private communications, and papers submitted but not yet accepted are not included in the reference list but instead must be included as footnotes or in parentheses in the text, giving all authors' names with initials; of a private communication, year of communication should also be given (e.g. J.S. Jones (personal communication, 1999)). If an unpublished book or article has been accepted for publication, include it in the reference list followed by the notation "In press". Do not in include volume, page number, or year in an in-press reference, as these are subject to change before publication.

## Presentation of the list

The reference list must be double-spaced and placed at the end of the text. References must be listed in alphabetical order according to the name of the first author and not numbered. References with the same first author are listed in the following order. (i) Papers with one author only are listed first in chronological order, beginning with the earliest paper. (ii) Papers with dual authorship follow and are listed in alphabetical order by the last name of the second author. (iii) Papers with three or more authors appear after the dual-authored papers and are arranged chronologically.

## General guidelines on references

References should follow the format used in current issues of the Journal. References to unpublished reports and to private communications are not included in the reference list (see Private communications and unpublished material, below). The names of serials are abbreviated the form given in Chemical Abstracts Service Source Index  (CASSI) (Chemical Abstracts Service 43210, U.S.A.) or in BIOSIS Serial Sources (BIOSIS, 2100 Arch Street, Philadelphia, PA 19103-1399, U.S.A.). In doubtful cases authors should write the name of the serial in full. References to **nonrefereed** documents (e.g. environmental impact statements contract reports) must include the address where they can be obtained. Following bibliographic citations illustrate the punctuation, style, and abbreviations for references.

*Journal article :* Redwood, R.G., and Jain, A.K. 1992. Code provisions for seismic design for concentrically braced steel frames, Can.J.Civ.Eng. **19**: 1025 – 1031.

*Book :* Healey, M.C. 1980. the ecology of juvenile salmon in Georgi Strait, British Columbia. In Salmonid ecosystems of the North Pacific. *Edited by* W.J. McNeil and D.C. Himsworth, Oregon State University Press, Corvallis, Oreg. Pp. 203-229.

*Paper in conference proceedings :* Whittaker, A.A., Uang, C.-M., and Bertero, V.F. 1990. Experimental seismic response of steel dual systems. Proceedings of the 4th U.S. National Conference on Earthquake Engineering, Palm Springs, Calif., Vol 2, pp. 655-664.

*Electronic citation :* Quinion, M.B. 1998. Citing online sources: advice on online citation formats [online]. Available from http://clever.net/uinion/words/citation.html [cited 20 October 1998].

## Tables

**Tables** must be typed on separate pages, placed after the list of reference, and numbered with Arabic numerals in the order cited in the text. The title of the table should be a concise description of the content, no longer than one sentence, that allows the table to be understood without detailed reference to the text. Footnotes in tables should be designated by symbols ( in the order *, ?, ?, §, êê,¶,# ) or superscript lowercase italic letters. Descriptive material not designated by a footnote may be placed under a table as a **Note.**

## Figures captions

Figures captions should be listed on a separate page and placed after the tables. The caption should informatively describe the content of the figure, without need for detailed reference to the text.

## Appendices

An appendix should be able to stand alone, as a separate, self contained document. Figures and tables used in an appendix should be numbered sequentially but separately from those used in the main body of the paper, for example. Figure A1, Table A1, etc. If reference are cited in an appendix, they must be listed in an appendix reference list, separate from the reference list for the article.

## Supplementary material

The National Research Council of Canada maintains a depository in which supplementary material may be placed either at the request of the author or at the suggestion of the Editor. In addition, supplementary material can now be made available in its native file format on the journal Web site. It will be linked from the Web page of the associated article. Such material may include extensive tables of data, detailed calculations, and maps not essential for understanding and evaluating the paper. Such material must be clearly marked when the manuscript is submitted. Tables and  figures should be numbered in sequence separate from those published with the paper (e.g. Figure S1, Table S1). The supplementary material should be referred to by footnotes. Copies of material in the depository may be purchased from the Depository of Unpublished Data, CISTI, National Research Council of Canada, Ottawa, ON K1A 0R6, Canada.

## Illustrations

**One original and two copies of figures must be submitted to the Editorial Office,** even if electronic versions are to be submitted for use in production. Each figure or group of figures should be planned to fit, after appropriate reduction, into the area of either one or two columns of text. The maximum finished size of a one column illustration is 8.6 × 23.7 cm (3.4 × 9.3 in.) and that of a two column illustration is 18.2 × 23.7 cm (7.2 × 9.3 in.). The figures (including halftones) must be numbered consecutively in Arabic numerals, and each one must be referred to in the text and must be self-explanatory. All terms, abbreviations, and symbols must correspond with those in the text. Only essential labeling should be used, with detailed information given in the caption. For hardcopy versions, each illustration must be identified by the figure number and the authors' names on the back of the page or in the left-hand corner, well away from the illustration area.

## Line drawings

All lines must be sufficiently thick (0.5 points minimum) to reproduce well, and all symbols, superscripts, subscripts, and decimal points must be in good proportion to the rest of the drawing and large enough to allow any of necessary reduction without loss of detail. Avoid small open symbols; these tend to fill in upon reproduction. Lettering produced by dot matrix printer or typewriters, or by hand, is not acceptable. The same font style and letter in sizes should be used for all figures of similar size in any one paper. Original recorder tracings of NMR, IR, ESR spectra, etc., are not acceptable for reproduction; they must be redrawn. For hard-copy versions, line drawings should be made with black ink or computer-generated in black on high-quality white paper or other comparable material; laser prints should be created at the highest resolution available.

## Photographs

Photographs should be continuous tone, of high quality, and with strong contrast. Only essential features should be shown. A photograph, or group of them, should be planned to fit into the area of either one or two columns of text with no further reduction. Electron micrographs or photographs should include a scale bar directly on the print. Hard-copy versions must be printed on glossy paper trimmed and mounted on thin flexible white Bristol board with no space between those arranged in groups.

## Colour illustrations

All colour files submitted must be as CMYK (cyan, magenta, yellow, and black). These colurs are used in full-colour commercial printing. RGB graphics (red, green, and blue; colours specifically used to produce an image on a monitor) will not print correctly. Colour illustration will be at the author's expense. Further details on prices are available from Angela Murphy, Managing Editor of the Journal (613-998-4485; fax: 613-952-7656; e-mail: angela.Murphy@nrc-cnrc.gc.ca).

## Preparation of electronic graphic files

### General

NRC Research Press prefers the submission of electronic illustration files for accepted manuscripts and will use these electronic files whenever possible.

If electronic files are not available or if those supplied are inadequate for reproduction, hard-copy originals of adequate quality, either previously supplied or requested from the author, will be scanned. Note that the scanner will easily reproduce flaws (e.g., correction fluid, smudges). Submission of noncontinuous (screened) photographs and scanned illustration printed out on laser printer is not recommended, as moirés develop; A moiré is a noticeable, unwanted pattern generated by rescanning of rescanning an illustration that already contains a dot pattern.

**If sending hard copies, please ensure that electronic files match the hard copies (i.e., figure number and figure content).** If sending a disk, on the disk label, identify, (*i*) the software application and version and (*ii*) file name(s), size, and extension. If you have compressed your files, indicate what compression format was used. PC or Macintosh versions of True Type or Type 1 fonts should be used. **Do not use bitmap or nonstandard** fonts. Electronic graphics can be accepted on the following disks: 3½ " disks, 100 MB Zip cartridge, and CD-ROM.

**The preferred graphic application of NRC Research Press is CorelDraw!** For other applications that can be used, see the electronic graphics list at *http:/ /pubs.nrc-cnrc.gc.ca/cgi-bin/rp/rp2_prog_e?gen_graphics_e.html.*

All figures should be submitted at the desired published size. For figures with several parts (e.g. a, b, c, d, etc.) created using the same software application, assemble them into one file rather than sending several files.

**Remember** that the more complex your artwork becomes, the greater the possibility for problems at output time. Avoid complicated textures and shadings, especially in vector illustration programs; this increases the chance for a poor-quality final product.

**Bitmap (raster) files** – Bitmaps are image files produced using a grid format in which each square (or pixel) is set to one level of black, colour, or grey. A bitmap (rasterized) files is broken down into the number of pixels or picture elements per inch (ppi). Pixels per inch is sometimes referred to as dots per inch (dpi). The higher the resolution of an image, the larger the number of pixels contained within the rectangular gird.

Proper resolution should be used when submitting bitmap artwork. The minimum requirements for resolution are 600 dpi for line art, 1200 dpi for finelines (line art with lines or shading), 300 dpi for halftones and colour, and 600 dpi for combinations (halftones with lettering outside the photo area).

## Manuscripts guidelines

### Style Guides

As a general guide for biological terms, *The CBE Manual for Authors, Editors, and Publishers: Scientific Style and Format* (6th ed., 1994) published by the Council of Biology Editors, Inc., Chicago, IL 60603, U.S.A., is recommended.

### Units of measurement

SI units (System international d' unites) should be used or SI equivalents should be given. This system is explained and other useful information is given in the *Metric Practice Guide* (2000) published by CSA International (178 Rexdale Blvd., Toronto, NO M9W 1R3, Canada). For practical reasons, some exceptions to SI units are allowed.

NRC Research Press

National Research Council of Canada

Ottawa, On K1A 0R6

Canada

Fax : 613-952-7656

E-mail : research.press@nrc.gc.ca

URL: http://pubs.nrc-cnrc.gc.ca

# CHAPTER - 46

# Indian Agriculturist

## Instructions To Contributors

The pages of Indian Agriculturist shall be open ordinarily only to members of the Society.

Manuscript should be sent in duplicate, typed with double spacing on one side of the paper, leaving 4 cm spacing on the top and left along with a floppy. Author will have to pay printing cost @ Rs. 150.00 per typed page of A 4 size and will have to bear 100 per cent of the block making charges. Short communications are also considered for publication. The accepted paper will be sent to the press after receiving payment in full. Authors will receive 50 reprints without cover, free. The cost of each extra reprint is Rs. 5.00 per pages per copy. The number of extra copies required should be stated while sending the manuscript. Manuscripts of the articles will not be returned, so the authors are requested to keep copies before sending for publication.

Paper should start with an Abstract, which should be a concise summation of the findings being reported, followed by *Introduction, Materials and Methods, Results, Discussion and Literature Cited.*

In short communications there should not be any subdivision of the text except *Abstract* (approximately 100 words) and *Literature Cited.*

*Literature Cited* should be prepared by quoting the name of the author(s), the year of publication, title of the paper, the volume and page numbers. Abbreviation of journals should be Literature Cited according to the *World list of Scientific Periodicals.* A specimen for quoting literature cited in the text is given below :

Datta. R.M. and Basak, S.L. 1959. Complex chromosome mosaic in two wild jute spices, *Corchorus trilocularis* linn. And C. *siliquisis* Linn. *India Agric.,*3, 37 - 42

Review articles are published only when invited.

Illustrations in line should be drawn boldly in Indian ink on Bristol board or smooth white card. All necessary shading must be done by well defined dots or lines. Colour, either in lines or wash, should be avoided. Due allowance for reduction must be made in the size of lettering, thinkness of lines and closeness of shading. Photographs should be  glossy print on black and white. The tables should be typed on separate sheets.

All correspondence should be made with the Editor, The Agricultural Society of India 35, Ballygunge Circular Road, Kolkata – 700 019 (India).

E-mail number of the corresponding author, if available should be, given in a footnote on the first page of the manuscript.

# CHAPTER - 47

# Indian Coconut Journal

Articles, research papers and letters on different aspects of coconut cultivation and industry are invited for publication in this Journal. All accepted material would be paid for. The Board does not accept responsibility for views expressed by contributors in this Journal. All remittances and correspondence should be addressed to the Chairman, Coconut Development Board, Kochi – 682 011.

## Publisher

**Coconut Development Board**

(Ministry of Agriculture, Department of Agriculture and Co-operation, Government of India )

Kera Bhavan, Kochi – 683 001, India

| | | |
|---|---|---|
| Phone | : | 0484-2377266,2377267, |
| | | 2376553, 2375266, 2376265 |
| Fax | : | 91 – 0484-2377902, |
| Grams | : | KERABOARD |
| E-mail | : | enk_cdrkochi@sancharnet.in |
| Website | : | www.coconutboard.nic.in |

# CHAPTER - 48

# Indian Farmer's Digest

## Attention Contributors

- Submit your neatly typed write up in duplicate with a certificate, that the article is original and unpublished.

- Please write full designation and mailing address of all the authors in the foot note of the article.

- Articles must be supported with line sketches, relevant data and photograph, preferably black and white. Please don't punch, paste or write at the back of the photographs and pack it safely for posting.

- Articles must be based on the research or practical experience of the author.

- Data and recommendations must be the latest. Articles, with obsolete Informations/ recommendations may not be published.

- The length of article should not exceed more than 2000 words.

- The articles on floppy (3 ½, 1.44 MB) is preferred. Please send two print alongwith the floppy.

- Mention your subscription number, if you are a subscriber of Indian Farmers Digest.

- Submit your article to:

  The Editor

  Indian Farmers' Digest,

  Department of Agricultural Communication,

  College of Agriculture,

  G.B. Pant University of Agriculture & Technology,

  Pantnagar-263145 (Uttaranchal)

# CHAPTER - 49

# Indian Farming

## Guidelines to Authors

**Indian farming, monthly, considers articles only on floppy in MS word 6, PM 5, WP 6** + two hard copies which conform to its style and meet the requirements given below:

1. Original articles, exclusive for the magazine, are considered on all aspects of farming, including those on corps (breeding, agronomy, soils, plant-protection, and agricultural engineering, biotechnology, agriculture environment, etc), livestock, poultry, fisheries, agricultural economics and information technology.

2. Articles should be based on recent experiments of the author or his/her practical experience. **Compiled matter or that of some survey, report or report, of extension education** as that of **theoretical nature or only of local importance is not considered.** Published matter need not be reproduced.

3. Give **Summary** mentioning purpose of study, duration/place of work, important methods and significant results. If required give further scope of research **in 3 / 4 lines in the end of the text.**

4. Text should be precise, giving complete relevant details of practical utility to the farmer in clear and simple English. Data should be correct, authentic and updated. Scientific names and technical nomenclature must be accurate. For plant-protection chemicals, trade names should be avoided; instead popular scientific names should be given. Dose of fertilizer may be in terms of N, P and K (not $P_2O_5$, $K_2O$), indicating clearly the exact amount of specific fertilizer (like superphosphate) to be used. The SI (metric) units may be used throughout. The text should be clear, **abiding abbreviations** (should be spelt out).

5. Tables and Figures, supporting text, should be kept to the minimum, and should not reproduce the same data. Histograms are preferable to Tables. Mean, relevant data should be given. Each Table should be typed separately (in double space), giving complete caption.

6. Relevant artwork should be sent in original, with two photocopies. Legend should be typed. Line drawing (including sketch) should be clearly drawn in black waterproof ink on smooth, tough white paper.

   - Photographs black-and-white as well as in colour are welcome. Theses should be large (minimum 10 cm × 15 cm), unmounted and using glossy paper of good quality. Subject should be focused, avoiding persons and irrelevant objects. Words (for identification or as label) or any extraneous matter should be introduced while taking a photograph, or pasted on it later on; each part should be labeled on the photocopy. Background when it is covered, should be white.

   - Colour photograph is preferable for identification or differentiation of different parts of an object. Colour should be sharp, with good contrast among various elements.

   - Clean photograph (do not paste or punch, or insert pin or clip) should be sent, packed flat between cardboards.

7. **Two copies of an article** (in double space) are required along with floppy and **relevant original figures** (+ two photocopies) for consideration. Designations and complete official addresses (with PIN) of all authors should be given, and any change in address must be communicated immediately.

8. Article Certificate (recent) should be sent, signed by all the authors. Subscription of *Indian Farming* is compulsory to all contributors.

9. Proofs are not sent to the authors for checking; hence the copy must be complete in all respects. For reprint, one copy of the issue would be supplied to each author.

10. There should not be more than two authors of one article. However, authors may be three or maximum four; if the article is an inter-sectoral package.

# CHAPTER - 50

# Indian Food Packer

## Instructions to Author(s)

1. Manuscripts of papers (in duplicate) should be typewritten in double space on one side of bond paper. **They should be complete and in final form. The paper should not have been published or communicated for publication anywhere else.** The typed or computer print should be bold, dark and easily readable.

2. The typescripts should be arranged in the following order: Titles (to be typed in bold capitals), Authors' names (in Upper/Lower bold type) and the institution's name and its (in italics/ordinary)

3. *Abstract :* The abstract to be given in the beginning should indicate the principal findings of the paper and typed in single space in upper/lower bold letters. It should not be more than 200 words and should be in such a form that abstracting periodicals can readily use it.

4. Use names of chemical compounds and not their formulae in the text. Methods of sampling, number of replications and relevant statistical analysis should be indicated. Foot notes especially for text should be avoided as far as possible.

5. *Tables :* Tables as well as graphs, both representing the same set of data, should be avoided. Tables should be short & typed on separate sheets. Nil results should be indicated as such and distinguished clearly from absence of data, which is indicated by '.......' Sign.

6. *Illustration :* Graphs and other line drawings should always be drawn in Indian ink on tracing paper or white drawing paper, preferably art paper, not bigger than 20 cm (OY axis) × 16 cm (OX axis). The lettering should be twice the size of the printed letter. Photographs must be on glossy paper and must have good contrast; two copies should be sent.

7. *References :* Names of the authors followed by the year of publication in brackets and the title of the paper should be cited. Abbreviations such as et. al. idem should be avoided. References should be serially numbered as superscripts while being cited in the text and the same order should be

maintained in the reference list. The titles of all scientific periodicals should be abbreviated in conformity with the World list of Scientific periodicals, Butterworths Scientific Publication, London, 1962.

List of References should be given exactly as follows with proper attention to punctuation and the journal's name in italics/ordinary:

(a) **Research Paper .** Sethi V and JC (1982). Keeping quality of whole tomoto concentrate Ind. Fd. Packer 35(6), 66.

(b) **Books :** Cruss WV (1958). Commercial Fruit and Vegetable Products, McGraw Hill Book Co. Inc., New York.

(c) **References to article in a book :** Hodge JE (1967): Origin of Flavour in  Foods-Non-enzymatic browning reactions in The Chemistry and Physiology of flavours by Scheetz HW, Day EA and Libbrey LM, Pub. Co., Inc. Westport  Connecticut pp. 465.

(d) **Proceedings,  Conferences and Symposia Papers :** Nambudiri ES and Lewis YS, Cocoa in on confectionery Proceedings of Symposium on the Status and Prospects of the Confectionery Industry in India, Mysore, May 1979, 27.

(e) **Thesis :** Sathyanarayan Y, Phytosociological Studies on the calcicolous Plants of Bombay, 1953, Ph. D. Thesis, Bombay University.

(f) **Unpublished Work :**  Roy SK, unpublished, Indian Agricultural Reseach Institute, New Delhi.

# CHAPTER - 51

# Indian Journal of Agricultural Economics

## Guidelines for Submission of the Papers by Author

1. **Subjects :** The objective of the Journal is to provide a forum for dissemination and exchange of findings of research on agricultural economics. Purely descriptive material is not appropriate for such a journal. Papers dealing with (i) new developments in research and methods of analysis, or (ii) which apply existing empirical research methods and techniques to new problems or situations, or (iii) which attempt to test new hypotheses. theoretical formulations or modifications of existing theories or policies to explain economic phenomena especially in the Indian context, or (iv) papers based on research done by the authors bringing out new fact or data presented in any analytical frame will be preferred. Younger economists are advised to seek guidance from their seniors in the preparation of the paper for the Journal. A one-page statement indicating the author's own assessment of the importance and relevance of the findings reported in his paper in the context of recent researches would be welcome.

2. **Size of Paper :** In view of the exorbitant increase in the cost of printing, it has become necessary to restrict the length of the papers accepted for publication to 20 (double space) typed pages (of the size 8 ½ by 11 inches) including tables and appendices (with margins on all sides of at least 1½ inches). Papers exceeding this size limit will be returned to the authors. The authors may however send along with their papers such "support material" as may be of the referees in evaluating them. Abstract not exceeding 100 words should be submitted in duplicate along with the papers. The authors may send their papers in electronic form at the following email address: isae@bom7.vsnl.net.in

3. **Copies :** Two copies of each paper should be submitted along with a processing fee of Rs.50/- either by M.O. or Demand Draft (US $ 15.00 or £ 10.00) drawn in the name of the Indian Society of Agricultural Economics, to meet partly the postage and other incidental charges.

4. **Author's identification** : To protect the anonymity of authors while referring the Papers for expert opinion on their merits, authors are advised to avoid  disclosing their identity in the text and to attach a separate page showing the name(s) and affiliations(s) along with any footnotes containing bibliographical information or acknowledgments.

5. **Mathematical notations** : Only essential mathematical notations may be used; it is prohibitively costly to typeset such notations. All statistical formulae should be neatly typed.

6. **Figure-Charts and Tables** : Professional assistance should be availed while presenting figures and charts. They should be drawn in black ink which can be easily reproduced by photographic process. Typescript is not suitable for reproduction. Roman numerals may be used for tables.

7. **Footnotes** : Footnotes should be numbered consecutively in plain Arabic superscripts.

8. **References and citations** : Only cited works should be included in the reference list. Please follow the style of citations as in previous issues of this Journal.

9. **General** : While sending papers the authors should state that the material has not been published elsewhere or is not being considered for publication elsewhere. It is editorial policy not to consider for publication more than one paper from the same author during a year.

   **Papers which are not found suitable for publication will not be returned unless accompanied by a self-addressed and stamped envelop. No correspondence will be entertained on the papers rejected by the Editorial Board.**

10. **Reprint** : Ten free reprints of the paper would be given to the author, additional reprints, if required, are supplied at cost which may be ascertained from the Editor.

# CHAPTER - 52

# Indian Journal of Agricultural Marketing

## Information for Authors

Indian Journal of Agricultural Marketing is to provide a forum for dissemination and exchange of original findings in various-disciplines of Agricultural Marketing of such a character as to be of primal interest to various functionaries in the marketing of different agricultural produce. It will also from time to time, contain articles summarizing the existing state of knowledge in the field of marketing of agricultural products.

Manuscripts sent for the Journal should be accompanied by a declaration of the senior author stating that the same paper has not been published or under consideration elsewhere and responsibility for the statement whether of facts or opinion rests with writers thereof.

In clearness, brevity and conciseness, punctuation, spelling and use of italics etc. the manuscript should conform to the best usage in leading journals published.

In case of data, reference to the source should be mentioned.

Manuscripts in duplicate should be typed in double space on one side of the bond paper, with sufficient margins. Paper should not normally exceed ten typed pages including tables and figures. Names of the authors should not be accompanied by their degrees, titles designations etc.

Name and address of the author along with the title of the paper should be typed on a separate sheet and the manuscript should bear only the title of the paper and at no place the name or any identity of the author.

Each paper should have a summary in 150 words, in the beginning. First and second copies of the typed manuscripts should only be submitted.

Drawings, graphs etc. drawn in black ink, which are good in quality and essential to a clear understanding of the paper should be submitted in original.

Each table should have a number hold followed by a concise heading. The references to the literature cited in the text of the paper should be serially numbered and given at the end of the paper under the heading REFERENCES. If the same reference is cited at different places in the text, common citation number should be used at all places. Only literature cited in the text should be given under References. Authors should uniformly follow the reference citation strictly in accordance to the examples given under –

Talukdar, K. C. and Muralidharan, M.A., 1989. Agrarian relation and marketing institutions for rice marketing in Assam : A case study. Ind. Jour.Agril. Mktg., 3(2) : 122-131

Pulses in India : Growth, Regional distribution and area responses. Ed. V. Satyapriya, Oxford IBH Publishing Co. (P) Ltd., new Delhi.1989.

All papers will be referred in anonymity. The comments of the referee should be complied by the authors within two months time. Rejected papers will be returned to the author, if claimed within a month, after the final communication is given in this regard. Authors may submit their papers through computer floppies specifying the name and version of word processor package used with one printout of the paper.

# CHAPTER - 53

# Indian Journal of Agronomy

## Guidelines to Authors

1. Indian Journal of Agronomy is published quarterly by the Indian Society of Agronomy, which includes the articles contributed by the members of the Indian Society of Agronomy only.

2. The journal publishes full-length research papers based on new approaches/ findings in English only.

3. The papers submitted for publications in the Journal must not carry material already published in the same form nor even a paper offered for publication elsewhere.

4. The articles submitted for publication in the Journal should contain data *not older than 5 years* on the *date of receipt of the article in the Society office.* The period shall be reckoned from the following *January* and *July* after the completion of the field experimentation in rainy (*kharif)* and winter (*rabi)* seasons respectively.

5. Articles On Original Research Completed, not exceeding 4,000 words (up to 15 double space typed pages, including reference, tables etc.) should be exclusive for the journal. They should present a connected picture of the investigation and should not be split into parts.

   Title Should be short, specific and phrased to identify the content of the article and include the nature of the study and the technical approach, essential for key-word indexing information retrieval.

   A Short Title not exceeding 30 letters should also be provided for running headlines.

   By-line should contain, in addition to the names and initials of the authors, the place where research was conducted. Change of address should be as a footnote and correspondence address must be indicated separately.

6. Abstract, should not have more than 150 words. It should contain a very brief account of the materials, methods, results, discussion and conclusion, so that the reader need not refer to the article except for details. It should have reference to literature, illustrations and tables.

7. Introductory part should be brief and limited to the statement of the problem or the aim of the experiment. The review of the recent literature should be pertinent to the problem and should be concise.

Relevant details should be given of the Materials And Methods, including experimental design and the techniques used. Where the methods are well known, citation of the standard work is sufficient. *Mean results* with the relevant standard errors should be presented rather than detailed data. The statistical methods used should be clearly indicated.

Results and Discussion  should be combined, to avoid repetition.

Results should be supported by brief but adequate tables or graphics or pictorial materials wherever necessary. Self-explanatory tables (up to 20% of text) should be typed on separate sheets, with appropriate titles. In no case the same data should be presented through both tables and diagrams.

The data should be so arranged that the tables would fit in the normal lay-out of the page. All weights and measurements must be in SI (metric) units.

The discussion should relate to the limitations or advantages of the author's experiment in comparison with the work of others.

The figures are expensive and should be kept to the minimum. The illustrations should be in 12.5 cm, width and drawn on *white art card paper* in black Indian ink and all letters must be written with suitable stencils (*not free hand*). Send illustrations in duplicate, one original and one photocopy. The legends must be clearly drawn and included for each figure. Figure number and caption should be given separately.

All reference in the text should appear at the end of the article and *vice versa* the spellings of names and dates at the 2 places should correspond. The recent issue of the Journal should be consulted for the method of citation of references in the text as well as at the end of the article. The references should include the names of all authors, year, full title of the article, full name of the journal (*no abbreviation is allowed*), volume number, also be given. The number of references should be kept at *minimum* possible.

8. All articles are sent to referees for SCRUTINY, and authors should meet criticism by improving the article.

9. Articles should be TYPE WRITTEN, and double spaced throughout (including by-line, references and table) on white, *durable bond paper* of 21 cm × 28 size, with a 5 cm margin at the top, bottom and left. Articles should be sent in duplicate after check-up of typographical errors.

10. For WRITING, authors are requested to consult *The Council of Biology Editors Style Manual, edn 4.* American Institute of Biological Sciences, Washington DC.

# CHAPTER - 54

# Indian Journal of Arecanut, Spices and Medicinal Plants

## Instructions to Contributors of Articles in the Journal

1. Articles may be sent in MS word format in 3.5 floppy diskette or CD along with two sets of hard copy, which should not exceed 12 pages. All pages (including tables, legends and references) should be numbered consecutively.

2. The matter is to be arranged in the following order :
   - Title in capital letters.
   - Name of the Authors.
   - Introduction highlighting the importance of the subject.
   - Subject matter.
   - Conclusion.
   - Recommendations.
   - Tables, Illustrations, Photographs etc. should be cited in the text appropriately. Line drawings must be in black colour. Photographs in colour or black & white with title indicated clearly on the back.
   - Authors full address to be given at the end of the first page

3. Preference will be given to articles in bilingual i.e., Hindi and English.

4. A certificate may also be furnished to the effect that the article submitted has not been published in any other journal in any form.

   Articles will be acknowledged immediately on receipt and acceptance or otherwise will be communicated with in a quarter.

The Director
Directorate of Arecanut and Spices Development
Ministry of Agricultural, Government of India
Department of Agriculture & Co-operation
West Hill P.O., Calicut – 673 005, Kerala
Phone: (0495) 2369877, 2765501, Fax: 0495-2765777
**E-mail: spicedte@nic.in**

# CHAPTER - 55

# Indian Journal of Biochemistry and Biophysics

## Instructions to Authors

The journal welcomes concise reports of original research in all areas of biochemistry and biophysics. Papers on nutrition or medical biochemistry will be considered only if they are of general biochemical interest.

## Submission of Manuscript

Hard copies of manuscripts (inclusive of illustrations) should be submitted in triplicate to the **Editor, Indian Journal of Biochemistry and Biophysics, National Institute of Science Communication and Information Resources (NISCAIR), Dr K S Krishnan Marg, New Delhi 110012.** Submission of a manuscript to this journal implies that it is not under consideration for publication elsewhere.

## Preparation of Manuscript

Manuscripts should be presented in as concise a form as possible and typewritten in double space on one side of the paper. Pages should be numbered consecutively, and the matter on Page 1 should be arranged in the following order: Title of the paper; Name(s) of Author(s), Department(s) and Institution(s); Foot Note containing address of Author for correspondence with e-mail id, telephone no., FAX no. followed by list of Abbreviations used in text. A short running title derived from the original title may also be given. Page 2 should contain the Abstract. Rest of the matter may be arranged in the following order: **Introduction; Materials and Methods; Results; Discussion; Acknowledgement; References. Abstract, Tables and Captions for Figures should be typed separately.**

*Title* : The title should be such as to be useful in indexing and information retrieval. If a paper forms part of a series, a sub-title indicating the aspects of the work covered in the paper should be provided.

*Abstract* : The abstract, not exceeding 200 words, should indicate the scope and significant content of the paper, highlighting the principal findings and conclusions. Keywords are to be given beneath the abstract.

*Introduction :* The introductory part should  bear one heading, should be brief and state precisely the objective in relation to the present status of knowledge in the field. Reviewing of the literature should be restricted to only essential background and the reason for the research undertaken.

*Materials and Methods :* The nomenclature, the sources of materials and the procedures should be clearly stated. New methods should be described in sufficient detail, but if the methods are already well known, a mere reference to them will do; deviations, if any, should, however, be given.

*Results :* Only such data as are essential for understanding the discussion and main conclusions emerging from the study should be included. The data should be arranged in a unified and coherent sequence for clarity and readability. The same data should  not be presented in both tabular and graphic forms. Only such tables and figures as are necessary should be given. Lineweaver-Burk plots, $p$H curves, substrate and enzyme concentration curves, and gel filtration and gel electrophoresis calibration curves for molecular weights of enzymes will be published only if they are non-standard.

In studies dealing with experimental animals, tests of statistical significance should be identified and references used should be cited. Statements about the statistical significance for the results should be accompanied by indications of the level of significance, preferably provided in respective Tables and Figure legends.

*Discussion :* Long rambling discussion must be avoided. The discussion should deal with the interpretation of results without repeating information already presented under Results. It should relate the new findings to the known and include logical deduction. In some cases, however, it may be desirable to combine results and discussion in a single section.

*Illustrations :* The number of illustrations should be kept to the minimum and numbered consecutively in Arabic numerals.. Simple linear plots or linear double reciprocal plots that can be easily described in the text should be avoided. Extension of graphs beyond the last experimental points is permissible only while extrapolating.

Captions and legends to the Figures should be self-explanatory and typed on a separate sheet of paper. Line drawing should be either laser prints of computer generated illustration or manually made in Indian ink on white drawing paper, and should be drawn to approximately twice the printed size. The drawings are reduced to the pager (width, 175 mm) or column (width, 85 mm) size, and care should be taken that the size of letters, numerals, dots and symbols is relatively uniform and sufficiently large to permit this reduction.

*Tables :* Each table should have an explanatory title and should be numbered in Arabic numeral. It should have sufficient experimental details, usually in the form of a paragraph immediately following the title. Units of measurement should be abbreviated and placed below the headings. Negative results should be indicated as 'Nil' and absence of a datum by a dash. Presentation of results should be limited to the accuracy of the method employed.

***Acknowledgement :*** Acknowledgement should be brief and for special assistance only, not for providing encouragement and routine facilities *et cetera.*

***References :*** Responsibility for the accuracy of bibliographic references rests with the author. Abstracts of papers presented at scientific meetings should be cited only after they appear in publications included in the Biological Abstracts. Such Abstracts may be cited in Foot Note. The Indian Journal of Biochemistry and Biophysics has accepted the recommendations of the Commission of Editors of Biochemical Journals of the International Union of Biochemistry, concerning presentation of reference lists. The recommended style is as follows:

## Periodicals

Upadhyay S N & Chattoraj D K (1972) *Indian J Biochem Biophys* 9, 17-20

In  case of mini review, title of the papers may be included.

## Books

Dixon M & Webb E C (1964) *Enzymes, 2ⁿᵈ* edn, pp.565-567, Logmans Green, London

Bracket J (1967) in *Comprehensive Biochemistry* (Florkin M & Stotz E M, eds), Vol 28, pp. 23-54, Elsevier, Amsterdam.

Laidler K J & Bunting P S (1973) The Chemical Kinetic of Enzyme Action, 2ⁿᵈ edn, Vol.2, Chapt. 4, pp.255-278, Clarendon Press, Oxford.

## Reports, non-serial symposia & proceedings

Harding B W, Whysner J, Cheng S C & Ramseyar J (1970) *Proceedings of the Third International Congress Hormonal Steroids,* Hamburg, September 1970, pp 294-300

*Technicon Autoanalyzer Methodology* (1971) method file N-24, pp. 1-4

Iverson H O (1969) *Homeostatic Regulators* (Ciba Foundations Symposium), pp. 29-30, J & A Churchill Ltd, London

## Thesis

Chandrasekharan K S (1956) *Studies on Crystal Structure and Absolute Configuration of Crystals,* Ph. D thesis, Madras University, Madras.

References should be cited in the text by number, and the reference list should be in the order of citation, not in alphabetical order. The first and last page numbers are to be included.

Unpublished work should  normally be avoided. If necessary it may not be listed in the references, but mentioned in the text as, for example, (Pande, A B, unpublished data).

Reference to patent should  include the names of patentees, the country of origin (underline) and the patent number, the organization to which the patent has been assigned (within circular brackets), the date of acceptance of the patent and the reference to an abstracting periodical where available [e.g. Trepagnier, J H, US Pat, 2, 463, 219 (to E I du Pont de Nemours & Co.,), 1 March 1949; Chem Abstr, 43 (1949), 7258].

***Nomenclature and Units :*** Standard terminology for biochemical compounds and uniform units in biochemistry and molecular biology  recommended by IUPAC-IUBMB joint committee should be used. (The website, *http://www.chem.qmw.ac.uk/iubmb/nomenclature*, contains updates and links to internationally agreed recommendations of use to biochemists).

***Enzyme Nomenclature :*** For enzymes, only the  trivial names recommended by the IUB-IUPAC Commission on Enzyme Nomenclature, 1992 (Academic Press, 1250, Sixth Avenue, San Diego, California 92101 – 4331, USA) should be used. In some cases, where the enzyme is the main subject of a paper, its code number and systematic name should  also be stated at its first citation in the paper.

***Spectrophotometric Data :*** While reporting Spectrophotometric data, the relation between the symbols used must be indicated. It is suggested that the symbols and terminology adopted by IUPAC [(1970) {*Pure Appl. Chem* 21, 1] may be adhered to. Beer's law be stated as

$$A = \log_{10} T = elc$$

Where A is the absorbance; T, the transmittance ($= I/I_0$); e, the molar absorption coefficient; *c,* the concentration of the absorbing substances in moles per litre; and 1, length of the optical path in centimeters.

**Molecular weight** being ratio is a pure number without any unit. It should be expressed as the relative moleclular mass ($M_r$).

**Molecular mass** (*m*) in contrast, is not a ratio, and can be expressed in Daltons (Da).

**Molecular mass** is the mass of one molecule of a substance.

***Abbreviations :*** Symbols and  abbreviations should conform to the recommendation of the Commission on Biochemical Nomenclature (CBN) of IUPAC-IUB as approved by the Commission of Editors of Biochemical Journals. Non-standard abbreviation, when used, must be defined in a footnote to the first abbreviation. Generic terms (virus, protein etc.), short trivial names (folate, adenosine, glycerol etc.) and enzyme names should not be abbreviated. However, an accepted abbreviation for the substrate may be incorporated in the enzyme name (e.g. glucose-6-P dehydrogenase, ATPase, but glucose 6-phosphatese, glutamate dehydrogenase). Rnase, for ribonuclease, and Dnase, for deoxyribonuclease, are permitted exceptions to this rule.

Authors are advised to see a recent issue of the journal to get familiar with the format and the practices adopted in respect of various elements of a paper.

## Instruction to Authors

*Accepted abbreviations :* The following abbreviations can be used without definition:

| | |
|---|---|
| ADP, CDP, | 5'-Phrophosphates of adenosing, cytidine, |
| GDP, IDP, UDP, | guanosine, inosine, uridine and |
| XDP | xanthosine. |
| AMP | Adenosine 5'-phosphate |
| ATP | Adenosine 5'-triphosphate |
| FMN | Flavin mononucleotide |
| GSH, GSSG | Glutathione, reduced and oxidized |
| CoA.Acyl-Coa | Coenzyme A and its acyl derivative |
| NAD, NAD | Nicotinamide adenine dinucleotide, its' |
| NADH | Oxidised and Reduced forms |
| NADP,NADP$^{\pm}$, | Nicotinamide adenine dinucleotide |
| NADPH | phosphate, its Oxidised and Reduced forms |
| NMN | Nicotinamide mononucleotide |
| Pi, Ppi | Orthophosphte, pyrophosphate |
| Q,QH,QH$_2$ | Ubiquinone, Ubisemiquinone, Ubiquinol |
| TDP | Ribosylthymine 5'-diphosphate |
| TMP | Ribosylthymine 5'-phosphate |
| UDPGlc | Uridine diphosphate glucose |
| UMP | Uridine 5'-phosphate |
| UTP | Uridine 5'-triphosphate |
| rRNA | Ribosomal RNA |
| nRNA | Nuclear RNA |
| mRNA | Messenger RNA |
| tRNA | Transfer RNA |
| cRNA | Complementary RNA |
| mtRNA | Mitochondrial RNA |
| hnRNA | Heterogeneous nuclear RNA |
| Tris | Tris (hydroxymethyl) aminomethane |
| EDTA | Ethylenediaminetetraacetate |
| CM cellulose | Carboxymethyl cellulose |
| DEAE | Diethylminoethyl |

***Proofs and Reprints :*** Gallely proofs are not sent to authors except to those who make specific request for the same. Authors should ensure that the data submitted for publication are error-free. Twenty five reprints, without cover, are supplied free of charges to the communicating author.

# CHAPTER - 56

# Indian Journal of Biotechnology

## Instructions to Contributors

The Indian Journal of Biotechnology, a quarterly journal, publishes original research papers, reviews, digests (biotechnology highlights), news-scan etc. The journal cover papers on Biotechnology in the following main areas: **(i) Agriculture; (ii) Animal husbandry; (iii) Environment; (iv) Industry; (v) Microbiology; (vi) Medicine; (vii) Bio-informatics; and (viii) Socio-legal and ethical aspects.**

Indian Journal of Biotechnology invites original research and review manuscripts not submitted for publication elsewhere. It is mandatory on the part of the corresponding author to furnish the following certificate at the time of submission of the manuscript:

*This is to certify that the reported work in the paper entitled "       " submitted for publication is an original one and has not been submitted for publication elsewhere. I/We further certify that proper citations to the previously reported work have been given and no data/tables/figures have been quoted verbatim from other publications without giving due acknowledgement and without the permission of the author(s). The consent of all the authors of this paper has been obtained for submitting the paper to the "Indian J Biotechnology".*

*Signatures and names of all the authors*

The copyright of the paper will be transferred from the author to publisher. One original and two copies of the manuscript should be submitted to the editor. The manuscript, after referees' acceptance, will be sent back to the author(s) along with referees' comments. For resubmission, two copies of the revised version of the manuscript, and a copy on floppy disk [3.5"(1.44 MB)] using word processing software such a MS Word (version 6 and onwards), or PDF files (version 4 and onwards), or as an attachment to e-mail should be submitted to the editor.

## Preparation of the Manuscript

Manuscripts should be typed in double space (11 pt, Times New Roman font preferred) on one side of the bond paper of $22 \times 28$ cm. All pages should be numbered consecutively. Use SI units, and give equivalent SI units in parenthesis when the use of other units is unavoidable. Symbols should conform to standard guidelines.

*Title :* It should be short & informative (15 pt), to be typed in only first letter of the first word capital; also, after colon or hyphen, first letter of the first word capital. Latin names are to be given in italics.

*Short Running Title :* Not in excess of 50 characters, to be all in capitals.

*Keywords :* Five or six keywords (in normal; 9 pt) indicating the contents of the manuscript.

*Authors :* Names of authors to be typed in first letters capital (10 pt).

*Addresses of Authors :* Addresses of the institution (s) where the work was carried out including telephone (office only),  fax number and e-mail address (9 pt). Author for correspondence should be indicated with an asterisk (*)

*Main Headings :* Each manuscript should be divided into the following main headings (typed in bold, first letters capital, on the left hand side of the page; 11 pt): Abstract, Introduction, Materials and Methods, Results, Discussion, Acknowledgement, References.

*Sub-Headings :* Typed in flush left, bold, first letters capital (9 pt).

*Sub-sub Headings :* Bold-Italics, first letter capital (9 pt).

*Abstract :* Should be brief not exceeding 200 words, typed in normal (9 pt).

*Introduction :* A brief and precise literature review with objectives of the research undertaken and essential background be given.

*Materials and Methods :* Materials and Methods should include the source and nature of material, experimental design and the techniques employed. New methods should be described in sufficient details, and others can be referred to published work.

*Results :* Results should contain data, which are essential for drawing main conclusion from the study. Wherever needed, the data should be statistically analyzed. Same data should not be presented in both table and figure form.

*Discussion :* Results should  deal the interpretation of the results. Wherever possible, results and discussion can be combined.

*Tables :* Tables should be typed in double space on separate sheets, numbered consecutively, and only contain horizontal cells. The table headings should be typed with the first letter capital.

*Figures :* The line drawings, illustrations, photographs, etc. will be accepted in TIFF files with hard copy. JPEG/GIF files will not be accepted. For each figure, a glossy print or original drawing may be submitted. Photomicrographs should have a scale bar. Line drawings should be roughly twice the final printed single column size of 7.5 cm width. The figures should be numbered in Arabic numerals. Lettering, numbering, symbols and lines in the graphs/illustrations should be sufficiently clear and large to withstand reduction up to 50%. Captions and legends to illustrations should be typed on a separate sheet of paper. Line drawings and photographs should contain figure number, author's name and the orientation (top) on the reverse with a soft lead pencil. Photostat copies and dot matrix prints will not be accepted.

***Reference* :** References should be cited in the text by the consecutive numbers of their occurrence; the  numbers are to be shown as superscript at the end of the statement related to that particular reference. e.g.  It also inhibits the activity of endogenous DNA polymerase of HBV[7].

Following the same sequence of the text, the list of references should be appended under the References heading. Each reference should provide names and initials of all the authors, giving coma in between the authors and '&' before the last author. In case, the authors are more than five, then use *et al.* after the 5th author. It should be followed by title of the paper, abbreviated title of journal (in italics), volume number, year of publication (within circular bracket), and the starting and closing page numbers. Abbreviated titles should conform to the international guidelines, e.g. The Chemical Abstracts Service Source Index (CASSI) or BIOSIS

The style of references should be:

## Research Papers

- Ghosh A C & Basu P S, Extracellular polysaccharide production by *Azorhizobium caulinodans*  from stem nodules of leguminous emergent hydrophyte *Aeschynomene aspera, Indian J Exp Biol,* 39 (2001) 155-159 [If accepted for publication, give (in press) in place of volume, year and pages].

- Newell C A, Lowe J M, Merryweather A, Rooke L M & Hamilton W D O, Transformation of sweet potato (*Ipomoea batatas*  (L) Lam) with *Agrobactericum tumefaciens* and regeneration of plants expressing cowpea trypsin inhibitor and snowdrop lectin, *Plant Sci,* 107 (1995) 215-227.

- Hoffman M P, Zalom F G, Smilanick J M, Malyj L D, Kiser J *et at,* Field evaluation of transgenic tobacco containing genes encoding *Bacillus thuringensis ä*-endotoxin or cowpea trypsin inhibitor: Efficacy against *Helicoverpa zea* (Lepidoptera: Noctuidae), *J Econ Entomol,* 85 (1991) 2516-2522

## Books & Proceedings of Conferences

- Truzuki  T & Irukayama K, *Minamata disease* (Elsevier, Amsterdam) 1977, 30-45.

- Roles O A & Mahadevan S, Vitamin A, in *The vitamins: Chemistry, pathology and methods,* 2nd edn, vol VI, edited by P Gyorgy & W N Pearson ( Academic Press, New York) 1967, 139-210.

- Allossp P G, Nutt K A, Geijsk R J & Smith G R, Transgenic Sugarcane with increased resistance to cenegrub, in *Sugarcane pest management in the new millennium, 4th Sugarcane Entomol Workshop,* held on 7-10 Feb, 2000 (Int Soc Sugarcane Technol, Khon-khon, Thailand) 2000, 63-67.

- Chaturvedi H C & Sharma A K, Citrus tissue culture, in *Proc Natl Semin Plant Tissue Cult* (ICAR, New Delhi) 1988, 36-46.

- Kapoor B C, 2000. *Managing in the face of not-so-developed and organized environment,* paper presented in *Natl Symp Manag* Dev, Institute of Public Administration, Jaipur, India, 21-23 July, 2000.

## Thesis & Dissertation

- Chaturvedi H C, *In vitro growth and controlled morphogenesis in callus tissue of Rauvolfia serpentina.* Ph D Thesis, Agra University, Agra, 1968.

## Patent

- Trepaginer J H, New  surface finishings and coatings, *US Pat 1276323* (to DuPont Inc, USA). 27 June, 2000; Chem Abstr, 49 (2000) 27689.

Manuscript along with referees' comments will be sent to the author identified for correspondence on the title page of the manuscript. It should be checked carefully and the modified manuscript should be returned within ten days of receipt. No page proofs will be sent to author(s).

*Reprints* : Twenty five reprints will be supplied gratis.

# CHAPTER - 57

# Indian Journal of Horticulture

## Guidelines To The Contributors

Indian Journal of Horticulture is the official publication of **the Horticultural Society of India.** It features the original research in all branches of Horticulture and other cognate sciences of sufficient relevance and primary interest to the horticulturists. The publication is generally open to the members of the Horticultural Societies of India but it also accepts papers from non-members on subjects related to Horticulture. All the authors have to become the annual member when a paper is accepted for publication intimated after a review.

The journal publishes three types of articles, i.e., **Review/Strategy paper** (exclusively by invitation from the personalities of eminence), **Research paper** and **Short communication.** The manuscripts should be submitted in duplicate in all respects to **the Editor, the Horticultural Society of India, C/0 Division of Fruits and Horticultural Technology, I.A.R.I. Pusa, New Delhi-110 012, India.** Each manuscript must be typed doubled spaced on one side of an A4 size page. Clearness, brevity and conciseness are essential in form, style, punctuation, spelling and use of English language. Manuscripts should conform to the S.1. systems for numerical data and data should be subjected to appropriate statistical analysis.On receipt of an article at the Editorial Office, an acknowledgement giving the manuscript number is sent to the corresponding author. This number should be quoted while making any future enquiry about its status.

*Review/Strategy paper.* It should be comprehensive, up-to-date and critical on a recent topic of importance. The maximum page limit is of 16 double spaced typed pages including tables and figures. It  should cite latest literatures and identify some gaps for future. It should have a specific **Title** followed by the **Name(s) of the author(s), Affiliation, Abstract, Key words,** main text with subheadings, Acknowledgements (wherever applicable) and **References.**

*Research paper :* The paper should describe new and confirmed findings. Should not generally exceed 12 typed pages including tables/figures etc. A paper should have the following features.

**Title** followed by **Author**(s) and **Affiliation** : Address of the institution (s) where the research was undertaken.

*Abstract :* A concise summary (200 to 300 words) of the entire work done along with the highlights of the findings.

*Key words:* Maximum five keywords to be indicated.

*Introduction :* A short introduction of the crop along with the research problems followed by a brief review of literature.

*Materials and methods :* Describe the materials used in the experiments, year of experimentation, site etc. Describe the methods employed for collection of data in short.

*Results and discussion :* This segment should focus on the fulfillment of stated objectives as given in the introduction. Should contain the findings presented in the form of tables, figures and photographs. As far as possible, the data should be statistically analyzed following a suitable experimental design. Same data should not be presented in the table and figure form. Avoid use of numerical values in findings, rather mention the trends and discuss with the available literatures. At the end give short conclusion. Insertion of coloured figures as photograph(s) will be charged from the author(s) as applicable and suggested by the printer.

**Acknowledgements** (wherever applicable).

*References :* Reference to literature should be arranged alphabetically and numbered according to author's names, should be placed at the end of the article. Each reference should contain the names of the author with initials, the year of the publication, title of the article, the abbreviated title of the publication according to the World List of Scientific Periodicals, volume and page(s). In the text,   the reference should be indicated by the author's name, followed by the serial number in brackets.


1.  Scaffer, B. and Guaye, G.O. 1989. Effects of pruning on light interception, specific leaf density and chlorophyll content of mango. *Scientia Hort.* **41**: 55-61.

2.  Laxmi, D.V. 1997. Studies on somatic embryogenesis in mango (*Mangifera indica* L.). Ph.D. thesis, P.G. School, Indian Agricultural Research Institute, New Delhi.

3.  Sunderland, N. 1977. Nuclear cytology. **In:** *Plant cell and Tissue Culture*, Vol. II. H.E. Street (ed.). University of California Press, Berkeley, California, USA, pp.171-2006.

4.  Chase, S.S. 1974. Utilization of haploids in plant breeding: diploid species. **In:** *Haploids in Higher Plants: Advances and Potential. Proc. Intl. Symp.* 10-14 June, 1974, University of Guelph. KJ.Kasha (ed.), University of Guelph, Canada, pp. 211-30.

5.  Panse, V.G. and Sukhatme, **P.V.** 1978. *Statistical Methods for Agricultural Workers*, Indian Council of Agricultural Research, New Delhi. 108p.

***Short communication :*** The text including table(s) and figure(s) should not exceed five pages. It should have a short title; followed by name of **author(s)** and **affiliation** and **References**. There should be no subheadings, i.e. Introduction, Materials and Methods etc. The manuscript should be in paragraphs mentioning the brief introduction of the topic and relevance of the work, followed by a short description of the materials and the methods employed, results and discussion based on the data presented in 1 or 2 table(s)/figure(s) and a short conclusion at the end. References should be maximum of seven in Nos. at the end.

## General instructions

- All the manuscript should be typed and double-spaced on one side of A4 size paper with proper margin.

- Generic and specific names should be italicized throughout the manuscript. Similarly, the vernacular names are to italicized.

- Each table should have a heading stating its content clearly and concisely. Place at which a table is to be inserted should be indicated in the text by pencil. Tables should be typed on separate sheets, each with a heading. Tables should be typed with first letter (T) only capital, table No. in Arabic numerals. All measurements should be in metric units.

- Data to be presented in graphical form should be sent on quality glossy contrast paper without folding. Each illustration must be referred to in the text and Roman numerals should be used in numbering. Photograph(s) of good contract must be mounted on hard paper to avoid folding and separate sheet must be given for the title for each photograph sent as figure.

- At the bottom of the first page present address of the corresponding author, **Tel. Fax. No. and E-mail ID,** etc. must be specified, if available.

- Revised manuscript is acceptable in duplicate along with a floppy/CD.

- No reprint of the published article will be supplied to the authors.

- Article forwarded to the Editor for publication is understood to be offered to INDIAN  JOURNAL OF HORTICULTURE exclusively. It is also understood that the authors have obtained a prior approval of their Department, Faculty or Institute in case where such approval is a necessary.

- Acceptance of a manuscript for publication in **Indian Journal of Horticulture** shall automatically mean transfer of copyright to the Horticultural Society of India. The Editorial Board takes no responsibility for the fact or the opinion expressed in the Journal, which rests entirely with the author(s) thereof.

# CHAPTER - 58

# Indian Journal of Marketing

## Guidelines to Authors

Indian Journal of Marketing – a Monthly Journal on Marketing welcomes original papers from both academics and practitioners on marketing. Papers should illustrate with empirical data and show the applicability of theory described.

***Text Preparation :*** Authors should send two copies of the manuscript. The text should be double spaced on A4 size paper with one-inch margin all around. The author's name should not appear anywhere on the body of the manuscript to facilitate the blind review Process. An abstract of 200 words and a brief biographical sketch of the author should accompany the manuscript on separate sheets. Research articles should be of 15-20 pages including tables, figures and exhibits. The authors should send a declaration stating that the paper is neither published nor under consideration for publication elsewhere.

All tables, charts and graphs should be typed on separate sheets. They should be numbered continuously in Arabic numeral as referred in texts. Wherever necessary, the source should be indicated at the bottom. Number and complexity of such exhibits should be as low as possible. All charts and graphs should be drawn cleanly and legibly. Tables and figures should contain self/explanatory titles. Footnotes, Italics, and quotation marks should be kept to the minimum.

Two or more referees review all contributions by following the 'double blind' system. The review process usually takes 2-3 months. The final draft is subject to editorial amendments to suit the journal's requirements.

An electronic version of the manuscript in MS Word would be required once the paper is accepted for publication.

The article must be sent to the :

Editor

Indian Journal Of Marketing

Y-21, Hauzkhas

New Delhi-110 016

# CHAPTER - 59

# Indian Journal of Nematology

**Information for Contributors**

The Indian Journal of Nematology is published twice a year. Original papers on all aspects of plant-parasitic, entomopathogenic and other soil-inhabiting nematodes will be considered for publication. Other publications will include short communications, papers discussing new concepts and abstracts of papers presented at the symposia sponsored by the Nematological Society of India. The author should be a member of the Nematological Society of India; a non-member, however, can communicate his papers through one of the members of the Society.

All manuscripts are to be typed in English, double-spaced throughout, including tables and citations, on white paper with 2.5 cm margin all around. A short title should be given at the beginning of each article. One original and one carbon copy of the manuscript should be sent to the Chief Editor, Indian Journal of Nematology, Division of Nematology. Indian Agricultural Research Institute, New Delhi – 110012, India, with the understanding that it is to be published exclusively in the Indian Journal of Nematology. Papers will be reviewed by the subject specialists and decision regarding acceptance will be communicated to the contributors in due time.

Normally, papers should not exceed 6 printed pages but longer papers will be considered on merits. However, contributors may be charged to pay extra cost of printing of the pages exceeding six. Metric and Celsius units are to be used. Title of the paper should be followed by author's name, institutional affiliation, postal and e-mail address, a condensed factual abstract, keywords, introduction, materials and methods, results, discussion, acknowledgement if any and references. The page number should be marked on the right hand corner of each page of the manuscript. For taxonomic papers, authors are advised to send the type / identified specimens along with manuscript for deposition in National Nematode Collection of India at Division of Nematology, I.A.R.I., New Delhi-110012.

Each table should be typed on a separate page. Figure legends are also to be typed on a separate sheet. Figures, illustrations and lettering should be large enough on final reduction, and in planning taxonomic plates, proportions of the journal page size (28.0 cm × 22.0 cm) should be kept in mind to allow for the figure legends. Photocopies of the original illustrations are to be sent at the time of sending manuscripts and original illustrations are to be sent only after acceptance of the paper is intimated. Author's name and serial number of the figure should be written on the back side of the illustrations or photographs lightly with pencil. Coloured illustrations will not be printed. Data are to be presented either by tables or figures, not by both. An identical copy of the manuscript may also be supplied on a 3.5" floppy diskette in MS-Word, as far as possible.

References cited in the text should be listed alphabetically and the citations should be complete including authors, year of publication, title of article, journal, volume and inclusive pages. Abbreviations should be as in the World List of Scientific periodicals. E.g. Linford, M B.& Yap, F. (1940). Some host plants of the reniform nematode in Hawaii. *Proc. helminth.Sco.Wash.* 7 : 42-44; White, P.R. (1943). *A handbook of plant tissue culture,* Jaques Cattel Press, Lancaster, Penn., pp.277.

International rules of nomenclature should be followed in naming of organisms. Binomials should be underlined.

The trade names of nematicides and other chemicals should be preceded by their chemical nomenclature.

# CHAPTER - 60

# Indian Journal of Plant Physiology

The Indian Journal of Plant Physiology is a quarterly publication which publishes records of original research in the fields of plant physiology, biochemistry, agronomy, soil science, horticulture, genetics, molecular biology and other cognate sciences that are of primary interest to plant physiology. The Journal also accepts critical review articles from the authors having substantial experience of research in that particular area of the subject. Review articles should indicate fruitful areas of further research. One review article in each issue may be published. The Journal is published quarterly, i.e. in March, June, September and December. However, the Society reserves the right to publish the Journal as a whole or in parts depending upon the material and the finances available.

**Manuscript** including illustrations should be sent induplicate to the Editor-in-Chief, Indian Journal of plant Physiology, Division of Plant Physiology, Indian Agricultural Research Institute, New Delhi 110 012, India. The papers forwarded to the Editor-in-Chief for publication are understood to be offered to **THE INDIAN JOURNAL OF PLANT PHYSIOLOGY** exclusively. It is also understood that the authors have obtained an approval the Head of their Department, Faculty or Institute in cases where such an approval is necessary. **Submission of a paper is taken to mean that the results reported have not been published and are not being considered for publication elsewhere.** The responsibility for statements, whether fact or opinion, rests entirely with the authors thereof.

On receipt of an article, an acknowledgement giving the registration umber of the paper is sent to the author who submits the manuscript. The number should invariably be quoted while making further enquiries about its publication. **All papers are refereed**. The author is informed of the article's acceptance/ rejection, as well as the comments of the referees, should a revision be thought necessary. The Editor-in-Chief decides on the final acceptance of the manuscript. He also decides in which issue of the Journal and in what position the paper is to be published.

Free reprints are not provided. Authors may however, order for the reprints while returning the proofs, which will be charged.

## Guidelines for the Preparation of Manuscripts

Manuscripts should be written in English. It should be as concise as possible and must not exceed 4 printed pages including tables and illustrations, extra pages will be charged. Short communications should not exceed 2 printed pages including figures and tables. **Manuscripts should be typed double spaced throughout on one side of the bond paper (21 × 29 cm)**, leaving 4 cm wide left hand margin. All pages (including the references, tables, figures and legends should be numbered consecutively.

**Title** should be short, specific and informative. It should identify the contents of the article. A short version of the title **(running title)** should also be given.

The names of authors should be written on the next line after the title of the paper followed by address of institution where the work was done. Change of address should be given as footnote.

**The contents** may be organized under : summary, key words, introduction, materials and methods, results, discussion, acknowledgements, if any and references. In short communication, separate titles for summary, introduction, materials and methods, results and discussion are not given. Please refer to recent issue of IJPP to note details of headings, tables, illustration, style and layout. **Manuscript not in the pattern of IJPP is returned to the author immediately.** Please also give telephone number/e-mail address of corresponding author in a separate sheet.

**Summary** should be brief and informative not to exceed 125 words. It should state the scope of work and the principal findings and should be complete enough for use by abstracting services.

**Key words** maximum of 6 in alphabetical order, suitable for indexing should be given

**Introduction** should be brief and limited to the statement of the problem or the aim of the experiment. The review of literature should be pertinent to the type of work.

**Materials and Methods** should include relevant details on the nature of material, experimental design, the techniques employed and the statistical methods used. For well known methods citation of reference will suffice. The **Results and Discussion** should preferably be combined to avoid repetition.

**References** should be arranged alphabetically, typed double-spaced. Make sure that all references in the text are listed and vice versa. Titles of periodicals should be abbreviated according to the world list scientific periodicals (1952). Refer to the latest issues of IJPP for the style used in references. Examples of common references are:

***Journal article***

Chinoy, J.J. Nanda, K.K., Sirohi, G.S. and Sawhney, K.L. (1959). Growth and phasic development of wheat. I. Vegetative period an photothermic requirement. *Indian J. Plant Physiol.* 2 : 29-45.

***Whole book***

Asana, R.D. (1976). Physiological Approaches to Breeding of Drought Resistant Crops. ICAR Publication, New Delhi.

***Chapter in a book***

Sirohi, G.S. (1965). Studies on growth and development of crop plant in relation to their genetic constitution. In: R.D. Asana and K.K. Nanda (eds.), Growth and Development of Plants, pp. 121-137. Today & Tomorrow's Book Agency, New Delhi.

**In the text, the references should be cited** by giving name(s) of the author(s) and year of publication. Do not use comma between the authors name and year of publication. Two or more references at one place in the text should be cited chronologically and are separated by comma. Use 'and' to link the name of two authors in the text and use *et al.* where there are more than two. Citation of 'personal communication' and 'unpublished data' should be avoided, unless absolutely necessary. Such citations should in text appear only as: (G.S. Sirohi, personal communication) and not in the reference list.

**Botanical genus and species names** should be set in italics or underlined. **Abbreviations** other than the standard should be explained within brackets when first used in the text. All units and measurements must conform to the international standard system (SI-system). However, non-SI units such as day and year are acceptable. In complex groupings of units, use the negative index system eg. mmol $m^{-2}s^{-1}$.

**Tables** should be typed on separate sheets. Each table should be numbered with Arabic numbers followed by a heading stating its contents clearly and concisely. Each table should be mentioned in the text and the places where tables are to be inserted, should be indicated.

**Illustrations** which are of good quality and are essential to a clear understanding of the paper can be accepted. Ensure that table and illustrations do not repeat data. Each illustration must be specifically referred to in the text. Text-figures should be used in preference to plates. Colour photographs are accepted if they are essential but the cost of production must be borne by the author.

**Line drawings** should be clearly drawn in black Indian ink on good tracing paper and should be of the size: width 12 cm × height (preferably, 9 cm to a maximum of 18 cm). Figures meant for reduction should have multiple dimension of the above size. Letters, dots, lines etc. in the drawing should be large enough to permit reduction without loss of details. Figures, diagrams etc. thus intended for reproduction as illustration must be of high contrast and sharp definition **(original drawings, no photostats)**. Text-figures should be numbered in arabic numbers, in order of their reference. Captions and legends to illustrations should be typed on a separate sheet of paper and attached at the end of the paper.

**Photographs** for publications should be of high contrast on glossy paper and should not be folded or creased. Photograph number and title of article with authors name should be given on the back of the photograph.

**Proofs** should be checked and returned as promptly as possible to the Editorial Office preferably with in a week. No editing or material changes at the proof stage will be permitted unless the extra cost involved is paid by the authors. Although every effort is made by the editors to correct proofs of all the paper, they assume no responsibility for any errors that may remain in the final printing.

# CHAPTER - 61

# Indian Journal of Plant Protection

Publication in this journal is open to the members of Plant Protection Association of India. All authors must be members. Original research papers on applied aspects of plant protection (insect pests, mites, nematodes, diseases, rodents, weeds and plant protection equipment) are considered for publication. Papers on routine laboratory / field studies are discouraged. Research articles based on Integrated Pest Management trials conducted for a minimum of two seasons shall be given due weightage. Field data especially on assessment of crop losses, control trails and host plant resistance should be for a minimum of two seasons.

Manuscripts not exceeding 6000 words may be submitted to:

> The Chief Editor
>
> Indian Journal of Plant Protection
>
> National Bureau of Plant Genetic Resources – Regional Station
>
> Rajendranagar, Hyderabad – 500 030
>
> Andhra Pradesh, India.

Manuscripts should be submitted induplicate, typed in double space, on standard A-4 size paper. Authors are advised to take special care while preparing the article to use clear and concise English to ensure consideration of the article by the editorial board. The first page of the typescript should contain a short and informative title with initial capitals for each word together with names, addresses and affiliations of the authors. A short running title of not more than 50 characters including spaces suitable for page headings should be given if the title is longer than this. The paper should start with a brief and informative Abstract of 200-250 words, containing all keywords. The keywords should be given immediately after the Abstract under a separate heading. The manuscript should be divided into four sections. Viz., Introduction, Materials and methods, Results and discussion, and References. The title (14 pt) and all headings should appear in Arial Black and the remaining text should be in Times New Roman font (10 pt)

The Introduction should be brief and state the subject of the experiment. The literature reviewed should be pertinent to the problem. Materials and methods should include the experimental design, treatments, replications, and techniques employed. Where standard methods / techniques are used, citation of reference is sufficient. Result and discussion should be combined and supported by Tables or Figures. The main body of the text should conform to the instructions given below. Abbreviations should be specified at first mention.

## Headings

All headings should be clearly indicated. The primary headings ( e.g. Materials and methods) should be in 12 pt whereas, the secondary headings should appear in 10 pt (e.g. Greenhouse studies). The tertiary headings in 10 pt should run on with para, full  stop and two spaces after the head before staring the text. (e.g., gramplasm screening. Genotypes consisting of....)

## References

References must follow the Harvard system and cited in the text as Singh *et al,*. (2002)or (Singh *et al.,* 2002). For several publications by the same author in a year, put a, b, c, etc., after the year of publication, e.g. Singh (2002a). The reference should be in chronological order in the text, and alphabetical order in the reference list at the end of the paper in the following format.

## Journal Articles

Jena, M. Dani, R.C. and Rajmani, S. 1992. Effectiveness of insecticides against the rice leaf folder. *Indian Journal of Plant Protection I20: 43-46.* (Name of the journal should be italicized and should not be abbreviated).

## Books/Proceedings

Matclf, R.L. 1975. Insecticides in pest management, *Introduction to Insect Pest Management* (eds. Metcalf. R.L. and Luckmann, W.H.). John Wiley and Sons, New York, USA. Pp 235-273. Where the citation refers to the whole book, the total number of pages should be specified. E.g. 274pp.

## Numbers and Units

Numbers and units are used for all specific units of measurement, e.g. 4 mm, 15 ml, and 5 g etc. Quantities of items up to nine are spelt out, e.g. six leaves, five fields, but 15 leaves and 140 fields etc. SI units should be used. Abbreviations for units should conform to British standard 1991 and are alike in singular and plural (thus g not gm or gms). Pesticide and fertilizer applications should be expressed as 1.5 kg a.i./ha

## Genera and species

The scientific names of genera and species should be in italics or underlined, followed by the author's name at first mention in the text (e.g. gram pod borer, *Helicoverpa armigera.* Hub.) and a subsequent mention, it should be cited as *H. armigera.*

## Application of  Pesticides

Pesticides and other industrial products should generally be referred to by their common names. If registered trade names are unavoidable, they should have a initial capital, e.g. Bavistin. Full details must be given of techniques used to apply pesticides (e.g. type of equipment, type of nozzle, pump pressure, volume of spray etc.) and of the amount of active ingredient applied per units area (250 g a.i./ ha).

## Tables

The table should be on a separate sheet, and intelligible without reference to the text. Table heading should be part of the table separated by a horizontal line between the head and table text. Vertical lines should not be drawn. Authors are required to prepare tables to fit in either single column width (83 cm) or 176 mm width. Appropriate statistical tests should be applied (SE, LSD, etc.,) to the data presented. Footnotes should be minimal. Authors must indicate on the margins where tables/figures should appear. Lengthy tables should be avoided.

## Figures

Figures should be computer generated or hand drawn in black ink and submitted in duplicate each on a separate sheet, numbered sequentially. They must be suitable for reduction to a width of 83 mm or 176 mm. Use a scale to indicate the size of the objects where necessary. Titles for all figures should be given along with the figures and kept to a minimum. For photograph figures only good quality black & white photographs are acceptable.

## Copyright

Manuscript will be accepted on the understanding that their content is original and that permission has been received in writing wherever necessary to reproduce previously published material (including quotations, data and illustrations), and that the manuscript has not been submitted/ accepted for publication elsewhere. Copyright resides with the Plant Protection Association of India.

## Softcopy

Authors whose articles pass through the preliminary review must submjit the revised manuscript along with a softcopy prepared in a word processor (Microsoft Word) in an IBM comparable floppy diskette ( 3 ½  inch). The graphs should be prepared in Microsoft Excel to fit in a single column width (83 mm) or two column width (176 mm) and enclosed as separate files only. The graphs should not be enclosed as pasted objects in the main article (Microsoft Word file). Authors may also email softcopies to *ijpp@epatra.com*.

## Page charges

There are no page charges.

# CHAPTER - 62

# Indian Journal of Social Research

## Guidelines of Contributors

Preparation of Copy

Submitted manuscripts should contain some definable and original contribution. Articles should be 15 to 20 typed double spaced pages including footnotes and references. Author should submit two copies of manuscript. The process of evaluating the manuscript normally takes about 6-8 weeks. Before submitting the paper, please prepare copy according to IJSR format which includes the following:

1. Type all copy, including indented matter, footnotes, and references, on white bond paper, double spaced, and on one side of the page. Lines should be six inches with one inch margin at top and bottom of the page.

2. All major headings should be capitalized and centered. All sub-headings should be on lefthand and underlined. Text immediately follows the subheading.

3. Draw each figure in black ink and type each table on separate pages. Send camera ready copy. Capitalize all letters of the title of tables and figures. Insert a location note in numerical order in the text, e.g. :Table 1 about here".

4. Footnotes should be used only for substansive comments not included in the main text. They should be numbered serially and typed double-spaced on separate sheets at the end of the text before the references.

## Format of References in Text

Identify all references to books, monographs, articles, and statistical sources at an appropriate point in the main text by author's last name, year of publication, and pagination where appropriate, all within parentheses.

1. If author's name is in the text, use only year of publication in parentheses, e.g. Lipset (1963). If the author's name is not in the text, include both author's name and year of publication separated by a comma within the parentheses e.g. (Dahrendorf, 1959).

2. Pagination follows year, separated by a colon, e.g. (Domhoff, 1970: 45-46).

3.  With dual authorship give both names; for three or more use "et al." e.g. (Lipset and Solari, 1967) and (Rosen et al., 1969).

4.  If more than one reference to the same author and year, distinguish between them by use of letter "a,b...." attached to year of publication;(Moskos, 1971a).

5.  Enclose within a single pair of parentheses a series of references, separated by semicolons, e.g. (Lang, 1971; Lopreato, 1972; Olsen, 1973).

6.  For institutional/Organisational authorship provide minimum identification from the beginning of the complete citation, e.g. (Am. Pol. Sc. Asso.1971:85) and (US Bureau of the Census, 1972).

# CHAPTER - 63

# Indian Journal of Weed Science

## Instructions to Contributors

Indian Journal of Weed Science is an international journal published quarterly by the Indian Society of Weed Science. Ordinarily, all authors must be members of the Society. Papers submitted by non-member may also be considered at the discretion of the Editorial Board. Papers submitted would be **peer reviewed**. Papers are accepted on the understanding that the work described is original and has not been published elsewhere and that the authors have obtained necessary authorization for publication of the material submitted. **Articles on all aspects of weeds and weed management will be considered for publication** in the journal.

## Manuscript

Manuscript should be written in clear concise English, typewritten in double space on one side of A4 size paper with at least 2.5 cm margin. The research papers should not exceed 15 typed pages including table and figures and should contain abstract, introduction, materials and methods, results and discussion, acknowledgement, if necessary, and references. Short communications based on new approaches or findings not exceeding 6 typed pages will also be considered for publication. Such articles should not exceed 6 typed pages and will not have abstract and are not divided into different sections. The review articles, as far as possible, should not exceed 30 typed pages. It should feature recent developments from the author's laboratory or his own field of specialization and include results of various workers in the given field. **Review article authors are advised to contact the Chief Editor before submission of the manuscript.** All articles are to be submitted in duplicate.

*Title :* The title must be short and specific. The by-line should include name(s) of the author(s) and address of the institution where the research has been carried out. Change of address or mailing address of the corresponding author should be given as a foot note.

*Abstract :* It should be brief and should include objectives, methodology and important findings and relevance of the research carried out. It should be followed by a maximum five key words for indexing purpose.

***Introduction :*** It should be brief, crisp and limited to the statement of the problem or objective of the experiment. Key references related to the work must be cited. It must end with proper justification for undertaking the work reported in the paper.

***Materials and Methods :*** All relevant details of the methodology including experimental design, treatment details, statistical application must be induded. Any modification of the original method must be duly highlighted. This section may be subtitled for clarity. All units must be in SI units and standard abbreviations of the various units wherever necessary may be used.

***Results and Discussion :*** This section may be subtitled for clarity. The salient results must be highlighted and discussed in relation to related works. Undue and vague claims and stretched explanations may be avoided. Data presented in figures should not be duplicated in Tables and vice versa.

***Figures and Tables :*** All figures and tables must be on separate pages. Figures should be drawn in black ink on  good quality tracing paper with numbers and point large enough to permit size reduction. Photographs must be on glossy paper. The data in tables should be so arranged that the table should fit in the normal layout of the page. Large tables which need to be arranged horizontally (landscape pattern) should be avoided. Discussion should be oriented towards explaining the significance and implication of the results obtained and should be supported with the previous work done in this direction. The results and discussion section may preferably be combined to avoid repetition.

***References :*** All references quoted in the text must be listed alphabetically. It should include names of all authors, year, full title of the article, name of the journal as per standard abbreviation, volume and page numbers. Authors are advised to consult a recent issue of the journal for correct style and format.

Manuscript complete in all respects should be submitted in duplicate to :

Dr. Govindra Singh

Chief Editor, IJWS

Department of Agronomy

G.B. Pant University of Agriculture & Technology

Pant Nagar – 263 145, India

# CHAPTER - 64

# Indian Perfumer

## Instructions for the Authors

*Indian Perfumer* is published quarterly, presenting research findings and reviews on the matters concerning cultivation and improvements of essential oil bearing plants and chemistry of essential oils as well as distillation, extraction, isolation, uses (fragrances, flavours and biological activity) and trade. All manuscripts must be typed in double space, on one side of the paper (A-4), with wide margins throughout. Do not underline any headings or any part of the text unless absolutely necessary. The language of publication is English.

1. Manuscripts should be submitted in triplicate (with original figures) to:

   Editor, *Indian Perfumer*

   Essential Oil Association of India

   301, 4832/24 Ansari Road, Daryaganj

   New Delhi 110 002, India

   Papers are accepted on the understanding that their content has neither been published not submitted for publication elsewhere.

   The lay-out of the paper will depend upon the content.

2. *Title :* should be concise and specific, describing the nature of work.

3. *Authors' Names :* should have initial(s) and surname (e.g. M.L. Maheshwari). Give full address of each author, indicating (with an asterisk) the author to whom the correspondence concerning the manuscript should be made.

   Footnotes should be kept to a minimum. They should be indicated by superscript number, e.g. change of address of an author.

4. *Abstract :* is a summary of the work done because it is often the only portion of the paper read by abstracting journals. Hence the abstract should be written so that it can be used verbatim in an abstracting journal. It should be a concise summary, giving essential information, data etc. and be intelligible without reference to the paper itself.

5. *Introduction :* should present the objectives or reason for the investigation. A summary of the pertinent literature should be included with latest references, in brief, pointing out lacunae that led you to initiate this work.

6. *Materials And Methods :* the section should describe clearly and in detail the materials and methods used, so that the experiment could be reproduced by others. New techniques need to be described fully, whereas the known methods must have only references.

7. ***Results And Discussion :*** should be presented in a clear and concise manner, using tables and illustrations for clarity. Do not list tabular data in text. Do not list significant figures of data for which the level of experimental error is unacceptable. The results of other studies of a similar or related nature should be compared with only the most pertinent data and can be listed in tabular form. Discussion should relate to your work in relation to previous work of similar nature or related aspects. Draw your conclusions.

8. *Tables :* should be typed in double space in the same form as the manuscript. However, they should not be formatted within the text, but should be attached at the end. Use tabs, not space bar, in typing.

9. *Figures* : Send original figures, properly packed, flat, without pins or clips. Do not write or mark on the main figure. Figures can be high-quality photographic prints or camera-ready original diagrams, graphs or drawings. Spectra or chromatograms should be presented as high-quality photographic prints or camera-ready computer-drawn representations. All figures descriptions can appear on the same page. The size of figures submitted should not exceed 14.5 × 20 cm. The figure number and the author's name should be written on the back of all figures in soft pencil.

10. *Acknowledgement :* Acknowledge the persons or institution that helped you personally or financially in conducting the experiment.

11. *References* : must be arranged alphabetically and should be typed in order. In the text, the reference should appear as: Gupta (1982), or (Smith, 1972; Jones&Brown, 1991; Atal *et al.,* 1980). All references must be full, using the surname, initials, year, title of paper, journal, volume, issue number and pages. The name of the journal should be abbreviated in accordance with *World List of Scientific Periodicals,* published by Butterworths, London.

For a book give title, volume, pages consulted, names of editors, publisher and place of publication. Give details of proceedings etc. similarly, leaving no ambiguity like-

1. Azzouz, M.A., Reinecius, G.S. and Moshonas, M.G. 1976. Comparison between cold pressed and distilled lime oils through the application of GC-MS. *J. Fd Sci.* 41: 324-328.

2. Guenther, E. 1972. *The Essential oils,* vol. 1, pp. 316-319. Robert E. Krieger Pub. Co., Huntington, New York.

3. Lawrence, B.M. 1988. A further examination of the variation of *Ocimum basilicum L. (In):" Flavours and Fragrances: a World Perspective,* pp. 161-170. Lawrence, B.M., Mookerjee, B.D. and Willis, B.J.(Eds). Elsevier Scientific Publishers, B.,V., Amsterdam.

4. Lincoln, D.E., Murray, M.J. and Lawrence, B.M. 1986. Chemical composition and genetic basis for the isopinocamphone chemotype of *Mentha citrata* hybrids. *Phytochemistry* 8; 1857-1863.

12. ***Reprints :*** The corresponding author will receive 25 complementary reprints for distribution among co-authors. The authors have permission to contact publisher to make further copies at their expense.

# CHAPTER - 65

# Irrigation Science

## Copyright

Submission of a manuscript implies: that the work described has not been published before (except in form of an abstract or as part of a published lecture, review or thesis): that it is not under consideration for publication elsewhere; that its publication has been approved by all co-authors, if any, as well as – tacitly or explicitly – by the responsible authorities at the institution where the work was carried out. The author warrants that his/her contribution is original and that he/she has full power to make this grant. The author signs for and accepts responsibility for releasing this material on behalf of any and all co-authors. Transfer of copyright to Springer becomes effective if and when the article is accepted for publication. After submission of the Copyright Transfer Statement signed by the corresponding author, changes of authorship or in the order of the authors listed will not be accepted by Springer.

The copyright covers the exclusive right (for U.S. government employees: to the extent transferable) to reproduce and distribute the article, including reprints, translations, photographic reproductions, microform, electronic form (offline, online) or other reproductions of similar nature.

All articles published in this journal are protected by copyright, which covers the exclusive rights to reproduce and distribute the article (e.g. as offprints), as well as all translation rights. No material published in this journal may be reproduced photographically or stored on microfilm, in electronic data bases, video disks, etc., without first obtaining written permission from the publisher.

The use of general descriptive names, trade names, trademarks, etc. in this publication, even if not specifically identified, does not imply that these names are not protected by the relevant laws and regulations.

An author may self-archive an author-created version of his/her article on his/her own website and his/her institution's repository, including his/her final version; however he/she may not use the publisher's PDF version which is posted on *www.springerlink.com*. Furthermore, the author may only post his/her version provided acknowledgement is given to the original source of publication and a link is inserted to the published article on Springer's website.

The link must be accompanied by the following text: 'The original publication is available at *www.springerlink.com*'.

The author is requested to use the appropriate DOI for the article (go to the Linking Options in the article, then to OpenURL and use the link with the DOI). Articles disseminated via www.springerlink.com are indexed, abstracted and referenced by many abstracting and information services, bibliographic networks, subscription agencies, library networks, and consortia.

While the advice and information in this journal is believed to be true and accurate at the date of its publication, neither the authors, the editors, nor the publisher can accept any legal responsibility for any errors or omissions that may be made. The publisher makes no warranty, express or implied, with respect to the material contained herein.

*Special regulations for photocopies in the U.S.A.* Photocopies may be made for personal or in-house use beyond the limitations stipulated under Section 107 or 108 of U.S. Copyright Law provided a fee is paid. All fees should be paid to the *Copyright Clearance Center,* Inc., 222 Rosewood Drive, Danvers, MA 09123, USA, Tel.: +1-978-7508400, Fax: +1-978-6468600, *http://www.copyright.com*, stating the ISSN of the journal, the volume, and the first and last page numbers of each article copied. The copyright owner's consent does not include copying for general distribution, promotion, new works, or resale. In these cases, specific written permission must first be obtained from the publisher.

*The Canada Institute for Scientific and Technical Information (CISTI)* provides a comprehensive, world-wide document delivery service for all Springer journals. For more information, or to place an order for a copyright-cleared Springer document, please contact Client Assistant, Document Delivery, CISTI, Ottwa K1A 0S2, Canada (Tel.: +1-613-9939251, Fax: +1-613-9528243, e-mail : cisti.docdel@nrc.ca).

## Subscription information

ISSN print edition 0342-7188

ISSN electronic edition 1432-1319

## Subscription rates

For information on subscription rates please contact:

Springer, Customer Service Journals

Haberstrasse 7, 69126 Heidelberg, Germany

Tel.: 49-6221-345-0, Fax +49-6221-3454229

e-mail: SDC-journals@springer.com

Business hours: Monday to Friday

8 a.m. to 8 p.m. local time and on Germen public holidays

## Electronic edition

All electronic edition of this journal is available at springerlink.com

## Science communication

Management Director: Beate Fruhstorfer

Advertising: Jens Dessin, anzeigen@springer.com

Springer, Heidelberger Platz 3

14197 Berlin, Germany

Tel.: +4-30-827875739, Fax: +49-30-827875300

www.springer.de/wikom

## Production

Springer, Sabine Taeger

Journal Production Life Sciences

Postfach 105280, 69042 Heidelberg, Germany

Fax: +49-6221-4878686

e-mail: sabine.taeger@springer.com

*Address for courier, express and registered mail:*

Tiergartenstrasse 17 69121 Heidelberg,

Germany

# CHAPTER - 66

# Journal of Agribusiness

## General Instructions

Please provide three clean, double-spaced hard copies of your final manuscript (using 1" margins, 12-point Times Roman font, *NO* right-margin justification, and *NO* end-of-line hyphenation) plus a disk file (standard 3.5" disk size) in WordPerfect (any version), Microsoft Word (any version), or as an ASCII file. (If you need to use an ASCII file, please contact the technical editor via e-mail for detailed instructions.) Be sure to include your e-mail address, fax/phone numbers so that the technical editor can contact you regarding any questions.

## Guidelines for Manuscript Components

1. **Title Page**

   (a)  Full title of manuscript

   (b)  All author names (exactly as you wish them to appear, i.e., first name/middle initial or just first/middle initials)

   (c)  Abbreviated manuscript title to be used as running head

   (d)  Author titles, affiliations, any acknowledgments, and any funding source ID if you wish to include it

2. **Abstract Page**

   (a)  Full title of manuscript

   (b)  Abstract (double-spaced, 12-point font), not to exceed 100 words

   (c)  *Key Words :* Provide up to 8 key words (or short phrases), in alphabetical order

   (d)  *Text :* All text (including abstract, endnotes, references, appendices) should be double-spaced, using a 12-point Times Roman font size and 1" top/bottom, left/right margins.

3. **Very Important**

   (a)  Please do not use right-margin justification (since it makes it very difficult for technical editor to intuit spacing, especially of math notations).

   (b)  Do *NOT* use end-of-line hyphenation feature.

(c) ***Footnotes/Endnotes :*** Do *NOT* use the WordPerfect (or Microsoft Word) footnote/endnote feature when preparing your text. Instead, insert superscript numbers within the text (numbering consecutively throughout the manuscript), and then prepare a full listing of your footnotes/endnotes (to be placed on a separate page immediately preceding the reference section). Do not place footnotes at the bottoms of manuscript text pages.

4. **Sequence of Manuscript Components**

   (a)  Title page (unnumbered)

   (b)  Abstract page (unnumbered)

   (c)  Text narrative (commence page numbering of text with page 1)

   (d)  Footnotes/Endnotes page

   (e)  Reference section

   (f)  Appendix (if more than one appendix, label Appendix A, Appendix B, etc.; assign titles to all appendices)

   (g)  Tables (each table should be on a separate page)

   (h)  Figures (each figure should be on a separate, *unnumbered* page)

5. **Heading Levels**

   So that the technical editor can clearly identify your heading levels, please use the following format:

   (a)    Level #1 = Centered, boldface, initial caps

   (b)    Level #2 = Flush left margin, italics, initial caps

   (c)    Level #3 = Flush left margin, plain typeface, initial caps

6. **Math/Equations**

   When numbering equations, use Arabic numbers enclosed in parentheses. Equation numbers should appear at flush left margin. Number only those equations that are referred to within the text, and number them consecutively—i.e., (1), (2), (3), etc.—throughout the manuscript. The math notations/equations should be centered between the L/R margins. Use *italic* typeface for all *variables*, and use boldface (no italics) for all vectors and matrices—both within equations and within narrative.*JAB* Author Guidelines Page 2 of 4

7. **Within-Text Citations**

   Citations may appear parenthetically or as part of the narrative. Within the text, use parentheses ( ) rather than brackets [ ] for citations.

   (a)  Spell out up to 3 author last names (i.e., use "et al." only for 4 or more authors).

   (b)  Include the year of publication for all within-text cites. If there is more than one work by the same author(s) in the same year, please designate, for example, as 1995a, 1995b, etc. Make sure the corresponding listings in the reference section also show the "a" and "b" designations.

(c)  *Very Important :* When citing a direct quotation, be sure to include the page number(s) from the author's work.

8.  Percent vs. %: Our journal style does not spell out "percent." Instead, use "%" throughout.

9.  **Tables**

Place each table on a separate page. For tables it is permissible to use single spacing and a smaller font size as needed. Tables should not be integrated into your text (but all tables should be introduced within the narrative discussion by table number). Tables should be numbered consecutively (1, 2, 3, etc.) and should be placed at the end of the manuscript. Footnotes within tables should be identified by superscript alphabetical letters (a, b, c, etc.) rather than Arabic numbers. When using asterisks (*, **, ***) to denote levels of significance/probability, a single asterisk is used for the lowest level, two asterisks for the next highest, etc. For example: * = .10 level (10%), ** = .05 level (5%), *** = .01 level (1%)

10.  **Figures**

Figures should be placed at the end of the manuscript (on unnumbered pages) immediately following tables. *Do not place figures within the manuscript text file.* Preferred (but not required) software for figure preparation is Microsoft Excel, Corel Quattro Pro, or Corel Presentations (through WordPerfect).

Regardless of the software you have used in preparing your graphics, you are requested to provide the following:

(a)  *Camera-ready* (i.e., run on laser printer . . . with a minimum of 600 dots per inch resolution required) copies of your figures, each on a separate, unnumbered page. The figure title can be included on the page with your graphic; however, *make certain the figure title does NOT appear within the graphic image itself* (since the title will be typeset in our precise journal font/format by our technical editor in the final layout). It is helpful (but not required) if the camera-ready hard copies of your graphics are sized to fit the margin constraints of the *JAB* journal page. (In the event our technical editor must cut/ paste your figures, this avoids loss of clarity through second-generation photocopies that must be scaled to fit.) < If the graphic is to appear in portrait format, make sure the graphic image is no more than 4.75" wide (height may vary as needed, but should not exceed 6"). < If the graphic is to appear in landscape (turned sideways on page) format, the image width should not exceed 6" and height should not exceed 4.25".

(b) *Important:* Provide a computer disk file(s) for your figure graphics, being sure to include spreadsheet data (i.e., the spreadsheet contains the values used in constructing the graphic image). The spreadsheet is critical, since it is frequently impossible to make even small cosmetic changes to the figure without the presence of the spreadsheet data file.

(c) Clearly label on your 3.5" disk the software used for preparation of your figures.

## The Reference Section

All citations within the manuscript must appear in the reference list . . . and all listings in the reference section must be cited somewhere within the manuscript. References should be in alphabetical order by author's last name. For clarity, *please do NOT use any abbreviations* (such as for journal names) in the references. Our journal style uses fully spelled-out journal titles.

***Special notes :*** Do not use "et al." (either for authors or editors) in reference list citations; all author (editor) names should be spelled out. Use only author/editor first and/or middle initials (we do not spell out first or middle names in reference section).

*You do not need to be greatly concerned with specific format/style requirements* when preparing your list of references. (The technical editor will handle this task for you.) The critical concern is that all reference *JAB* Author Guidelines Page 3 of 4 components are present. The *JAB* reference style is adapted from the *Publication Manual of the American Psychological Association* (APA). A number of sample reference citations are provided below (as generally fictitious illustrations only), showing the components needed for various reference types.

## Book

Cremlyn, R. J. (1991). *Agrochemicals: Preparation and Mode of Action*, 3rd ed. Chichester, England: John Wiley and Sons.

## Chapter In Book

Green, R. E., J. M. Davidson, and J. W. Biggar. (1980). "Methods for determining adsorptiondesorption of organic chemicals: An assessment." In A. D. Banin and U. F. Kafkafi (eds.), *Agrochemicals in Soils* (pp. 273S282). Elmsford, NY: Pergamon Press.

EDITED BOOK (citing entire book rather than individual authors)

Bredahl, M. E., P. C. Abbott, and M. R. Reed, eds. (1994). *Competitiveness in International Food Markets*. Boulder, CO: Westview Press.

## Journal Article

(Note: *FULLY SPELL OUT* name of journal; be sure to include volume number and inclusive page numbers.) The first two examples shown below use 1985"a" and "b" designations to illustrate more than one work by same authors in same year; the third illustration shows inclusion of the journal issue number as well as volume number:< Addiscott, T. M., and R. J. Wagenet. (1985a, March). "Concepts of solute leaching in soils: A review of modeling approaches." *Journal of Soil Science* 36, 411S424. < Addiscott, T. M., and R. J. Wagenet. (1985b). "A simple method for combining soil properties that show variability." *Soil Science Society of America Journal* 49, 1365S1369. < Anderson, J. L. (1995, Winter). "The environmental revolution at twenty-five." *Rutgers Law Journal* 26(2), 395S430.

## Papers, Reports, Bulletins

Alam, A., and S. Rajapatirana. (1993). "Trade policy reform in Latin America and the Caribbean in the 1980s." Policy Research Working Paper No. 1104, International Trade Division, The World Bank, Washington, DC. < Schatzer, R. J., M. Wickwire, and D. Tilley. (1986). "Supplemental vegetable enterprises for cow-calf and grain farmers in southeastern Oklahoma." Research Report No. T-874, Agricultural Experiment Station, Department of Agricultural Economics, Oklahoma State University, Stillwater. < Grise, V. (1990, February). "The world tobacco market: Government intervention and multilateral reform." Unnumbered staff report, U.S. Department of Agriculture, Economic Research Service, Washington, DC. < U.S. Department of Agriculture. (1983, December). "Food consumption, prices, and expenditures: 1962S1982." Statistical Bulletin No. 702, USDA, Economic Research Service. Washington, DC: U.S. Government Printing Office. < Huang, K. S. (1985, December). "U.S. demand for food: A complete system of price and income effects." Technical Bulletin No. 1714, U.S. Department of Agriculture, Economic Research Service, Washington, DC. < U.S. Department of Agriculture, Agricultural Marketing Service. (1985S96). *Fresh Fruit and Vegetable Arrival Totals for 23 Cities* (various issues). USDA/ AMS, Market News Branch, Washington, DC.

## Paper Presented At Meeting

Eginton, C., and L. Tweeten. (1982, February 12). "Impacts of national inflation on entrance and equity growth—Opportunities on typical commercial farms." Paper presented at the annual meetings of the Southern Agricultural Economics Association, Atlanta, GA.

## Proceedings

Badger, D. D. (1981). "Economics of manure management." In *Livestock Waste— A Renewable Resource: Proceedings of the Fourth International Symposium on Livestock Wastes* (pp. 275S291). Held in Amarillo, TX, July 10S13, 1980.

Lubbock, TX: Texas Tech University Press. < Aguilar, G. C., H. Medina, and F. Carranza. (1993). "Pesticide poisonings among agricultural workers in Bolivia." In G. Forget, T. Goodman, and A. de Villiers (eds.), *Impact of Pesticide Use on Health in Developing Countries* (pp. 86S97). Proceedings of a symposium held in Ottawa, Canada, September 17S20, 1991. Ottawa, Ontario, Canada: Agricultural Research Center of Canada. *JAB* Author Guidelines Page 4 of 4

## Trade Magazine or Publication

Bickers, C. T. (1994, March 1). "The 75% domestic content law." *Tobacco International* 196(4), 16S21. < Tweeten, L. G. (1995, 2nd Quarter). "The twelve best reasons for commodity programs: Why none stands scrutiny." *Choices*, pp. 4S7, 43S44. < Schoenberg, T. (1995, October 20). "Nebraska regents seek to redirect focus of research." *Chronicle of Higher Education*, p. A37. < Kuchler, F., and K. Ralston. (1993, January/February). "Impacts of the Delaney Clause ruling." *Agricultural Outlook*, pp. 29S32. < Terry, S. J. (1993, September 26). "Drinking water comes to a boil." *New York Times Magazine*, pp. 42S48, 62S65.

## Newspaper

Falgout, C. G. (1993, May 7). "Louisiana's trade with Mexico up sharply." *The* [Baton Rouge] *Morning Advocate*, pp. B6, B17. < "Manufacturers decry a shortage of workers while rejecting many." (1995, September 8). [Editorial]. *Wall Street Journal*, p. A4.

## Thesis / Dissertation

Rowland, W. W. (1996). "A nonparametric efficiency analysis for a sample of Kansas swine operations." Unpublished M.S. thesis, Department of Agricultural Economics, Kansas State University, Manhattan. < Knowles, G. J. (1980, June). "Estimating utility of gain functions for southwest Minnesota farmers." Unpublished Ph.D. dissertation, Department of Agricultural and Applied Economics, University of Minnesota, St. Paul.

## Internet / Online Citations

Reinhardt, A. (1995, March). "New ways to learn." *Byte*. 65 paragraphs. Online. Available at http://www.enews.com.magazines/byte/archive. [Retrieved September 7, 1995]. < U.S. Department of Agriculture, National Agricultural Statistics Service. (1999). NASS historical data. Online. Available at http://www.usda.gov/nass/pubs/histdata.htm. [Retrieved March 2000].

## Other

Cox, T. (1995, May 23). "Assessing the regional impacts of alternative proposals for reform or elimination of federal milk market orders." Testimony before the House Committee on Agriculture, Subcommittee on Livestock, Dairy, and Poultry, Washington, DC. < U.S. Congress. (1994, January). *The Total Costs of Cleaning Up Nonfederal Superfund Sites.* Unnumbered special publication, Congressional Budget Office, Washington, DC. * * * * * * * *

Return your final manuscript package to: If there are any questions about these instructions,

**Contact :**

Chung L. Huang, *JAB* Editor please contact our technical editor:

Department of Agricultural & Applied Economics

313-E Conner Hall Judy Harrison

The University of Georgia e-mail: WordDoctor@aol.com

Athens, GA 30602-7509

Fax : (706) 542-0739

Phone : ((706) 542-0747

e-mail : jab@agecon.uga.edu

# Journal of Agricultural Economics

## Guidelines of Submission

*Subjects :* Contributors to the *JAE* should note that, in choosing subjects for articles, the objects of the Society are to promote the study and teaching of agricultural economics and relevant disciplines, and their application to issues in the agricultural, food and related industries, rural communities and the environment. The relevant disciplines include economics, statistics, marketing, business management, politics, history and sociology'. No rigid rules are applied but preference will be given to articles, which either deal with new developments in research and methods of analysis, or apply existing methods and techniques to new problems and situations, which are likely to be of general interest to the *Journal's* international readership. The *Journal* is generally not a suitable medium of publication for articles dealing with subjects of purely local interest. Articles should be clearly and concisely written and should normally not exceed 5,000 words. Articles greatly in excess of this limit cannot be considered for publication.

*Note and Comments :* Shorter papers and comments, up to 1,000 words, will also be considered for publications. Such notes might appropriately deal with the economic aspects of current questions of agricultural policy, with the results of small research projects not justifying a full-length article, or comment briefly on articles previously published.

*Editorial Process :* The Editor-in-Chief and Associate Editors assess the initial suitability of articles submitted. Submissions considered to be suitable for the JAE will be sent out to two reviewers who will referee the article for the Editor. Unsuitable articles are returned to the authors with a short note of explanation from the Editor-in-Chief. The refereeing process in "double-blind" : the identity of the author remains anonymous to the referee and *vice versa.* Authors may be asked to re-submit their article in revised form, but in all but exceptional circumstances, the Editor-in-Chief will consider only one re-submission. Upon completion of the refereeing and editorial processes, authors and referees will be notified of the Editor-in-Chief's decision about publication with explanatory feedback including referees' reports. The lag between receipt of submission and first response varies, but a four-month turnaround should be expected. Every effort is made to ensure that this target is achieved, but it cannot be guaranteed.

***Preparation of Manuscript :*** There are no specific requirements for the style or presentation of articles submitted to the *JAE* except that they should be appropriate for the purposes at hand. A 12 point type face with line spacing of 1.5 is satisfactory. Diagrams and tables should be arranged, as far as possible, in portrait with figures, headings and labels orientated horizontally, using black and white only. Equations should appear on a separate line with equation numbers (where necessary) aligned at the right hand margin. Where appropriate, full mathematical workings should accompany the articles in order to assist the referees. These workings will not be published. References should be cited according to the Harvard System,  i.e. author(s) names and date of publication to be given in the body of the text and the references collected alphabetically at the end of the paper. The title of the journal or source should be given in full. See a recent issue of the JAE for guidance. In the case of articles quoting statistics (such as regression coefficients) based on unquoted data, authors should be prepared to send to the Editor a copy of the data and details of the methodology used (possibly in the form of a self-explanatory computer printout) so that referees may test the derivation of such statistics.

***Submission :*** Submissions should be sent in electronic (pdf) form to: Professor David Harvey Editor-in-Chief, e-mail: jae.editor@ncl.ac.uk, with a copy to david.Harvey @ ncl.ac.uk.  We need a title page, with abstract and authors names, affiliations and addresses, and a separate anonymous version of the full paper, complete with title and abstract, for review purposes. Please also send a cover letter confirming acceptance of our submission criteria:

(i)    Papers are submitted exclusively to the *Journal of Agricultural Economics;*

(ii)    If published, copyright belongs to the *Agricultural Economic Society;*

(iii)    At least one author is a member of the *Agricultural Economics Society* in the year of submission, the Society's web site: *http://www.aes.ac.uk/ membership.htm.*

***Publication :***  Authors of articles accepted for publication will be asked to supply the final version in *JAE* house style. A template in Microsoft Word and accompanying guidance notes will be forwarded to authors of articles accepted for publication, and are available on the AES web site. Authors will be required to complete and sign a copyright assignment form for Blackwell Publishing, and will receive free PDF offpritns of their contribution.

***Page Charges and Submission Fees :*** There are no page charges or submission fees at the *JAE.* Non-members wishing to submit an article are required to join the Society, and confirm that they have done so when they submit.

***Book Reviews :*** The *JAE* publishes reviews, typically around 1500 words in length, of relevant books listed in the publications received section of each issue. Anyone wishing to act as reviewer should contact, Dr. L.J. Hubbard Book Review Editor, Email: Lionel.hubbard@ncl.ac.uk

***General Information :*** General information and enquiries should be addressed to the JAE Office at jae.editor@ncl.ac.uk or Ms. Nicola Forster, JEA Office, Centre for Rural Economy, School of Agriculture, Food and Rural Development, University of Newcastle upon Tyne, NE1 7RU, UK. Telephone:44(0)191 222 6623, Facsimile: 44(0)191 222 5411. Further details are available on the Agricultural Economics Society's website: *http://www.aes.ac.uk.*

# CHAPTER - 68

# Journal of Agricultural Engineering

## Information and Instructions for Authors

The Journal of Agricultural Engineering contains original papers only and submission of a paper will be taken to imply that no similar paper has been or is being submitted elsewhere. Papers (three copies) in English, typed in double space on one side of A-4 size paper, should submitted to the Chief Editor.

The subject matter reflects the wide range and interdisciplinary nature of research in engineering and physics for agriculture. Papers describing engineering innovations for field crops, horticulture, livestock, and fisheries and those reporting research seeking to understand and model the physical processes underlying agricultural systems are welcome.

The following three types of papers are invited for publication. Research papers are the normal type of papers published and make up the main bulk of the Journal. They should not normally exceed ten Journal pages, that is, about 5000 words. Research notes enable important findings to be speedily communicated and facilitate the reporting of work not meriting a full length paper. They should not exceed three journal pages, that is, about 1500 words. Review papers are intended to be in-depth studies of the state-of-the-art in the chosen subjects. They should not normally exceed 15 journal pages, that is, about 7500 words.

Your paper should be arranged as follows.

- *Title, authors, names, affiliations and addresses.* Please necessarily include e-mail address of the corresponding author.

- *Abstract.* The Abstract of not more that 2000 words should state the problem investigated, outline the methods summarize the main conclusions.

- *Main body text.* The main body of the text should be divided under suitable headings, numbered to show the hierarchical order. Some or all of the following headings may be appropriate.

  - *Introduction.* This should explain the problem investigated, review earlier work and state the intentions of the present paper.

  - *Literature review.* A separate section may be justified for this, depending on length.

- *Theoretical considerations*. It may be appropriate to explain or develop a theory to shed light on a problem or to correlate experimental results.

- *Materials and methods or Procedures Experimental details*.

- *Results*. These should be presented with reference to appropriate figures and tables. The most appropriate presentations should be chosen and tabulated data should not duplicate that is shown graphically.

- *Discussion*. This may merit a separate section, although Results and Discussion can sometimes be presented together.

- *Conclusions*. Research and Review papers must have this final section but it is optional in Research Notes. Conclusions emerging from the work should be succinctly summarized. These conclusions will have been discussed in the paper and new material must not be introduced at this stage.

- *List of symbols*

- Acknowledgments. These are optional

- *References*

- *Appendices*. It may be appropriate to place detailed mathematical derivations in appendices, rather in the main body of the text.

- *Figure captions*. These should be listed, together with any 'key' information necessary to aid interpretation.

- *Tables*. Each table must be placed on a separate sheet and numbered consecutively throughout the text (with Arabic numerals),and referred to as Table 1, table 2, etc. There must be a caption at the top of each table.

- Figure 1, Figure 2, etc. Figure captions should be placed beneath each figure and should also be listed on a separate sheet.

## References

References in the text should be in the format of surnames of authors followed by the year of publication of the appropriate reference, separated by a comma. The artifice "Leading author et al." should be used in the text for multiple authorship papers, where there are more than two authors. Each reference should give the names of all authors (or, if anonymous, the name of the organization), the year of publication, the title in that order. Some examples of references are given below.

In the text: "Previous work has shown (Hammer & Langhans, 1972; Federer, 1955)..."

In the references list (to journals):

**Hammer P A; Langhans R W.** 1972. Experimental design considerations for growth chamber studies. Hortscience, 7, 481-483.

In the reference list (to books):

**Federer W** T. 1955. Experimental Design. Macmillan, New York.

### *Figures*(on first submission)

All illustrations, whether line drawings, graphs or photographs, are presented as figures and are given numbers (eg. Figure1) in ascending numerical order as reference is first made to each in the text.

### *Figures*(on acceptance)

When the paper is accepted for publication, line drawings without lettering will be required. The publisher will insert the lettering to maintain a common style but it greatly helps if the figures provided for guidance (as for refereeing) are in the required style. Line drawings should preferably be in Indian ink on tracing paper, Bristol board should be of faintly lined graph paper but without numerals and letters. They should be numbered in soft pencil for ready identification with the complete figures attached to the paper.

# Journal of Applied Horticulture

## Guideline for Authors

Contributions must be original, clearly and precisely presented in English, and submitted in duplicate exclusively to the *Journal of Applied Horticulture.* Observational, experimental or applied papers on any aspect of horticulture are welcomed. Original articles should not normally exceed 18 typed pages. Reviews should not exceed 24 typed pages and Short Communications must be restricted to five pages. Address for the submission of manuscripts is given at the first inner cover of the journal. Authors should provide e-mail and postal addresses, fax and telephone numbers. Papers text on floppy disk/ CD, processed in MS Word/ WordStar/ Word Perfect/ will be helpful in early publication and omission of errors. It is also presumed that the authors have obtained the approval of their department, faculty or institute where such approval is essential. Journal or Editorial Board takes no responsibility for the facts or opinion expressed, which entirely rests with the authors of the paper.

Text must be typed on one side of white opaque paper, using double spacing throughout, with at least 30 mm of margin on all sides. Elaborately wordprocessed or typeset layouts should be avoided. On all points of style regarding text and tables, follow a current copy of the *Journal.* Words to be italicized should be underlined or typed in italics. The title should be concise, specific and informative. Please suggest a short running title of six to ten words. The abstract must not exceed 250 words. A list of up to 12 key words, including the complete botanical name and common name (if any) of the plant material, must be supplied. Other key words should include the topic investigated and any special techniques used. The list of key words should be complete in itself. Title, abstracts and key word list should be informative without reference to the remainder of the paper. The complete scientific name (Genus, species and authority), and cultivar or strain where appropriate, must be cited for every organism on first mention. The generic name may be abbreviated to its initial thereafter, except where reference to other genera could cause confusion. Vernacular names may be added, but should be used alone only when they are unambiguous. Introduction should be brief, limited to the statement of problem or the aim of the study. Materials and methods should include relevant details of materials, experimental design, techniques employed and statistical methods used. Results and discussion should be relevant, critical and supported by published evidence. Last paragraph of the Results and discussion should contain the conclusion drawn from the investigation. Tables should be typed on separate sheets with title stating its contents clearly and concisely.

Metric measurements in SI units should be used. If non-SI units have to be used, the SI equivalents should be added in parentheses on first mention. Units should be spelled out except when preceded by a numeral, when they should be abbreviated g, mg, ml, d, h, ha etc. (not followed by full stops). Use the minus index (m-1, l-1, h-1) except in such cases as 'per plant or 'per pot'. Numbers up to ten should be given in words except when referring to measurements: give 11 and upwards as numerals, except at the beginning of a sentence. Fractions should be expressed as decimals. Use %' not  'per cent' in the text. Dates should be as : 1 Jan, 1992.

In acknowledgements, please be brief, 'We      thank...' (not 'The present authors would like to express their thanks to....'). Citations in the text should take the form: Hamson and Noirot (1974) or (Jones, 1990; Smith, 1998 a, b). When papers are by two authors, use all names. For papers by three or more authors use et al. throughout. Publications in the text should be cited by- author, year of publication and multiple citations should be in chronological order. Underline only Latin and generic names and the titles of books / journals.

*Journal article* :  **Stassen, P.J.C., H.G Grove and S.J. Davie, 1999. Tree shaping strategies for higher density mango orchards. J. Appl. Hort., 1(1): 1-4.**

*Chapter in book* : **Ferguson, A.R., A.G.Seal, M.A. McNeilage, L.G.Fraser, C.F.Harvey and R.A.Beatson, 1996. Kiwifruit. In: J. Janick and J.N.Moore (Eds.), *Fruit Breeding*, Vol. II: Vine and Small Fruits. Wiley, New York, pp. 371-417.**

*Thesis* : **Marsh, H. 1999. *Phenotypic variation of kiwifruit in a factorial mating design.* M.Appl.Sc. Thesis, Massey University, Palmerston North, New Zealand.**

*Symposium/workshop/meeting proceedings* : **Wample, R.L., and T.K. Wolf. 1996. Practical considerations that impact vine cold hardiness. In: *Proceedings for the Fourth International Symposium on Cool Climate Enology and Viticulture/* T. Henick-Kling et al. (Eds.), pp. 23-38 N.Y. State Agricultural Experiment Station, Geneva, NY.**

Citations such as papers 'in press' may appear in the list, but not papers 'submitted' or 'in preparation'. In the text refer to a paper 'in press' by the expected year of publication. Avoid 'unpublished work'. A 'Pers. Comm.' may be cited in the text, but not in the list.

Only scientifically necessary illustrations should be used. Illustrations must be sent flat and unfolded. Good figures and black and white photographs are encouraged where they enhance significantly the clarity of scientific information. Only original line drawings will be reproduced. If computer drawn, these must be of the highest quality. Please comply exactly with the following instructions. No lettering, shading or hatching should be done on original drawings; indicate these on accompanying photocopies. Draw in black ink on white card, white paper, tracing film. Supply two photocopies with each figure. Do not exceed twice the linear dimensions desired in final reproduction. Draw small figures to reproduce within the width of one text column, which is 80 mm, or group several together into labeled parts for two-column reproduction (see *Journal* for style). Be sure to draw all lines and symbols boldly. For reduction to 50% linear, a suitable thickness for axes is 0.35 mm and for other lines 0.5 or 0.6 mm.

Black and white photographs must be of excellent quality in copies of the submitted MS and printed on glossy paper. Black and white prints may be submitted up to twice linear size of final reproduction, several photographs may be grouped together by mounting them on white card with a narrow gap between them to produce a pleasing symmetrical layout. The same requirements apply as for line drawings regarding type area, column width and the labeling of groups. No lettering, arrows, scales, etc, should be done on original prints; indicate these on mounts or on accompanying photocopies. Each legend should be explanatory and meaningful without reference to the text.

Book Reviews on recent publications pertaining to or of interest to Horticulture will also be accepted.

# CHAPTER - 70

# Journal of Biotechnology

## Guide for Authors

*Journal of* Biotechnology provides a medium for the rapid publication of both full-length articles and short communications on all aspects of biotechnology. The Journal will accept papers ranging from genetics or molecular biological positions to those covering biochemical, chemical or bioprocess engineering aspects as well as computer application of new  software concepts, provided that in each case the material is directly relevant to biotechnological systems. Papers presenting information of a multidisciplinary nature that would not be suitable for publication in a journal devoted to a single discipline, are particularly welcome. The following is an outline of the areas covered by the Journal: Nucleic acids/Molecular biology; Physiology/Biochemistry; Biochemical engineering/ Bioprocess engineering; Industrial processes/New products.

**Submission of Manuscripts.** Submission of a paper to the *Journal of Biotechnology* implies (1) that it is not being submitted for publication elsewhere; (2) the transfer of the copyright from the author to the Publisher.

One original plus two copies of the paper and the figures (copy sets of photographs must be original prints) should be submitted to the Chief Editor:

*Professor Dr A. Piihler,* Universitat Bielefeld, Universitatstraâe 25, D-33615 Bielefeld, Germany. Tel.: +49 521 1065607; Fax: +49 1065626. E-mail: jbiotech@genetik.uni-bielefeld.de

All questions arising after acceptance of the manuscript, especially those relating to proofs, should be directed to : *Journal of Biotechnology,* Elsevier Ireland, Elsevier House, Brookvale Plaza, East Park, Shannon, Co. Clare, Ireland, Tel. (+353)61 709644; FaxL+353) 61 709109.

**Electronic manuscripts :** Electronic manuscripts have the advantage that there is no need for the rekeying of text, thereby avoiding the possibility of introducing errors and resulting in reliable and fast delivery of proofs.

For the initial submission of manuscripts for consideration, hardcopies are sufficient. For the processing of accepted papers, electronic versions are preferred. After final acceptance, your disk plus three, final and exactly matching printed version should be submitted together. Double density (DD) or high density (HD) diskettes ($3\frac{1}{2}$ or $5\frac{1}{2}$ inch) are acceptable. It is important that the file saved is in the native format of the wordprocessor program used. Label the disk with the name of the computer and wodprocessing package used, your name, and the name of the file on the disk. Further information may be obtained from the Publisher.

***Author in Japan please note :*** Upon request, Elsevier Japan will provide authors with a list of people who can check and improve the English of their paper *(before submission)*. Please contact our Tokyo office: Elsevier Japan, 4F Higashi-Azabu, 1-Chome Bldg, 1-9-15 Higashi-Azabu, Minato-ku, Tokyo 106-0044, Japan, Tel.: (+81) (3) 5561 5037; Fax: (+81) (3) 5561 5047.

## Types of papers

1.  Full-length papers, generally not exceeding 20 typewritten pages.
2.  Short Communications, not exceeding 1500 words or equivalent space including figures and tables. These must be brief definitive reports and not preliminary findings.
3.  Reviews will be published following invitation from Review Editors.

Letters to the Editor and announcements, meetings and courses will be included at the discretion of the Editors and the Publisher.

## Manuscripts

The manuscript should be typed with double spacing and wide margins, on one side of the paper only, and should be accompanied by a separate title page giving the authors' names and affiliations, as well as an address for correspondence. Words to be printed in italics should be underlined.

**Full-length articles** should :

(a)  be divided into sections (Abstract, Introduction, Materials and methods, Results, Discussion):

(b)  contain an Abstract, not exceeding 200 words, at the beginning of the paper, followed by 3-6 keywords;

(c)  not exceed 12-15 printed pages (approximately 20 typewritten pages) including the space required for figures. Longer papers will be considered, but may be subject to delayed publication.

**References** should be assembled on a separate sheet. In the text they should be referred to by name and year (Harvard System). More than one paper from the same author in the same year must be identified by the letters a,b,c,etc., placed after the year of publication. In the text, when referring to a work by more than two authors, the name of the first author should be given followed by et al. and year in brackets. Literature references must consist of names and initials of all authors, year, title of paper referred to, abbreviated title of periodical, volume number and first and last page numbers of the paper. Periodicals, books and multi-author books should accord with the following examples:

Ponti, C.,Sonneleitner, B.,Fiechter, A., 1995. Aerobic thermophilic treatment of sewage sludge at pilot plant scale. 1. Operating conditions. J. Biotechnol. 38, 173-182.

Walter, H.,Brooks,D.E.,Fisher,D.,1985. Partitioning in aqueous two-phase systems.Academic press, Inc., Orlando, FL.

Hamer, G.,1989. Fundamental aspects of aerobic thermophilic digestion. In: Bruce, A.M., Colin, F., Newman, P.J. (Eds.), Treatment of Sewage Sludge: Thermophilic Aerobic Digestion and Processing Requirements for Landfilling. Elsevier Applied Science, London, pp 2-19.

Abbreviations of journal titles should conform to those adopted by *List of Serial Title Word Abbreviations*. International Serials Data System, 20 rue Bachaumont, 75002 Paris, France. ISBN 2-904938-02-8.

**Tables** should be typed double-spaced on separate sheets, numbered consecutively with Arabic numerals, and only contain horizontal lines. A short descriptive title should appear above each table, with possible legend and footnotes (identified with a,b,c, etc.) below.

**Figures** should be line drawings in black ink or very sharp, well-contrasting prints on glossy paper suitable for immediate reproduction. Half-tone figures should be black-and-white, very sharp, well contrasting and on glossy paper. Figures should be completely lettered, the size of the lettering being appropriate to that of the drawing, taking into account the necessary reduction in size. All legends should be typed double-spaced on separate sheets. If figures are not to be reduced their format should not exceed 16.0 × 20.2 cm. Tables, figures and photographs should be clearly marked on the reverse side with the number, author's name, and orientation (top), using a soft pencil. Use negative powers of the dimension.

**Nomenclature :** A list of symbols and abbreviations (e.g. of enzyme) should be provided. 'Fermentation' has become a very ambiguous expression and therefore should not be used in this Journal. Preferably use:

- *For microbial subjects:* growth, microbial growth (batch, continuous) cultivation, reaction, etc.

- *For biochemical subjects:* glycolysis, Embden Meyerhof pathway, repression, derepression, oxidation, oxido-reduction, etc.

- *For industrial, practical or technical subjects:* bioprocess, microbial process, industrial process, reaction, biochemical engineering, biotechnology, bioproduct, bioethanol (instead of fermentation ethanol), biosynthesis,…

Similarly: it is frequently preferable to use the term 'Reactor' or more specifically 'Bioreactor'. 'Fermenter' should be used only for very unspecific devices like jars, pots, etc. 'Bioreactor' is used for the physical containment of a bioreaction but never for biocatalyst enzyme. 'Broth' should be replaced by medium, spent medium, reaction mixture, etc.

*Units and Dimensions* should be expressed according to IUPAC nomenclature, e.g. Time, s, min, h, d, a; Mass, ng, ìg, mg, g, kg; t; Length, nm, ìm, mm, cm, m, km, Volume, ìm, ml, 1; Dalton, Da,  for molecular mass. Molecular weight has no dimension.

Negative powers should be used instead of fractions, e.g. g $1^{-1}$ $h^{-1}$,  nmol $ml^{-1}$ etc. and dimension should not be mixed with specifications, e.g.  protein per biomass (g $g^{-1}$) instead of g  protein/g biomass.

**Scientific and Engineering Symbols**

***Growth kinetics and cultivation:*** As recommended by the International Commission at the $2^{nd}$ Int. Symposium on Cont. Cultivation of Microorganisms, Prague 1962 (Proceedings published by Academic Press, New York, p 379, 1962)

***Other symbols :*** as per 'Perry's Chemical Engineering Handbook'.

Instructions of authors regarding GenBank/DNA sequence linking

DNA sequences and GenBank Accession numbers Many Elsevier journals cite "gene accession numbers" in their running text and footnotes. Gene accession numbers refer to genes or DNA sequences about which further information can be found in the databases at the National Center for Biotechnical Information (NCIB) at the National Library of Medicine. Elsevier authors wishing to enable other scientists to use the accession numbers cited in their papers via links to these sources, should type this information in the following manner:

For each and every accession number cited in an article, authors should type the accession number in bold, underlined text. Letters in the accession number should always be capitalized. (See Example 1 below). This combination of letters and format will enable Elsevier's typesetters to recognize the relevant texts as accession numbers and add the required link to GenBank's  sequences.

***Example 1:*** "GenBank accession nos. **AI31510, AI63511, AI632198,** and **BF223228)** a B-cell tumor from a chronic lymphatic leukemia (GenBank accession no.**BE675048),** and a T-cell lymphoma (GenBank accession no **AA361117)**".

Authors are encouraged to check that the accession numbers are used very carefully. An error in **a letter or number can result in a dead link**

In the final version of the printed article, the accession number text will not appear bold or underlined (see Example 2 below).

***Example 2:*** "GenBank accession nos. AI631510, AI631511, AI632198, and BF223228), a B-cell tumor form a chronic lymphatic leukemia (GenBank accession no. Be675048), and a T-cell lymphoma (GenBank accession no. AA361117)".

In the final versionof the Electronic copy, the accession number text will be linked to the appropriate source in the NCBI databases enabling readers to go directly to that source from the article (see Example 3 below)

***Example 3 :*** "GenBank accession nos. AI631510, AI631511, AI32198, and BF223228), a B-cell tumor from a chronic lymphatic leukemia (GenBank accession no. BE675048), and  a T-cell lymphoma (GenBank accession no. AA361117)".

**Proofs** will be sent to the first-named author of an article, unless an alternative is requested on the title page of the manuscript. They should be checked carefully and returned by fax and airmail within 2 days of receipt . Only printer's errors may be corrected: no changes in or additions to the edited manuscript will be allowed at this stage.

**Offprints**  may be ordered by filling in and returning to the Publishers the order form sent to the author with the proofs. Twenty five free offprints per contribution will be made available.

# CHAPTER - 71

# Journal of Central European Agriculture

## Guide to Authors

The Journal publishes original articles, review articles, rapid communications and proceedings of scientific meetings in English, Bulgarian, Croatian, Czech, Hungarian, Polish, Romanian, Slovak and Slovenian. If you want to submit an article to be published in JCEA in your own native language, contact your National Editorial Board (NAB). Otherwise contact the Main editor, Editor in chief, NAB or Technical Board.

## General Presentation

The complete set of the manuscript should be submitted in electronic format. The manuscript should be typed in Times New Roman or Arial fonts, double-spaced with margins of at least 3.5 cm at top, bottom and sides, on paper and CD (diskettes) or E-mail attachment. The Editorial Board maintains the option of returning, before evaluation, manuscripts to authors who do not comply with these recommendations. The author is advised to keep one copy of the manuscript and a set of figures. After the article has been accepted for publication, the authors are encouraged to forward the revised version on CD (diskettes), or E-mail attachment to the editor (RTF format is preferred). The publication of the text and coloured figures is free of charge.

## Original Articles

Original articles should be no longer than ten pages i.e., about 30 000 characters, including tables and figures. The manuscript should be arranged as follows: title page, short abstract, detailed abstract, introduction, materials and methods, results, discussion, acknowledgements, references, tables, figures. You can write your article in English or in your own native language.

## Title page

The title page should include the following: the title of the paper/article (in English and native language), the surname and forenames (in full) of each author (forenames should be written in capital letters), the department and institution where the study was carried out (international names), telephone and fax numbers and e-mail address of the corresponding author (this author being identified by an asterisk). The title of the article should not contain abbrevations, chemical formulas, trade marks (instead of generic) or slang.

## Abstract

The abstract (less than 850 characters) should be written in both English and native language, in a form suitable for abstracting services. Paragraphs, footnotes, references, cross-references to figures and tables and underlined abbreviations should be avoided.

## Keywords

Up to five keywords should be supplied, both in English and native language of the author. Keywords may be taken from the title, abstract or text. The plural form and uppercase letters should be avoided.

## Detailed abstract in English

A detailed abstract (less than 3000 characters) should be written. If article is written in English language, detailed abstract must be written in native language and vice versa. Detailed abstract should provide sufficient information about methods and results (with reference to the tables and figures), the discussion and conclusion. The purpose of this detailed abstract in English is to give the reader full understanding of the work in all essential respects.

## References

In the reference list, the references should appear in alphabetical order, preceded by an Arabic numeral enclosed in square brackets. The authors' names are listed in alphabetical and chronological order for each author. The references are cited in the text by the corresponding number enclosed in square brackets. All entries in the reference list must correspond to references in the text and vice versa. When authors are cited in the text (within a phrase and not just between brackets), the spelling of the authors' names must be exactly the same as in the reference list. The Harvard system of references should not be used.

The style and punctuation of the references should follow the format illustrated in the following example:

### *Article in a journal*

[14] Stopar M., Lokar V., The effect of ethephon, NAA, BA and their combinations on thinning intensity of 'Summerred' apples, J. Cent. Eur. Agric. (2003) 4: 399-404.

### *Book*

[5] Brown M.B., Cotton: History, Species, Varieties, Morphology, Breeding, Culture, Diseases, Marketing and Uses, McGraw-Hill, New York, 1938.

*Article in a book*
[17] Kubkin S., Measurement of radiant energy, in: Sestak Z., Catsky J., Jarvis P.G. (Eds.), Plant Photosynthetic Production. Manual of Methods, Junk, The Hague, 1971, pp. 702–765.

*Illustrations (figures and tables)*
As this part is electronically subject to change and mishaps, figures and especially tables demand extra care and safety.

We recommend sending illustrations also in separated files. Preferred formats are TIFF, EPS and BMP in resolutions more than 300 bpi. JPG and GIF are optional. The description of tables, figures and illustrations should be written in both English and native language.

## Rapid Communications

Brief accounts of particularly interesting results can be published as rapid communications. Their length is limited to four printed pages, i.e., about 10 000 characters (including figures and tables). They have the same structure as original articles. Authors do not receive proofs.

## Review Articles

The length of review articles is limited to 18 printed pages, i.e., about 65 000 characters. The usual division into 'materials and methods, results and discussion' may be replaced by a more adapted structure.

## Proceedings of Scientific Meetings

Summaries of communications are limited to 1700 characters. They should have no chapters, bibliographic references, or acknowledgements and are published by prior arrangement with the Editorial Board.

## Scientific Notes

The Journal will publish brief notes of scientific interest to disseminate information and observation of preliminary nature. The length of such notes will be strictly restricted to two pages (approx. 5 000 characters) and publication will depend on the availability of space in the Journal and general interest of the readers. This form is intended to augment scientific communication and all submissions must include 'scientific note' in the title. The title should be written in English and native language.

## Proofs and Reprints

Proofs will be sent to the author indicated on the title page. They should be carefully corrected and returned to the publisher within 48 hours of reception. If this period is exceeded, the galleys will be proved only by the editorial staff and published without the authors' corrections. Should substantial changes in the original manuscript be requested (other than typographical errors), they will be made at the author's expense. The final version of the paper is placed on the homepage of the Journal as PDF files, and they can be downloaded and printed if required. The downloaded PDF files are available for the authors at JCEA web page [http://www.agr.hr/jcea] and can be used instead of reprints.

## Copyright

As soon as the article is published, the author is considered to have transferred his rights to the publisher.

## Contact Groups

### Main Editor

Prof. Dr. Nikola Keziæ; Faculty of Agriculture, Svetošimunska 25, 10000 Zagreb, Croatia

E-mail: nkezic@agr.hr

### Technical Board

JCEA Technical Board; Faculty of Agriculture, Svetošimunska 25, 10 000 Zagreb, Croatia

E-mail: jcea@agr.hr

### Bulgaria

Prof. Iordanka Kouzmanova, DSc; Agricultural University Plovdiv, 12, Mendeleev Str., 4000 Plovdiv

E-mail: rector@au-plovdiv.bg

### Croatia

Prof. Dr. Nikola Keziæ; Faculty of Agriculture, Svetošimunska 25, 10000 Zagreb

E-mail: nkezic@agr.hr

### Czech Republic

Doc. Ing. Miroslav Maršalek; University of South Bohemia, Faculty of Agriculture, Studentska 13, 370 05 Èeske Budejovice

E-mail: marsalek@zf.jcu.cz

### Hungary

Dr. Dr. Zsuzsanna Bacsi, PhD; University of Veszprem, Georgikon Faculty of Agriculture, Deak F. u. 16, 8361 Keszthely Pf. 71

E-mail: h5519bac@ella.hu

### Poland

Dr. Hazem M. Kalaji; Marszalkowska Street 56/4, 00 545 Warsaw

E-mail: kalaji@bluewin.ch

**Romania**

Prof. Dr. Doru Pam. 1, PhD; University of Agricultural Sciences and Veterinary Medicine, 3-5 Manastur Str., 3400 Cluj - Napoca

E-mail: dpam. l@usamvcluj.ro

**Slovakia**

Prof. Ing. Ondrej Debreceni; Slovak Agricultural University, Faculty of Agrobiology and Food Resources, A. Hlinku 2, 94976 Nitra

E-mail: ondrej.debreceni@uniag.sk

**Slovenia**

Dr. Vladimir Megliè; Agricultural Institute of Slovenia, Hacqutova 17, p.p. 2553, 1001 Ljubljana

E-mail: jcea@kis-h2.si

**How original articles should be organised:**

- Title (English)

- Title (Native language)

- Author(s)

- Department(s) and institution(s) where the study was carried out, telephone and fax numbers and e-mail address of the corresponding author (this author being identi. ed by an asterisk)

- Abstract and Keywords (English)

- Abstract and Keywords (Native language)

- Detailed abstract - if the rest of the article is in English, detailed abstract should be in native language and vice versa.

**The rest of the article can be written in English or in native language.**

**Usual form:**

- Introduction

- Materials and methods

- Results

- Discussion

- Acknowledgements

- References

- tables, figures and illustrations (must have descriptions in both English and native language)

If you have any questions, please contact Technical Board: jcea@agr.hr JCEA web site: *http://www.agr.hr/jcea*

# CHAPTER - 72

# Journal of Cereal Science

## Guide for Authors

*The Journal of Cereal Science* publishes papers originating in any country. Papers dealing with topics of only restricted local interest will not be accepted, however, unless the information presented can be demonstrated to be of general applicability.

The journal exists to advance scientific concepts in cereal science, and the content of papers published within it must be consistent with this goal. Papers that are essentially archival will not be accepted.

## Submission of Manuscripts

Submission for all types of manuscripts to *Journal of Cereal Science* proceeds totally online. Via the Elsevier Editorial System website for this journal, *http://ees.elsevier.com/yjcrs*, you will be guided step-by-step through the creation and uploading of the various files. When submitting a manuscript to Elsevier Editorial System, authors need to provide an electronic version of their manuscript. For this purpose only original source files are allowed, so PDF files are not permitted. Once the uploading is done, the system automatically generates an electronic proof, which is then used for reviewing. All correspondence, including the Editor's decision and request for revisions, will be by e-mail. Authors may send queries concerning the submission process, manuscript status, or journal procedures to the Editorial Office at jcs@elsevier.com.

If online submission is unavailable, manuscripts may be submitted by sending the source files on disk together with a matching hard copy (both text and figures and tables), by registered mail to the editor. (Please note that this is not the preferred way of submission and could cause a delay in publication of the article). Manuscripts can be submitted to:

Journal of Cereal Science

Elsevier Editorial Services Office

The Boulevard, Langford Lane

Kidlington, Oxford OX5 1GB, UK

Tel.: +44(0) 1865 843530; E-mail: jcs@elsevier.com

## Submission checklist

It is hoped that this list will be useful during the final checking of an article prior to submitting it to the journal for review.  Please consult this Guide for Authors for further details of any item.

Ensure that the following items are present:

- One Author designated as corresponding Author.
- E-mail address
- Full postal address
- Telephone and fax numbers
- All necessary files have been uploaded
- Keywords
- An alphabetical list of abbreviations
- All figure captions
- All tables (including title, description, footnotes)
- Manuscript has been "spellchecked"
- References are in the correct format for this journal
- All references mentioned in the Reference list are cited in the text, and vice versa
- Permission has been obtained for use of copyrighted material from other sources (including the Web)
- Colour figures are clearly marked as being intended for colour reproduction on the Web (free of charge) and in print or to be reproduced in colour on the Web (free of charge) and in black-and-white in print.
- If only colour on the Web is required, black and white versions of the figures are also supplied for printing purposes

For any further information please contact the Author Support department at authosupport@elsevier.com

## Types of Papers Published

### Research Papers

The main activity of the Journal in fulfilling its mission is the publication of original and innovative research papers of a high scientific standard.  These papers should: (a) report a specific identifiable advance in knowledge that has not been published elsewhere; (b) claim no more than can be substantiated by the results; (c) be logically consistent both within themselves and within the existing body of knowledge; (d) give enough information to allow the research to be tested and repeated by competent researchers elsewhere; and (e) give due reference to previously published work relevant to the research described.

*Rapid communications :* These are intended as vehicles for conveying news of advances in cereal science, the scientific importance of which merits preferential treatment. Scientific importance and novelty of the information will be the key criteria in judging their acceptability.

*Research notes :* These are intended as a means of publishing the results of studies of limited size that do not merit high-priority treatment.

*Reviews :* These should present critical appraisals of the current status and future directions of specific areas of topical interest. They are not intended as exhaustive, archival literature surveys over a broad front. They should aim to give balanced, objective assessments by giving due reference to relevant published work and not merely represent the prejudices of individual authors or summarise only work carried out by the authors or by those with whom the authors agree. They should also avoid undue speculation.

*Letters to the Editor* concerned with issues raised by articles recently published in the journal or by recent developments in cereal science are welcomed. These may be submitted informally to the Editor at any time. Letter should not exceed 750 words.

*Book reviews :* Please contact the Editor-in-Chief if you wish to submit a book review.

## Review System

Papers are peer-reviewed by independent reviewers with appropriate expertises in the subject area of the paper. The review process is anonymous, although the reviewers' recommendations and comments are usually transmitted to the authors to help them in revising their manuscripts (which is almost invariably required). The Editors and reviewers attempt to make the review system as constructive and sympathetic as possible, although they must, at the same time, attempt to ensure that only papers of a high standard are published. Many contributors acknowledge the help they receive from the review process in improving their papers. No revision of Rapid Communications will be allowed in order to ensure rapid publication.

As well as advising on the paper's acceptability, the reviewers are also asked to give a priority rating, which will help to give the highest priority to papers that represent important new advances. Papers recommended for publication will be categorized as a) being of outstanding scientific standard and representing and Important advance in the particular subject area; (b) being of high Scientific standard but representing a logical or predictable extension of previous research; (c) presenting necessary information and of good scientific standard but being essentially confirmatory in nature.

*Please note :* Authors may suggest the name of appropriate reviewers for their papers or may identify individual reviewers whom they would prefer not to review the manuscript; provided that valid reasons are given. In the latter case, the Editors will respect the author's wishes.

## Originality of Research

Submission of a paper for consideration for publication in the *Journal of Cereal Science* will be held to imply that the material represents the results of original research or of an original interpretation of existing knowledge not previously published, that it is not under consideration for publication elsewhere, and that, if accepted for publication in the *Journal of Cereal Science,* it will not be published elsewhere in the same form, in English or any other language without the consent of the Editorial Board and the Publisher.

## Resubmission of Revised Manuscripts

If a manuscript returned to the author for revision is not resubmitted within 6 weeks (making due allowance for postage times), it may on re-submission be deemed a new paper and the date of receipt altered accordingly.

## Preparation of Manuscripts

The standard of preparation of the manuscript determines to a  considerable extent the speed of processing and publication. Authors are advised in their own interests to read these notes carefully and to ensure that their manuscript meets the requirements; they are also urged to ensure that their manuscript meets the requirements; they are also urged to ensure that the manuscript does not contain superfluous material.

Manuscripts should meet the obvious criteria of relevance, originality and scientific validity.  Two other important attributes should also be considered: first, papers should be intelligible to an international readership, many of whom may not be experts in particular specialist fields with their attendant assumptions and jargon.  Second, papers should contain adequate and concise information to enable a competent research worker to reproduce the work. Authors are urged to read their own manuscripts objectively and with the same critical approach that they would employ in reading the work of others.

All categories of manuscript should be typed on standard-sized (preferably A4) paper on one side of the paper only, triple-spaced with two wide margins (at least 3 cm all round). **Pages should be numbered, and each line on the page should also be numbered.** The first page should contain: the Title indicating the subject matter as briefly as possible and, in any case, in not more than 250 characters, including spaces; names of authors; address(es) of the laboratoryies where the work was done; full postal and e-mail address for correspondence; an alphabetically arranged list of al abbreviations used; current addresses of authors, if different from above. The texts of different types of paper will differ and guidelines for each are set out below.

## Key Word Index

To assist in the preparation of a key word index, authors should provide a list of up to four key words on the title page of the manuscript.

## Language

The language of the journal is English (Concise Oxford Dictionary, Webster's New Collegiate Dictionary). To expedite publication and to avoid ambiguities and errors, authors whose first language is not English, are strongly advised to have their manuscript checked by an English-speaking colleague who is knowledgeable about written English and English grammar. Contributors from North, Central and South America may use American spellings if they so wish. Authors from all other countries should use English spellings. For the latter, 's' spellings are preferred in words such as 'summarise', 'hydridise'. Care should be taken over the use of a- and b- as prefixes for carbohydrates. Greek letters a- and b- should be used only for enzymes that have specificity for glycosidic linkages with particular configuration at the anomeric carbon atom, e.g., a-glucanases, a and b-glucosidases. In the particular case of alpha- and beta-amylases, both enzymes are specific for the (1n4)-a-linkage between glucose residues instarch polymers (and glycogen) and in this case 'alpha' and 'beta' should be spelled out in full in italics.

## Nomenclature

Abbreviations and symbols should, wherever possible, follow the IUBMB recommendations on Biochemical and Organic Nomenclature, Symbols and Terminology, at *http://www.chem.qmul.ac.uk/iubmb/*

Non-standard abbreviations should be kept to a minimum. The words to be abbreviated should be spelled out in full on the first citation and the abbreviation given in parentheses. All abbreviations used should be listed and their meanings given on the title page (this list will be included on the front page of the published article). Enzyme nomenclature should follow the IUBMB Enzyme commission recommendations (http://www.chem.qmul.ac.uk/enzyme/) (relevant EC numbers should be given).

The International System of units (SI) should be followed (see "Quantities, Units and Symbols in Physical chemistry", Mills, Ian; Cvitas, Tomislav; Homann, Klaus; Kallay, Nikola; Kuchitsu, Kozo, C R C Press Blackwell Science (UK), ISBN: 0632035838, 1995, or "Specification for quantities, units and symbols, Physical chemistry and molecular physics" BS 5775-8:1993 ISO 31-8:1992, ISBN:0580221954, 1993). You may also wish to consult the website of the Bureau International des Poids et Mesures, http://www 1.bipm.org/en/si.

Non-standard, but conventional, units may be accepted if unambiguous and where there is no SI unit. Non-standard, but conventional, units may be accepted if unambiguous and where there is no SI unit.

## Research Papers

An ideal paper would probably contain a maximum of 6000 words of text (approx. 4-5 printed pages), no more than six tables or figures and up to 30 references. The second page of the manuscript should contain the Abstract only. The text should then follow the sequence: Introduction, Experimental, Results, Discussion, Acknowledgements, References. Tables, Figure Captions and, finally, Figures. Please number the pages, and each section should also be numbered

*The Abstract* should be clear and concise with a maximum of 200 words.

*The Introduction* should be brief and contain sufficient information to provide the background to the research reported in the paper, but should not present a complete historical review. The objectives of the work (but not the results or conclusions) should be stated clearly at the end of this section.

*The experimental section* should contain sufficient information on material and methods to enable a competent worker to repeat the work. Details of methods published in commonly available journals need not be given at length; instead, appropriate references should be quoted and a brief summary of the method given.

*The Results* section should present concisely the experiments done and the results obtained. Discussion of the results should not appear in this section.

*The Discussion section* should interpret the findings in the context of current knowledge but should not reiterate material in the Results section. Authors should be careful to distinguish between interpretation and speculation and should avoid the latter.

For conciseness and clarity, it may be convenient to combine the Results and Discussions section, in which case a brief concluding paragraph would be necessary.

*Acknowledgements* should be brief. The References should be checked carefully before submission. Responsibility for the accuracy of bibliographic citations lies entirely with the authors.

*Text :* All citations in the text should refer to:

1. *Single author :* the author's name (without initials, unless there is ambiguity) and the year of publication;

2. *Two authors :* both authors' names and the year of publication;

3. *Three or more authors :* first author's name followed by 'et al.' and the year of publication.

Citations may be made directly (or parenthetically). Groups of references should be listed first alphabetically, then chronologically.

*Examples :* "as demonstrated (Allan, 1996a, 1996b, 1999; Allan and Jones, 1995). Kramer et al.(2000) have recently shown ..."

*List :* References should be arranged first alphabetically and then further sorted chronologically if necessary. The letters "a", "b", "c", etc. must identify more than one reference from the same author(s) in the same year, placed after the year of publication. Journal titles should be given in full.

*Examples :* Reference to a journal publication:

Cuvelier, G., Launay, B., 1986. Concentration regimes in xanthum gum solution deduced from flow and viscosity properties. Carbohydrate Polymers 6, 321-333,

***Reference to a book :*** Strunk Jr., W., White, E.B., 1979. The Elements of Style, third ed. Macmillan, New York.

***Reference to a chapter in an edited book :*** Mettam, G.R., Adams, L.B., 1999. How to prepare an electronic version of your article, in: Jones, B.S., Smith, R.Z. (Eds.), Introduction to the Electronic Age. E-Publishing Inc., New York, pp.281-304.

It is important that the references cited should be accessible to the general reader. References to unpublished materials should not appear in the reference list. References to papers 'in press' or in obscure sources should also be avoided, as should references to proceedings of conferences/conference abstracts available only to the conference attendees. References to papers in private publications, e.g. a report appearing in a publication directed to the membership of a private research organisation, must not be used.

## Reviews

The layout for reviews is flexible, and will be dictated to a large extent by the points that the author is attempting to discuss. An Abstract should be included, however, and the background should be contained in an Introduction. Details on citation and listing of references, preparation of figures and tables, abbreviations and units, etc., are as for conventional research papers.

### *Rapid communications and Research Notes*

The format for these papers is flexible. No Abstract is required, and there is no specification as to number of tables, figures or references. The paper should not be split into sections, although it should begin with a few sentences to introduce the subject area and to indicate the nature of the problem being examined. Likewise, at the end of the paper the conclusions drawn from the work should be summarized.

Rapid Communications and Research Notes will be strictly limited to two printed pages in the journal (equivalent of approx.2000 words) in total, i.e. including title, references, tables and figures, etc. Where figures or tables are used, the number of words must be reduced to compensate for these, giving due regard to the size of such tables and figures. Other details on preparation are as for conventional research papers.

For Rapid Communications, authors are required to justify in a covering letter why the paper should be accorded priority treatment.

## Preparation of Illustrations

Photographs, charts and diagrams are all to be referred to as 'Figure(s)' and should be numbered consecutively in the order to which they are referred. They should accompany the manuscript, but should not be included within the text. All figures are to have a caption that should always indicate the source of the figure. Captions should be supplied on a separate sheet. A maximum total of 5 figures, tables and boxes are allowed. Boxes are useful to highlight a concept that is central to the article, or to set aside necessary explanatory material that might otherwise impede the flow of the text. Boxes may contain only text, or a mixture of text and figures. Tables should be numbered consecutively and given a suitable caption and each table typed on a separate sheet. Arabic numerals should be used. Footnotes to tables should be typed below the table and should be referred to by superscript lowercase letter. No vertical rules should be used. Titles should be brief but unambiguous. Any explanatory material should be itemized in footnotes and indicated with symbols or superscript, lower case letters (a, b, c). Standard abbreviations and units should be used wherever possible. Experimental values should be qualified by indications of statistical significance (standard deviation, standard error of the mean, number of determinations, P-value, etc.) or range. Tables should not duplicate results presented elsewhere in the manuscript, (e.g. in graphs).

If, together with your accepted article, you submit usable colour figures then Elsevier will ensure, at no additional charge, that these figures will appear in colour on the web (e.g. Science Direct and other sites) regardless of whether or not these illustrations are reproduced in colour in the printed version. For colour reproduction in print, you will receive information regarding the costs from Elsevier after receipt of your accepted article. For further information on the preparation of electronic artwork, please see http://authors.elsevier.com/artwork.

***Please note :*** Because of technical complications which can arise by converting colour figures to grey scale (for the printed version should you not opt for colour in print) please submit in addition usable black and white prints corresponding to all the colour illustrations. As only one figure caption may be used for both colour and black and white versions of figures, please ensure that the figure captions are meaningful for both versions, if applicable.

### *Preparation of electronic illustrations*

Submitting your artwork in an electronic format helps us to produce your work to the best possible standards, ensuring accuracy, clarity and a high level of detail.

### *General points*

- Make sure you use uniform lettering and sizing of your original artwork.
- Save text in illustrations as "graphics" or enclose the font.
- Only use the following fonts in your illustrations: Arial, courier, Helvetica, Times, Symbol.
- Number the illustrations according to their sequence in the text.
- Use a logical naming convention for your artwork files, and supply a separate listing of the files and the software used. Provide all illustrations as separate files.
- Provide captions to illustrations separately.
- Produce images near to the desired size of the printed version.

A detailed guide on electronic artwork is available on our website: http://authors.elsevier.com/artwork.

### You are urged to visit this site.

*Preparation of supplementary date.* Elsevier now accepts electronic supplementary material to support and enhance your scientific research. Supplementary files offer the author additional possibilities to publish supporting applications, movies, animation sequences, high-resolution images, background datasets, sound clips and more. Supplementary files supplied will be published online alongside the electronic version of your article in Elsevier web products, including Science Direct. In order to ensure that your submitted material is directly usable, please ensure that data is provided in one of our recommended file formats. Authors should submit material in electronic format together with the article and supply a concise and descriptive caption for each file. For more detailed instructions please visit our Author Gateway at http://authors.elsevier.com.

### Proofs

When your manuscript is received at the Publisher it is considered to be in its final form. Proofs are not to be regarded as 'drafts'. One set of page proofs in PDF format will be sent by e-mail to the corresponding author, to be checked for type-setting/editing. No changes in, or additions to, the accepted (and subsequently edited) manuscript will be allowed at this stage. Proofreading is solely your responsibility. A form with queries from the copy editor may accompany your proofs. Please answer all queries and make any corrections or additions required. The Publisher reserves the right to proceed with publication if corrections are not communicated. Return corrections within two working days of receipt of the proofs. Should there be no corrections, please confirm this. Elsevier will do everything possible to get your article corrected and published as quickly and accurately as possible. In order to do this we need your help. When you receive the (PDF) proof of your article for correction, it is important to ensure that all of your corrections are sent back to us in one communication. Subsequent corrections will not be possible, so please ensure your first sending is complete. Note that this does not mean you have any less time to make your corrections, just that only one set of corrections will be accepted. Proofs are to be e-mailed to proofcorrections @else-vier.com.

## Offprints

Twenty-five offprints will be supplied free of charge. If colour has been paid for within the article, 100 extra offprints will be supplied free of charge. Additional offprints and copies of the issue can be ordered at a specially reduced rate using the order form sent to the corresponding author after the manuscript has been accepted. Orders for reprints (produced after publication of an article) will incur a 50% surcharge.

***Please note :*** Papers published in the *Journal of Cereal Science* do not incur page charges or any manuscript processing fee.

## Copyright

Upon acceptance of an article, authors will be asked to transfer copyright (for more information on copyright see http://authors. Elsevier.com). This transfer will ensure the widest possible dissemination of information. A letter will be sent to the corresponding author confirming receipt of the manuscript. A form facilitating transfer of copyright will be provided. If excerpts from other copyrighted works are included, the author(s) must obtain written permission from the copyright owners and credit the source(s) in the article. Elsevier has preprinted forms for use by authors in these cases: contact Elsevier Ltd., Global Rights Department, The Boulevard, Langford Lane, Oxford, OX5 1GB, UK; phone: (+44) 1865 843830, fax: (+44) 1865 853333, e-mail: permissions@elsevier.com

## Author Enquiries

Authors can keep track of the progress of their accepted article, and set up e-mail alerts informing them of changes to their manuscript's status, by using the "Track a Paper" feature of Elsevier's Author Gateway. Other questions or queries will also be dealt with via the website http://authors.elsevier.com. Contact details for questions arising after acceptance of an article, especially those relating to proofs, are provided when an article is accepted for publication.

## Journal of Cereal Science incurs no page charges

Visit the journal website at *www.elseveir.com/locate/jcs* for further information and access to abstracts and tables of contents.

# CHAPTER - 73

# Journal of Ecobiology

## Instructions to Authors

Submission of a paper to the Journal of Ecobiology/ J. Ecotoxicol. Environ. Monit. Implies that it has not previously been published and not being submitted for publication elsewhere. The published papers will become the copyright of the publisher. The journal will include full length research papers, short research communication, book reviews and news (Information about symposia, conferences etc;).

**Types of papers :**

1.  Research papers-4 to 12 printed pages including figures and tables
2.  Brief communication-1 to 3 printed pages
3.  Mini-reviews

The manuscripts (2 copies) should be type written with double spacing and wide margins on one side of the paper(A4). Manuscripts can also be submitted in 1.44 floppy disk indicating the name of the file along with printed copies. Words to be printed in italics should be underlined. The research paper should be arranged as follows: Title, Address, Abstract, Key words, Introduction, Materials & Methods, Results, Discussion (Results and Discussion may be combined), Acknowledgements (if any) and References.

**Brief Communications** should be arranged as Title, Address, Abstract, Body of the Text (Methods, Results and Discussions as separate para without heading), Acknowledgements and References. Metric units should be used throughout. A running title not more than 40 letters should be provided.

**Tables** should be compiled on separate sheets. A title should be provided for each table and all tables should be referred to in the text. Illustration should be numbered consecutively and referred to in the text. Drawings should be lettered; the size of the letterings being appropriate to that of the drawings. The photographs must be of good quality paper. Figure captions should be in a separate sheet.

**References** should be cited  as the name of the author, year, title, journal, volume number and pages. The reference should be in alphabetical order.

Arunachalam S and Palanichamy  S 1982 Toxic and sublethal effects of carbaryl on survival, food utilization and oxygen consumption in the air breathing fish *Mocropodus cupanus*. **Physiol.Behav.29**:23-27.

Odum E P 1972 Fundamentals of ecology. W B Saunders Co: Philadelphia. P.574.

Bryan G W 1972 Adaptation of estuarine polychaete to sediments. In: Pollution physiology of marine organisms. Ed: Vernberg  F  J  Acadmic Press, New York p.123-315.

Proof reading will be undertaken by the Editor-in-Chief or sent to the corresponding authors, 50 reprints will be supplied to the contributions on orders of a minimum of 50 reprints.

The management does not accept  responsibility for any type of damage or loss of manuscript. Keep a copy for your record. The authors are responsible for the opinions expressed in the journals. All authors must be subscribers of the journal. Manuscript will be accepted only by the advice of a member of the advisory board or an expert in the field.

All editorial correspondence should be addressed to

Prof. Dr.S. Palanicharmy,

57, Anna Nagar, Palani 624 602 India.

All trade enquire should be addressed to:

Palani Paramount Publications

57, Anna Nagar, Palani 624 602 India.

Back volumes of both Journals are available. Request rates and other details.

# CHAPTER - 74

# Journal of Eco-Friendly Agriculture

## Guidelines to the Authors

The Journal of Eco-friendly Agriculture is a biannual journal published in English by the *Doctor's Krishi Evam Bagwani Vikas Sanstha, Lucknow*, India. This journal will be published half yearly in the month of January and July each year. Journal is devoted to basic and applied aspects of Agriculture, Horticulture, Environmental Science and Human Ecology and welcomes original research papers in these areas. The articles can be published as full research paper or as short communication. The Editorial Board may invite important short reviews from eminent scientists working in the respective fields. Authors may note that the articles submitted to the Journal of Eco-friendly Agriculture are not submitted simultaneously to any other journal for publication. All research papers and review articles submitted for publication will be reviewed by referees. Authors may provide the names of at least five referees with complete postal address, who can be approached to review the paper. However, final decision will rest with the Editorial Board.

## Content

The articles published in this journal should be related to eco-friendly agriculture viz bio-pesticides, bio-agents, bio-fertilizers, IPM, IDM, INM and other allied eco-friendly areas. Special emphasis must be laid on the qualitative aspects of modern agro practices interventions vis-a-vis environmental sustainability at large.

## Preparation of manuscript

Articles (3 copies) should be computer typed in double space in font size 12 (Arial or New Times Roman) on one side of good quality bond paper in A 4 size with one inch margin on all four sides. After acceptance of the paper, the revised article in final computerized typed format (one original and two hard copies) should be sent along with a CD. Use a new CD and ensure that the electronic version is free of virus. Mention file name, name of the software and version used and name of the corresponding author. File format of the text should be MS Word.

## Full research paper should be divided into following headings

Title (bold, upper lower case font size 16), author's name (font size 14, upper lower case), complete address of the author(s) (italized, font size 12) and e-mail address (if available). Short communications should not exceed four typed pages and should contain title, author's address, e-mail address (if available), text, table, figures, acknowledgments and references. In full paper title should be short, brief and clear. It should be written in bold. Title will be followed by author's full name with initials. If there are more than one author use 'and' before the last name. Title, author(s) name and address should be written in sentence with left margin aligned. Two years field data is the minimum requirement for publication of Full paper. However for laboratory trials sufficient replications of the treatments are required.

*Title :* It should normally be short running into not more than 6-8 words.

*Abstract :* It should cover the main findings of the research paper and should not exceed 150 words.

*Key words :* Maximum of 6 words. Separate the words with comma (,).

*Text :* It should be divided into Introduction, Materials and Methods, Results and Discussion, Acknowledgements and References. Footnotes should be avoided. New paragraphs should be indicated by clear indentation. Metric units of weights and measures should be used. Tables, figures and, legends should be given on separate pages.

## Introduction

It should be brief with clear objectives and justification for taking up the study. This section should have a very short review on work done in the concerned area.

## Materials and Methods

All procedures followed in the experimentation should be indicated in brief. Methods adopted from other references should be indicated by quoting proper references.If a new method is included, it should be given in detail so that other workers may be able to use it.

## Results and Discussion

This should be combined to avoid repetition. Do not describe the results already indicated in table or graph but blend it with the discussion with supporting references. All the data presented in the paper should be statistically analysed.

## Acknowledgments

This should appear after the main text.

## References

Literature cited should be arranged alphabetically (names of authors) with year, title of paper, name of the journal, volume and pages. Citation of personal communication should be avoided but if necessary, it may be cited in the text as: (G.S. Fraenkel-personal communication). This should not be included in the reference list. Write full name of the Journal. For e.g., references should be quoted as follows:

Mathur, A.C., Krishnaiah, K. and Tandon, P.L. 1974. Control of tomato fruit borer (*Heliothis armigera* Hub.). *Pesticides*, 8 : 34-35.

Pasricha, N. S. 1998. Integrated nutrient and water management for sustainable crop production. In: Ecological Agriculture and Sustainable Development, Vol. I. (eds. G. S. Dhaliwal, N. S. Randhawa, R. Arora and A.K. Dhawan). Indian Eco. Soc. and CRPID, Chandigarh, p. 521-535.

## Book by one author

Cressie, N.A.C. 1993. Statistics for Spatial Data. John Wiley, New York, 900 p.

## Book edited by more than one author

Pedigo, L.P. and Nuntin, G.D. (eds.). 1994. Handbook of Sampling Methods for Arthropods in Agriculture. CRC Press, Boca Raton, Florida, 714 p.

## Page charges

All authors should be the member of this Sanstha (society) before or during the submission of research paper/ review article. Responsibility of membership fee would rest with senior author/next author, in case any author(s) has left the institutions, where the studies were conducted. It should be sent in form of DD in favour of Doctor's Krishi Evam Bagwani Vikas Samsthe payable at Lucknow. Research paper/Review will not be printed unless the dues are cleared.

The Society will bear the cost of publication. However, due to increase in postal charges and printing costs, a uniform amount of Rs. 300/- (US $ 20) for each full article and Rs. 200/- (US $15) for the short communications will be charged.

## Reprints

Fifty reprints may be made available to authors at Rs 200/- (US $ 5) per printed page.

## Declaration

Interpretations and data presented in the research paper and reviews will be the sole responsibility of the author(s) and not the Journal of Eco-friendly Agriculture published by the *Doctor's Krishi Evam Bagwani Vikas Sanstha*, Lucknow. Mention of the proprietary products does not necessarily mean its endorsement by the Sanstha (Society).

## Use of Materials Published in the Journal

Persons interested to use the illustrations published in the Journal of Eco-friendly Agriculture should obtain permission from Chairman, *Doctor's Krishi Evam Bagwani Vikas Sanstha* and must provide a line crediting author(s) and Journal of Eco-friendly Agriculture as a source of material.

## Corresponding Address for Submission of the Manuscript

Dr. R.P. Srivastava, Chairman, *Doctor's Krishi Evam Bagwani Vikas Sanstha*, A-601, Sector-4, Indira Nagar,Lucknow-226 016 (U.P.) India, Phone No. 91522-2342815 (0), 3251389 (R), Fax 91-522-2351389, e-mail: vrsystem@sify.com. Website : ecoagrijournal.com

**Please Provide E-mail address in All Correspondence**

# CHAPTER - 75

# Journal of Economic Entomology

## Information for Contributors

*The Journal of Economic Entomology* is published in February, April, June, August, October and December.  Contributions report on the economic significance of insects and are divided into categories by subject matter: forum; apiculture and social insects; arthropods in relation to plant disease; biological and microbial control; commodity treatment and quarantine entomology; ecology and behavior; ecotoxicology; field and forage corps; forest entomology; horticultural entomology; household and structural insects; insecticide resistance and resistance management; medical entomology; molecular entomology; plant resistance; sampling and biostatistics; stored-products; and veterinary entomology.

All manuscripts are reviewed by scientists qualified to judge the validity of the research.  Acceptance or rejection, however, is the decision of the subject editor.  Acceptance of manuscripts is based solely on their scientific merit. Appeal of a rejected manuscript should be made to the Editorial Board, via the managing editor. Accepted manuscripts are published approximately in the order they are received.

In addition to research papers, the *Journal of Economic Entomology* publishes *Letters to the Editor,* interpretive articles in a *Forum* section, and *Book Reviews.* The same guidelines apply to these as to research articles.  For more detailed instructions concerning the preparation and submission of manuscripts, consult *Publish with* ESA located on our web site (http://www.entsoc.org/pubs/ PUBLISH/esa_publish.htm).

**Manuscript Submission and Preparation.** As of December 2, 2002, all manuscripts for the journal should be submitted electronically using our new web-based Rapid Review system.  Go to http://www.rapidreview.com and select the *Author Log on* button next to the ESA journals. All authors submitting through Rapid Review for the first time are required to set up an account by filling in their contact information and determining their user name and password. Once you have done so, please follow the instructions for submitting your manuscript online. Tables and figures should be included as part of the manuscript file. Authors can subsequently log on to Rapid Review and see the status of their manuscript(s) in the "My Manuscripts" screen. If you have questions, try using the online Help button first. Or, you can e-mail pubs@entsoc.org or call 301-731-4535, ext. 3020.

Submit manuscript as an MS Word or WordPerfect file (PC version recommended) with a page size of 8.5 × 11". Type all as double-spaced, with 1-inch margins, and do not justify text. Use the font Times (New) Roman (12 point). Use italicization only to indicate scientific names, symbols for variables, and words that are defined. Use quotation marks for quoted material only. Use American English spelling throughout and follow *Merriam-Webster's New collegiate /Dictionary,* 10the ed., for guidance on spelling. Number pages consecutively, beginning with the title page. Note that ESA now requires that line numbers be used on your submitted manuscript. Begin each of the following on a separate page and arrange in the following order: title page, abstract and key words (three to five words), text, acknowledgments, references cited, footnotes, tables, figure legends, and figures. Type all captions on a separate page and put each figure and table on a separate page.

**Title Page.** In the upper right corner, type the complete name, address, Telephone number, fax number, and email of the author who is to receive and approve page proofs. In the upper left corner, type *Journal of Economic Entomology,* the section of the journal in which the manuscript should appear, and the running head. Use a brief and informative title that includes order and family names in parentheses. Four lines beneath the title, type in capital letters the complete names of all authors. Four lines beneath the authors' names, type the department, institution, city, state, zip code, and country (other than the United States) of the institution where the research was done. Use numbered footnotes for authors not affiliated with the research institution, and give their addresses or affiliations on the footnote page.

**Abstract.** Abstract should be an informative digest of the significant contents and the main conclusions of the research.

**References Cited.** Citations in the text should be by last name of the author and date and refer to a reference listed alphabetically by author in References Cited. For further formatting instructions, see the ESA *Style Guide* at www.entsoc.org/Pubs/PUBLISH/esa_ publish.html

**Tables.** Keep tables to a minimum. Type all parts of each table double spaced, with starting a new page.

**Figures.** Include all figures, in black and white or color, as part of the manuscript file, with each figure on a separate page. Figures should be inserted in the manuscript file in one of the following formats: TIFF, EPS, WMF, or Postscript (GIF and JPEG formats produce unacceptable laser printouts because of low resolution, even for peer review purposes). Image resolution should be 300 pixels per inch and figures should be one or two columns wide (i.e., 72 or

148 mm).  Separate parts of the same figure must be grouped together and arranged to use space efficiently.  Wherever possible, it is best to avoid using a full page for a set of illustrations.  That is, authors should attempt to have each figure appear separately from the others and should consider numbering illustrations as separate figures rather than as multiple parts of the same figure. Authors should indicate a preference for one- or two-column reproduction, and are asked to account for proportionate reduction or enlargement in the choice of lettering size, type style, and the format of the artwork itself.  Final lettering size should be 8 or 9 point, using the font Arial or Helvetica.  Authors with questions concerning the preparation of artwork can refer to the Cams web site at http://cjs.cadmus.com/da/or contact the publications office staff at 301-731-4535 or fpubs@entsoc.org. Once a paper is accepted for publication, ESA can use original drawings and photographs if a suitable electronic version is not available.

**Style Conventions.**  When first mentioned, a plant or animal should include the full scientific name and the author of the zoological name unabbreviated-except for Linnaeus (L.) and Fabricius (F.).  Use only common names approved by the Entomological Society of America *(Common Names of Insects & Related Organisms 1997)*.  Otherwise, the manuscript should conform to the guidelines established by the latest edition of the *Council of Biology Editors Style Manual.*

**Proofs and Reprints.**  Page proofs and reprint order forms are sent to authors after manuscripts are typeset.  All changes in proof, except for printer and editorial errors and answers to copyeditor queries addressing style or format, will be charged to the author.

**Editorial Charges.** Editorial review charges are assessed on all accepted manuscripts to cover some of the expenses of publishing. Contact the publication office at 301-731-4535 or the ESA web site for current rates.  Reprints (printed and electronic) are ordered at the time page proofs are returned to ESA; see the reprint order form for rates.  Author alterations are billed at $6 per alteration (alterations to figures are billed at a higher rate). Member authors who are unemployed or retired and not affiliated with any institution can publish five pages gratis each year.  Member authors who are employed but who have no grants or funds for publication can request a partial waiver of editorial charges. Such requests must be made at the time of manuscript submission.  For more information on waivers, publication charges, and policies, see the ESA web site at www.entsoc.org/Pubs/PUBLISH/esa_publish.htm.

# CHAPTER - 76

# Journal of Education and Psychology

## The Journal Requests

- Contributors may send their articles, research papers, abstracts of research and gleaning comments for publication. Two copies of each article must be sent.

- The manuscript should be presented in typescript with wide margins and double spacing.

- Name(s) of authors and their organization should be accompanied with the article, so readers may contact if necessary.

- Title of the research article should be in capital letters only.

- Each article must be proceeded by an abstract in English and succeeded by a list of references cited. The references should follow standard pattern.

- Authors may be asked to amend, rewrite or shorten articles, or accept amendments made by the editors. No guarantee of publication can be given and articles may be held over to a later issue.

- Manuscripts of accepted and not accepted articles are not returned.

- Author(s), whose articles are accepted will be communicated; if they have enclosed self addressed envelope with adequate stamps.

- The author(s) will receive five reprints of their article and one copy of the issue in which their article is published.

- All books and publications for review, exchange of periodicals and all business communication should be addressed directly to the Editor, Journal of Education and Psychology, Sardar Patel University, Vallabh Vidyanagar – 388 120, Gujarat, India.

- Subject to Anand Jurisdiction.

# CHAPTER - 77

# Journal of Entomological Research

## Guidelines to the Contributors

Journal of Entomological Research is the official publication of the **Malhotra publishing House**. It features the original research in all branches of Entomology and other cognate sciences of sufficient relevance and primary interest to the entomologists. The publication is generally open to all members but it also accepts papers from non-members on subjects related to Entomology. All the authors have to become the annual member when a paper is accepted for publication intimated after a review. The journal publishes three types of articles, i.e. **Review/ Strategy paper** (exclusively by invitation from the personalities of eminence), **Research paper** and **short communication**. The manuscripts should be submitted in duplicate complete in all respects to **the Editor, Journal of Entomological Research, C/o Mlahotra publishing House, B-6 DSIDC Complex, Kirti Nagar, New Delhi-110 015, India**. Each manuscript must be typed doubled spaced on one side of a A4 size pages. Clearness, brevity and conciseness are essential in form, style, punctuation, spelling and use of English language. Manuscripts should conform to the S.I. system for numerical data and data should be subjected to appropriate statistical analysis. On receipt of an article at the Editorial Office, an acknowledgement giving the manuscript number is sent to the corresponding author. This number should be quoted while making any future enquiry about its status.

*Review/Strategy paper :* It should be comprehensive, up-to-date and critical on a recent topic of importance. The maximum page limit is of 16 double spaced typed pages including tables and figures. It should cite latest literatures and identify some gaps for future. It should have a specific Title followed by the **Name(s) of the author(s), Affiliation, Abstract, Key words**, main text with subheadings, Acknowledgements (wherever applicable) and **References.**

*Research paper :* The paper should describe a new and confirmed finding. Should not generally exceed 12 typed pages including tables/figures etc. A paper must have the following features.

**Title** followed by **Author(s)** and **Affiliation** *:* Address of the institution(s) where the research was undertaken.

*Abstract :* A concise summary (200 to 300 words) of the entire work done along with the highlights of the findings.

*Key words :* Maximum five keywords to be indicated.

*Introduction :* A short introduction of the crop along with the research problem followed by a brief review of literature.

*Materials and methods :* Describe the materials used in the experiments, year of experimentation, site etc. Describe the methods employed for collection of data in short.

*Results and discussion :* This segment should focus on the fulfillment of stated objectives as given in the introduction. Should contain the findings presented in the form of tables figures and photographs. As far as possible, the data should be statistically analyzed following a suitable experimental design. Same data should not be presented in the table and figure form. Avoid use of numerical values in findings, rather mention the trends and discuss with the available literatures. At the end give short conclusion. Insertion of coloured figures as photograph(s) will be charged from the author(s) as applicable and suggested by the printer.

## Acknowledgements (wherever applicable).

*References :* Reference to literature should be arranged alphabetically and numbered according to author's names, should be placed at the end of the article. Each reference should contain the names of the author with initials, the year of the publication, title of the article, the abbreviated title of the publication according to the World List of Scientific Periodicals, volume and page(s). In the text, The reference should be indicated by the author's name, followed by the serial number in brackets.


1.  Scaffer, B.and Guaye, G.O. 1989. Effects of pruning on light interception, specific leaf density and chlorophyll content of mango. *Scientia Hort.,* **41**: 55-61.

2.  Laxmi, D.V. 1997. Studies on somatic embryogenesis in mango (*Mangifera indica* L.). Ph.D. thesis, P.G. School, Indian Agricultural Research Institute, New Delhi.

3.  Sunderland, N. 1977. Nuclear cytology. **In:** *Plant Cell and Tissue Culture,* Vol.II.H.E Street(ed). University of California Press, Berkeley, California, USA, pp.171-206.

4.  Chase, S.S. 1974. Utilization of haploids in plant breeding: breeding diploid species. In: *Haploids in Higher Plants: Advances and Potential. Proc.Intl. Symp.* 10-14 June, 1974, University of Guelph. K.J. Kasha (ed.), University of Guelph, Canada, pp. 211-30.

5.  Panse, V.G.and Sukhatme, P.V.1978. *Statistical Methods for agricultural Workers,* Indian Council of Agricultural Research, New Delhi. 108p.

***Short communication :*** The text including table(s) and figure(s) should not exceed five pages.  It should have a short title; followed by name of **author(s)** and **affiliation** and **references**. There should be no subheadings, i.e. Introduction, Materials and Methods etc. The manuscript should be in paragraphs mentioning the brief introduction of the of the topic and relevance of the work, followed by a short description of the materials and the methods employed, results and discussion based on the data presented in 1 or 2 table(s)/figure(s) and a short conclusion at the end.  References, should be maximum seven in Nos. at the end.

## General instructions

- All the manuscript should be typed double spaced on one side of A4 size paper with proper margin.

- Generic and specific names should be italicized throughout the manuscript. Similarly, the vernacular names are to italicized. Each table should have a heading stating its content clearly and concisely. Place at which a table is to be inserted should be indicated in the text by pencil.  Tables should be typed on separate sheets, each with a heading.  Tables should be typed with the first letter (T) only capital., table no. in Arabic numerals.  All measurements should be in metric units.

- Data to be presented in graphical form should be sent on quality glossy contrast paper without folding.  Each illustration must be referred to in the text and Roman numerals should be used in numbering. Photograph(s) of good contract must be mounted on hard paper to avoid folding and a separate sheet must be given for the title for each photograph sent as figure.

- At the bottom of the first page present address of the corresponding author, **Tel.Fax No. and E-mail ID** etc. must be specified, if available.

- Revised manuscript is acceptable in duplicate along with a **floppy/CD**.

- No reprint of the published article will be supplied to the authors.

- Article forwarded to the Editor for publication is understood to be offered to JOURNAL OF ENTOMOLOGICAL RESEARCH exclusively.  It is also understood that the authors have obtained a prior approval of their Department, Faculty or Institute in case where such approval is a necessary.

- Acceptance of a manuscript for publication in **Journal of Entomological Research** shall automatically mean transfer of copyright to the **Malhotra Publishing House**. The Editorial Board takes no responsibility for the fact or the opinion expressed in the Journal, which rests entirely with the author(s) thereof.

# CHAPTER - 78

# Journal of Food Science and Technology

## Instructions to Authors

Journal of Food Science and Technology publishes original research papers and reviews in all branches of science, technology and engineering of foods and food products.

## Submission of papers

Authors are requested to submit their manuscript and figures in triplicate to Editor-in-chief, Journal of Food Science and Technology, AFST(I), CFTRI Campus, Mysore-570 020, India.

Submission of paper implies that it has not been published previously, is not under consideration for publication elsewhere, and if accepted, it will not be published elsewhere in the same form, in English or any other language, without the written consent of the publisher and its submission has the approval of the all co-authors and the authorities of the host institute where work has been carried out. All papers should be written in English. Manuscripts may be sent enclosing a covering letter providing name, address and email of three competent referees who would be reviewers for the manuscript.

## Manuscript preparation

*General :* Manuscript must be type written (12 pt font size), double spaced on one side of A4 size bond paper with 2.5 cm margin on all sides. Each page should have line numbers. Authors should consult the recent issue of the journal for style and format. If the manuscript is not according to the journal format and style, and incomplete with any respect including tables, figures and references, it will be returned to the author to resubmit it after attending to all changes. Two referees and an editorial board member will review all submissions and the points raised will be intimated for action. The authors are expected to revise the manuscript as per the suggestions of referees and editor. The revised manuscript with serially numbered pages upto the last page should be submitted in disks along with two print copies of the revised and the original manuscript.

***Arrangement of manuscript :*** The manuscript should be arranged in the following order:

***Title page :*** The title is to be typed in sentence case letters. Authors' name(s)should be followed by their institutional affiliation. The corresponding author should be indicated by an asterix and full address including e-mail details should be provided.

***Abstract :*** Each paper should have an abstract not exceeding 250 words, reporting concisely the objective, methodology and results. Abstract should be followed by at least 6 key words reflecting the main topics of the paper.

***Text :*** The text should start on a new page with introduction (without heading), materials and methods, results and discussion, acknowledgement (if any) and references. Materials and methods should include details of sampling methods, replicates and relevant method of statistical analysis. Scientific names, and all vernacular names like *pulav, kabab* etc., must be in italics and briefly described when first used. Varietal names such as 'IR 150', 'Pb 56' must be in single inverted commas. Results and discussion may be combined. All abbreviations and symbols should be as per the journal style.

***References :*** In the text reference to publication should be with the author (s) name and year of publication in bracket e.g. (Smith 1990, Smith and Gibbons 1990, Smith et al. 1992). All publications cited in the text should be presented in the list under Reference in alphabetical order. The titles of all scientific periodicals should be abbreviated as in chemical Abstracts, Biological Abstracts and Annual BIOSIS. Unpublished data including thesis and dissertation or personal communications should not appear in the list but can be indicated in the text.

References should be cited in the following form:

Tairu AO, Omotosu RA, Bamiro FO 1991. Studies on oxidative stability of crude and processed yellow nutsedg tuber and almond seed oil. J Food Sci Technol 28:8-11

Hacking AJ 1986. Economic aspects of biotechnology. Cambridge University Press,  Cambridge

AOAC 1984. Official Methods of Analysis. 14$^{th}$ edn. Association of Official Analytical   Chemists, Washington DC

Kurtzman CP, Phaff HJ, Myer SA 1983. Nucleic acid relatedness among yeast. In: yeast Genetics. Fundamental and applied aspects. Spencer JFT, Spencers DM, Smith ARW (eds). Springer-Verlag, New York, p 139-166

Gross E 1975. Subtilin and nisin: The Cher and biology of peptides and amino acid Peptides, chemistry, structure and bio Walter R, Merenhoper J (eds). Proceeding the Fourth American Peptide Symposium Arbhor, Michigan, USA. P 31-42

Bhalerao SD, Mulmulay GV, Potty VH 1989 fluent management in food industry.  In: venir, National Symposium on Impact of lution in and from Food Industries and Management.  Association of Food Science and Technologists (India), Mysore, p 1-

Schmidt GR, Means WJ 1986.  Process of prepare algin/calcium gel-structures meat product Patent 4 603 054

***Tables*** *:* Tables, numbers and Arabic numerals should be indicated in order in text and should be typed on separate sheets and placed after the Reference section. No vertical lines in between columns should be drawn and no table should have more than 12 columns. Nil results should be indicate by ND (not detected), while absence of by the sign, '-'.  The mean±S.D/SEM should be indicated as footnote.

***Illustrations*** *:* All illustrations should be provided in a form and size suitable for production and each figure should not exceed  half the size of the A4 page and letters should not be less than 10 pt size as it should not obtain legibility when reduced to column.  All photographs, charts and diagrams should be indicated as "Figure(s)" and number consecutively as per the order in which are referred in the text. They should accompany the manuscript, but should not be included within the text.  All figures should have a caption.  Original photographs must supplied.  Line drawings and graphs should preferably computer generated in MS Dos Xerox copies of graphs/drawing and photographs are not acceptable.

***Proofs*** *:* Corresponding author will receive page-proofs for correction of typewriting errors, which should be returned immediately after attending to all corrections major changes, or additions in the paper ending the additions of new authors names allowed at this stage.

## Copy right

The author(s) should fill and signed copyright form while submitting manuscript, which automatically transfer copyrights from authors to the journal.

---

### Copyright Assignment Form/Undertaking

Journal Name :  Journal of Food Science and Technology

Manuscript Title ___________________________________________________

Authors ___________________________________________________________


I/We hereby confirm the assignment of all copyrights in and to the manuscript named above in all forms and media to the publishers journal namely, Association of Food Scientists and Technologists (India) effective if and when it is accepted for publication by the editor-in-chief of the journal.

Date :                                      (Corresponding author's name and signature)

---

# CHAPTER - 79

# Journal of General Plant Pathology

## Instruction for Authors

The *Journal of General Plant pathology* welcomes all manuscripts dealing with plant diseases or their control, including pathogen characterization, identification of pathogens, disease physiology and biochemistry, molecular biology, morphology and ultrastructure, genetics, disease transmission, ecology and epidemiology, chemical and biological control, disease assessment, and other topics relevant to plant pathological disorders.

## Types of papers published

Contributions should fit one of the following categories: (i) Full length articles, (ii) Short communications, (iii) Disease notes, (iv) Techniques, (v) Sequence records, (vi) Letters to the editor, and (vii) Reviews. **Full-length article and short communications:** These should be original research reports that have not been submitted elsewhere. **Disease notes:** Authors should describe symptoms, the hosts, when and where the disease occurred, and pathogen identification (or proof of pathogenicity). They should also state the significance of the disease. **Techniques:** JGPP accepts only reports of techniques that are unique and useful in the plant pathology field. Authors should explain in a cover letter why publication is important. **Sequence records:** JGPP accepts only new sequence information that is unique. The report should give information on the provenance of the material (the origins of the materials together with a reference if available), a reference to the sequence, an annotated diagram of the sequence information (ORFs, promoters, etc.), and any biological information. The authors should explain in the text why the information is important. **Letters to the editor:** A letter to the editor is a comment on research published in the journal or elsewhere. **Reviews :** Reviews should be discussed with the Editor-in-Chief prior to submission.

## Submission of manuscripts

All manuscripts should be submitted in triplicate, one original and two copies including table and figures, with a completed Manuscript Submission Form and Copyright Transfer Statement Form. The forms are published regularly in the journal or can be obtained from http://www.springeronline.com. Send all materials to the Editor-in-Chief at the following address:

Dr. Ichiro Uyeda

Graduate School of Agriculture, Hokkaido University, Sapporo 060-8589, Japan

Tel. + 81-11-706-2473; Fax +81-11-706-2479

e-mail : uyeda@res.agr.hokudai.ac.jp

Manuscripts should be written in English. If the manuscript conforms to the guidelines specified in the instructions, the date received will be the date the manuscript arrived at the editorial office.

The Editorial Committee reserves the right to accept or reject the manuscript for publication.

The Committee may advise the author to revise the manuscript according to suggestions by reviewers. A manuscript written in poor English may not be accepted regardless of its content. When revision of a manuscript has been requested, the revised manuscript should be returned within one month after notification. Otherwise, the manuscript will be processed as one withdrawn from submission. The accepted date will be the day when the Editor-in-Chief has judged it to be publishable after the completion of the reviewing process.

New nucleotide data must be deposited in the DDBJ/EMBL/GenBank databases and an accession number obtained before a paper can be accepted for publication. Submission to any one of the three collaborating databanks is sufficient to ensure data entry in all. The accession number should appear as a footnote on the title page: The nucleotide sequence data reported are available in the DDBJ/EMBL/GenBank databases under accession number(s) ——. The accession number should also be included in the text, tables, or figure legends, as  appropriate.

## Electronic submission of manuscripts

Please send your manuscript in Electronic form as an e-mail attachment to jgpp@res.agr.hokudai.ac.jp. The data should be sent as a single PDF or MS Word file. Prepare the content of your paper in exactly the same way as for conventional submissions (following the Instructions for Authors), ensuring that all illustrations, tables, etc. are included in a single file with the main text. Please add line numbers as well as page numbers in the text. Half-tone or color figures in the file should be ? 150 dpi to keep the file size suitable for sending by e-mail for reviewing. Please try to ensure that the size of your file does not exceed 1 MB. Name the file xxx.pdf or xxx.doc, where xxx is the surname of the first author (no more than 15 characters, no spaces). When you make subsequent submissions to the system, add digits to the end of the file names. In the e-mail massage sent together with the file, please indicate the name of the author to whom correspondence should be directed. Please do not send hard copies by postal or express mail if you send you manuscript in electronic form.

## Page limits and page charges

Reviews and full-length articles should not exceed 8 and 6 printed pages, respectively, and short communications and letters to the editor should be no longer than 3 printed pages. The papers for disease notes and sequence records should not exceed 2 pages, and techniques should not exceed 4 pages. Pages beyond these limitations will be subject to an excess page charge. Reviews and full-length articles cannot exceed 12 and 10 printed pages, respectively, and short communications or letters to the editor, 5 printed pages. Disease notes cannot exceed 2 printed pages, without exception.

Because of fluctuation in printing costs, page charges are subject to change without notice.

Currently there is no page charge for members of the Phytopathological Society to Japan; for nonmembers the charge is $15 per printed page, for members and nonmembers alike. Japanese researchers will be charged at a different standard and should pay in yen as specified in the general rules in the *Japanese Journal of Phytopathology* ( ¥6000 per printed page for nonmembers; ¥36 000 per excess page.)

## Copyright

Authors transfer the copyright to their articles to The Phytopathological Society of Japan and Springer-Verlag Tokyo effective when the articles are accepted for publication. The copyright covers the exclusive and unlimited rights to reproduce and distribute an article in any form of reproduction (printing, electronic media, or other form); the copyright also covers translation rights for all languages and countries. For U.S. authors, the copyright is transferred to the extent transferable.

## General

Manuscripts should be double-spaced with 3 cm margins, 25 lines per page on either A4 (21.0 × 29.5 cm) or 8 ½ × 11 inches, nonerasable bond paper. All pages, including tables, figures, and legends, should include the author's name and the page number at the top right corner for identification. Italic and boldface type should be specified using the features of standard word-processing software. If such features are not available, please indicate italic and bold by single underline or wavy underline, respectively.

### Arrangement of the manuscript

Pages should be numbered consecutively and arranged in the following order.

### Page 1

**Title page** (including title of paper; the names and affiliations of all authors; total text pages; numbers of tables and figures; address to which reviewed manuscripts and proofs should be sent, including e-mail address, telephone, and fax number)

**Page 2**

**Abstract** (no more than 350 words for reviews and full-length articles, 100 words for short communications, techniques, and letters to the editor, and one to two sentence(s) for disease notes). A maximum of six Key words that include the name of organisms (common name or scientific name), method or other words or phrases that represent the subject of the study, such as fungistasis, *Fusarium oxysporum,* phytoalexins, late blight, *Solanum tuberosum.*

**Page 3**

**Text,** divided into the following sections: **Introduction, Materials and methods, Results, Discussion, Acknowledgments, References, figure legends, and tables**. Authors should consult recent issues for details of style and presentation. The positions of tables and figures should be marked in the left margin. A short communication, letter to the editor, or disease note should not be divided into section, except for References.

Mathematical equations should be written in a form such as (RT/nF). In (b/a) instead ofRT/n unless it is confusing to do so.

## Tables and figures

Tables and figures must be mentioned in the text and should be numbered consecutively with Arabic numerals. A unit of measure should be reported as the actual quantity multiplied by a power of 10 to give the reported quantity (the unit may be changed by the use of m or μ).

Tables must be typed on separate sheets with short, informative titles. All parts of each table should be double-spaced.

Figures should match the size of either the column width (8.6 cm) or the printing area (17.6 × 23.6 cm). If figures must be mounted to create one composite figure, they should be mounted on lightweight, flexible cardboard. Composite parts of figures should be labeled with letters (a, b, c).

Each figure must be on a separate sheet, the author's name, and the figure number written on the back in pencil. Figure legends should be grouped together on a separate sheet.

Figures should be good-quality prints in the desired final size. If reduction is necessary, the alternative scale desired should be stated. Capital letters should be about 2 mm high in the final version. The publisher reserves the right to reduce or enlarge figures.

Color illustrations will be accepted. However, authors will be expected to make a contribution toward the extra costs (approx. ¥110 000 for the first and ¥60 000 for ech additional page).

Original artwork should not be sent until the manuscript has been accepted for publication.

## References

References should be cited in the text by the author and year. The reference list at the end of the paper should include only works cited in the text and should be arranged alphabetically by the name of the first author. Citations of "unpublished results" or papers "in preparation" should be included in the text but not in the reference list.

References should be cited as follows: journals papers – names and initials of all authors, year in parentheses, full title, journal as abbreviated in accordance with international practice, volume number, first and last page numbers; books – names and initials of all authors, year, chapter title, names and initials of all editors, full title, edition, publisher, place of publication.

## Example of a journal paper

Virtudazo EV, Nakamura H, Kakishima M (2001) Phylogenetic analysis of sugarcane rusts based on sequences of ITS, 5.8 S rDNA and D1/D2 regions of LSU Rdna. J Gen Pathol 67:28-36 If available the Digital Object Identifier (DOI) of the cited literature should be added at the end of the reference in question. Kischner R, Braun U, Chen Z-C, Oberwinkler F (2002) Pleurovularia, a new genus of hyphomycetes proposed for a parasite on leaves of Microstegium sp. (Poaceae). Mycoscience 43:15-20 DOI 10.1007/s102670200003

## Example of a book

Kempken F (ed) (2002) The mycota XI. Agricultural applications. Springer, Berlin

## Example of a chapter in a book

Waterhouse PM, Upadhyaya NM (1999) Genetic engineering of virus resistance. In: Shimamoto K (ed) Molecular biology of rice. Springer, Berlin, pp 257-281

Responsibility for the accuracy of bibliographic data rests entirely with the author.

## Units of measurement

Authors should use C.G.S. units in the text, table and figures as follows:

*Length* : km, m, mm, mm, nm, etc

*Area* : $km^2$, $m^2$, $cm^2$, etc. a, ha are acceptable.

*Capacity* : kl, 1 (liters in the text), ml, ml, etc. Do not use lambda and italic .

*Volume* : $km^3$, $m^3$, $cm^3$, (not cc), $mm^3$, etc.

*Mass* : kg, g, mg, mg (not gamma), ng, pg, etc.

*Time* : s, min h, day(s), week(s), month(s), year(s)

*Concentration* : M, mM, N, % (only after numbers and in table and figures), g/l, mg/1, mg/l, ppm, ppb

*Temperature :* °C

*Gravity :* Xg

*Molecular weight :* mol wt

*Others :*  Radiosotopes: $^{32}$P

Radiation does : Bq

Oxidation-reduction potential: rH

Hydrogen ion concertration: pH

## Final version of manuscripts in electronic form

After manuscripts have been accepted for publication, authors should sent the files for manuscripts to the Editor-in-Chief, following the Technical Instructions for Manuscripts and Illustrations in Electronic Form, published in most issues. Please note that for these final files, the preferred data format is RTF; PDF files are not acceptable. Text and illustrations should be prepared in separate files.

*Templates :* To help authors prepare their manuscripts, Springer offers a template that can be used with Microsoft Word. The template is available:

*via ftp :* ftp.springer.de

*User name :* anonymous or ftp, Password: your own e-mail address

*Directory :* /pub/Word

*file names :* sv-journ.zip of sv-journ.doc and sv-journ.dot

via   www:ftp://ftp.springer.de/pub/Word

*file names :* sv-journ.zip or sv-journ.doc and sv-journ.dot

## Proofs and offprints

Proofs must be corrected upon receipt and immediately returned to the publisher. Correction should be limited to misprints and other mistakes for which the printer is responsible. A change in sentences can be accepted only with the approval of the Editor-in-Chief and payment of a compensation fee.

Offlprints may be ordered when the proofs are returned. Original manuscripts will not normally be returned. Original artwork will be returned only at an author's request.

# CHAPTER - 80

# Journal of Information Technology in Agriculture

## Submission Guidelines

The Journal of Information Technologies in Agriculture is the peer-reviewed publication of INFITA. It is available exclusively on the World Wide Web at www.jitag.org.

To assist researchers, extension professionals, students, and the general public in their quest for information, the JITAg web site offers a full site search and archives of back publications. Also, we are in the process of converting 20 years' worth of print publications to web format in order to provide a comprehensive JITAg database.

We invite you consider submitting your work to JITAg for review and publication. The links below offer helpful advice for those interested in submitting articles to JITAg .

JITAg expands and updates the research and knowledge base for Agricultural IT professionals and other adult educators to improve their effectiveness. In addition, JITAg serves as a forum for emerging and contemporary publications affecting food, fibre production, and natural resources.

JITAg is written, peer-reviewed, edited, and published by agricultural IT professionals, sharing with their colleagues successful educational applications, original and applied research findings, scholarly opinions, educational resources, and challenges on publications of critical importance to agriculture.

Authors submitting articles to JITAg must follow the guidelines in this document. Submissions that deviate from these guidelines will be returned to the corresponding authors for changes. Because the guidelines are updated as appropriate, authors should check them again before they submit their articles.

JITAg is published on the World Wide Web. Authors should prepare their articles with the Web and on-screen reading in mind. This means, among other things, shorter paragraphs and more bullet and numbered lists than are conventional in more traditional, on-paper journals.

Help for JITAg Authors offers additional information on writing for JITAg. Authors are strongly encouraged to visit this site.

## Journal Sections/Article categories

JITAg accepts submissions in the following categories. Authors should note the differences among the article categories, and the corresponding authors should indicate the category of the article they are submitting.

**Feature** (reviewed by three reviewers): Discuss concepts and research findings of particular interest and significance to agricultural IT professionals and to the knowledge base, methodology, effective practice, and organization. Emphasize implications of IT. Maximum length: 3,000 words, plus tables, graphics, and abstract.

***Research in Brief*** (reviewed by three reviewers): Summarize research results of importance to U.S. Extension professionals. Maximum length: 2,000 words, plus tables, graphics, and abstract.

***What's the Difference?*** A Feature focuses on the implications of the data or concepts for as wide an audience of agriculture professionals as possible (hence, the "extra" 1,000 words). A Research in Brief focuses more on the data, itself, and the methods used to gather it. A Feature is broader in scope and implication. A Research in Brief is more specific and localized.

***Ideas at Work (reviewed by one reviewer):*** Describe novel ideas, innovative programs, and new methods of interest to agricultural professionals. Maximum length: 1,000 words (including abstract), plus tables and graphics.

***Tools of the Trade (reviewed by the editor)*** : Report on specific materials, books, techniques, and technology useful to agricultural professionals. Maximum length: 1,000 words (including abstract), plus tables and graphics.

***What's the Difference?*** An Ideas at Work focuses on what is novel. A Tools of the Trade focuses on what is useful. An Ideas at Work focuses on an idea. A Tools of the Trade focuses on a thing.

***Commentary (reviewed by the editor)*** : Offer a challenge or present a thought-provoking opinion on an publication of concern to agricultural IT professionals. Initiate discussion or debate by responding to a previously published JITAg article. Maximum length: 1,500 words, plus abstract.

***What's the Difference?*** The difference between a Commentary and the other types of JITAg articles is challenge, immediacy, and conviction.

## Submission Formats & Procedure

Articles for JITAg must be written in English.

JITAg accepts submissions in electronic format only and from a single designated corresponding author. Submissions must be on a single file using the JITAg template. A description of the formats used in the template can be found here.

Submitting a paper requires that the author first register to JITAg. Once an author is registered with JITAg and logged into the submission web page follow the "STEP ONE OF THE SUBMISSION PROCESS" link. Step by step instructions will be provided until successful completion of the process.

Go register and/or login to the JITAg submission web page.

Authors who have questions can contact the editor.

## Review Procedures

### Editorial Review

JITAg employs a two-tiered review system. That is, the editor first reviews each submission to determine whether or not it is suitable to be sent out to the peer reviewers on the JITAg Manuscript Review Committee.

If the submission is not suitable for review, the editor either rejects the submission or returns the submission to the author with (often substantive) revision suggestions.

## Peer Reviewers

The Journal of Information Technology in Agriculture (JITAg) is a peer-reviewed publication. A Manuscript Review committee composed of agricultural IT professionals with backgrounds in a variety of subject areas and from different parts of the world serve on the committee. Reviewers are appointed for two-year terms by the editor, with the approval from the JITAg's Board of Directors. A list of current Manuscript Review Committee members can be found in each publication of JITAg . Instructions for retrieving that listing are given in each publication's Table of Contents.

## Peer Review Process

JITAg uses a blind review process. That is, all references to the author(s) are removed before the manuscript is sent out to reviewers.

A set of criteria is used by reviewers to evaluate manuscripts submitted to the JITAg . Reviewers are asked to assign a numerical rating from 1 (weak) to 10 (strong) for each criterion and provide comments on a rating sheet and/or on the manuscript itself. This process is handled through e-mail.

Feature and Research in Brief manuscripts are reviewed by three committee members. Ideas at Work manuscripts are reviewed by one committee member. Tools of the Trade and Commentary manuscripts are reviewed by the editor.

Reviewers are asked to make a disposition on each manuscript they review and submit it to the editor. They can recommend:

- Publish manuscript
- Publish with minor revisions
- Publish with major revisions
- Use ideas and start over
- Reject manuscript

The editor weighs the reviewers' comments and recommended disposition for each manuscript in making the final publication decision. When authors are asked to revise and resubmit manuscripts, the revision may be sent for another round of reviews by the Manuscript Review Committee members or reviewed by the editor. That decision is made at the discretion of the editor.

The two tiers in the JITAg review system add up to a unique combination of academic rigor and professional development. JITAg both "keeps the bar high" and helps authors get published.

## Criteria for Evaluation

Criteria vary somewhat depending upon the review category (Feature, Research in Brief, etc.). The criteria for each category are listed below.

## Feature Article

*Content Criteria*

***Contribution :*** Expands or updates agricultural ITresearch and knowledge base. Important enough to give space in JITAg.

***Audience :*** Of broad interest to agricultural IT professionals in general.

***Usefulness :*** Helps Extension educators improve their effectiveness. Specifically suggests applications.

***Rigor :*** Based on valid and reliable information, documentation or sound concepts; content is empirically, logically and/or theoretically supported.

***Clear Focus :*** Central ideas, findings and conclusions control the article. Has a clear main point.

## Readability Criteria

***Interest :*** Captures and holds readers' attention.

***Understandable :*** Uses easy-to-understand language and flows smoothly.

***Development :*** Appropriately sequences and constructs paragraphs and sentences to support the central idea and conclusions.

***Mechanics :*** Uses acceptable standards of spelling and grammar.

## Research in Brief

*Content Criteria*

*Audience :* Of interest to agricultural IT professionals.

*Importance :* Important enough to give space in JITAg . Research uses unique methods and/or produces interesting or unusual findings.

*Methods :* Offers a clear statement of the research problem and methods used.

*Rigor :* Based on valid and reliable information, documentation or sound concepts; content is empirically, logically and/or theoretically supported.

*Findings :* Describes the research findings with emphasis on their implications.

*Usefulness :* Indicates the usefulness of the methods or findings to Extension educators.

*Readability Criteria*

*Interest :* Captures and holds readers' attention.

*Understandable :* Uses easy-to-understand language and flows smoothly.

*Development :* Appropriately sequences and constructs paragraphs and sentences to support the central idea and conclusions.

*Mechanics :* Uses acceptable standards of spelling and grammar.

## Ideas at Work

*Content Criteria*

*Importance :* Important enough to give space in JITAg. Something that agricultural professionals need to know or would want to know.

*Innovative :* Describes something not already widely known or done. New enough to be considered innovative in some way.

*Audience :* Of interest to agricultural or closely associated professionals involved in the use of IT.

*Usefulness :* Provides suggestions for practical applications.

*Repeatability :* Provides information or resources so that program or idea can be repeated or adapted.

*Clear Focus :* Clearly describes one idea.

*Readability Criteria*

*Interest :* Captures and holds readers' attention.

*Understandable :* Uses easy-to-understand language and flows smoothly.

*Development :* Appropriately sequences and constructs paragraphs and sentences to support the central idea and conclusions.

*Mechanics :* Uses acceptable standards of spelling and grammar.

## Tools of the Trade

Tools of the Trade articles report on specific techniques, materials, books and technology that can be useful to agricultural IT professionals. They are reviewed by the editor for appropriateness and relevance for the Journal of Information Technology in Agriculture, and for readability according to the criteria applied to other articles.

## Commentary

Commentary articles state an opinion, offer a challenge, or present a thought-provoking idea on an publication of concern to Extension, including a published article in JITAg . They are reviewed by the editor for appropriateness and relevance for the Journal of Information Technology in Agriculture, and for readability according to the criteria applied to other articles.

## Contact

Department of Agricultural Economics

Prof. Dr. Gerhard Schiefer

Meckenheimer Allee 174

D-53115 Bonn

Germany

E-mail : info@infita.org

Fax +49 228 73 3431

Phone +49 228 73 3500

For **submission of publications** please contact our e-mail address and send the following informations:

**Last Name, First Name, Address, Title, Source, Keywords, pdf-file**

# CHAPTER - 81

# Journal of Maharashtra Agricultural Universities

## Preparation of Manuscript

Authors are urged to have one or more colleagues, read the manuscript critically prior to submission. Submission implies non-submission elsewhere and if accepted, no future publication without the consent of the Editorial Board

Submit with a covering letter one original and one copy of the manuscript typewritten on one side, no bond paper ($22 \times 28$ cm) and double space including title, abstract, text, literature cited, tables, headings and figure legends. One copy should be without author's name, address and acknowledgement. Footnotes should be avoided. Liberal margins should be left for Editorial marking. Number the pages of title, text, literature cited, tables and figures in that order. Tables and captions of figures should be separate and every sheet of manuscript numbered. A serial number, assigned to each new manuscript accompanies its acknowledgment to the author who submitted it. Refer to that number in all subsequent correspondence.

Mode of Submission – Authors should submit the review / research article / note along with prescribed "Article Certificate" directly to the Journal with a copy of it to the Director of Research / Dean / Head of the Institution and after its publication a copy of reprints of the paper for his record.

Title – The title should be unique and concise description of the paper. It should not exceed 10 to 15 words. Names of the authors centered directly below the title. The name and location of the senior author/institution (The institution where most of the research work is conducted) is centered below the author's name. The designations and addresses of the author/s are foot noted.

Abstract – The abstract is placed on the next line after Institution's name on page,1. It should be a concise summation of the findings and should include names of organisms, effect of major treatments and major conclusions. It should not exceed 5 percent of the length of the paper.

Key Words – A list of additional Key words may be included below the abstract.

Headings – Centre and capitalize letter of main headings of the paper i.e. ABSTRACT, MATERIALS AND METHODS, RESULTS AND DISCUSSION and LITERATURE CITED. Do not centre secondary headings. Place the secondary headings with the first letter of the word capitalized at the beginning of the paragraph.

Style – Manuscripts must conform to current standards of English style and usage. The first person pronoun is accepted and often preferred for clarity. Main clauses should usually be stated first. The introduction without a heading should  state clearly why the research was conducted with reference to earlier work. Give enough information on methods to indicate how the research was conducted so that the reader will have confidence in the validity of the result. Omit details unless it is method paper. Present results succinctly with emphasis on main effects. Use tables and figures to illustrate the text, interpret the results in discussion. Brief conclusion can terminate the discussion. Complex conclusions should  form a separate section. Underline the botanical/scientific names.

Literature Cited – List citations alphabetically by names of authors and type them after the text. The form should be in the order of names of the authors, year, subject title, name of publication (In Roman). Vol. and inclusive page numbers.

**Examples**

Dorsey. M.J. and J.W. Bushnall. 1920. The hardiness problem. Proc. Amer. Soc. Hort. Sci. 17:210224.

Snedecor, G.W. 1956. Statistical Methods. 5$^{th}$ Edn. Iowa State Univ. Press, Ames, Iowa, pp.534.

Whiteside, W.F. 1973. A study of light as influenced by time and planting date on growth of onion (*Allium cepa* L.) in the glasshouse and the field. Ph.D. Thesis, Univ. of Illinois at Urbana-Champaign, pp.53.

Tables – Type tables in double-space on separate pages, each one referred to in the text and numbered with Arabic numerals (e.g. Tables 1). The title of each table should identify its contents so that reference to the text is not necessary. Title and column heading should be brief. Unnecessary descriptive matter should be avoided. Column and row heading with footnotes should be self explanatory. Capitalize only the first letter of the first word of each column and row heading. Use lower case letter from the end of alphabet (x,y,z) to identify tabular footnotes. Use lower case letters from the beginning of alphabet (a,b,c) or asterisks with explanatory footnotes at 5% level, capitals for significance at 1 % level.

Figures – Identify graphs, line drawings and photographs with consecutive Arabic numbers, as Fig. 1, Fig. 2. Place figures after tables. Photographs must be clear, glossy prints. Original graphs and drawings must be in Indian ink or equivalent on plain, white drawing paper. Letter in Indian ink with a lettering guide large and bold enough to permit reduction. Each figure should be referred to in the text. The title of paper and fig. number should be placed in pencil on the back of each fig. for identification.

Trade or Brand Names – Trade or brand names are not used in permanent literature. State trade or brand names only in parentheses. The active ingredient, chemical formula, purity and diluent or solvent may be stated in the text. Capitalize the first letter of trade or brand names.

Abbreviations – In both text and tables except for the first word in a sentence long words or terms used repeatedly should be abbreviated usually without a period (e.g a.m. ppm. m. 0C, N, P, K, K2SO4). Use standard abbreviations. Names of organic chemicals and other terms, abbreviated for the reader's convenience should be spelled out first with parenthetical abbreviation used.

Arabic Numerical – Use Arabic numericals in all cases except the beginning of sentence or where the use would be unclear.

Separates / Reprints – Ten copies will be supplied free of cost.

Advertisement – Advertise of Agro-based industries will be published on Cover Page No. 2 and 3 full page at the cost of Rs.4000/- and 3000/- respectively.

Correspondence – Business correspondence, remittances, subscriptions, change of address etc. should be addressed to The Editor-in-Chief, Journal of Maharashtra Agricultural Universities, College of Agriculture, Pune – 411 005, INDIA.

# CHAPTER - 82

# Journal of Nuclear Agriculture and Biology

### Hints for the Preparation and Submission of Papers

Full length articles, short communications and book reviews are accepted for publication in the Journal. The articles or short communication should concern *original investigations* on the use of nuclear techniques in various branches of Agriculture, Biology and Animal Sciences. At least one of the authors (in case of joint authorship) should be a member of the Indian Society for Nuclear Techniques in Agriculture and Biology and should not be in arrears of subscription.

For contributing an article or a short note for publication in the Journal it is advisable for the contributors to consult current issue of the Journal of Nuclear Agriculture and Biology for style of presentation and other minute details. The intending contributors are also requested to follow rigorously the *Suggestions to the Contributors* for the preparation and submission of papers and notes as printed in the March issue of the Journal.

Abbreviations of periodicals should be in accordance with those given in *A World List of Scientific Periodicals, Butterworth,* London.

Illustrations, which can stand sufficient reduction to the size of the Journal should be drawn with Indian ink (black) on good quality white art card/tracing paper. Line should be of uniform thickness and numbers and letters printed with the help of suitable stencils. Legends should be clearly drawn and included for each figure. At least one set of original figures/diagrammes (with one set of xerox or photo copies) must accompany the article.

Careful observance of the points indicated above will facilitate quick handling and printing of the paper.

The original typescript and one carbon copy of the manuscript complete in all respects, should be sent under *Registered Cover* to the:

**Secretary**

Indian Society for Nuclear Techniques in Agriculture & Biology,

Nuclear Research Laboratory, I.A.R.I.,

New Delhi – 110 012

# CHAPTER - 83

# Journal of Oilseeds Research

## Information for Contributors

Contributions from the members only on any aspect of oilseeds research/ extension will be considered for publication. Articles for publication (in triplicate) and subject reviews should be addressed to

The Editor,

Journal of Oilseeds Research

Directorate of oilseeds Research

Rajendranagar

Hyderabad-500 030, A.P., India

**(A Floppy Diskette containing the manuscript should also be sent along with the revised article)**. Manuscript should be prepared strictly according to the pattern of Journal of Oilseeds Research 18(1) (June, 2001) and should not exceed **15 and 5 types pages** for articles and short communications, respectively (including tables and figures).

# CHAPTER - 84

# Journal of Ornamental Horticulture

## Guidelines to Authors

Journal of Ornamental Horticulture is a Quarterly periodical published in English by the Indian Society of Ornamental Horticulture, for the presentation of records of research papers and review articles on different aspects of Ornamental Horticulture. Book Reviews on recent publications pertaining to area of interest of Ornamental Horticulture will also accepted. Publication in the Journal of Ornamental Horticulture is open to the members of the Indian Society of Ornamental Horticulture. All the authors must be members (Life or Annual) of the Society at the time of publication.

For format of Journal, the recent issue of the publication should be consulted. The way the aspects (Title of the paper), Name of the Author(s), Affiliations, Abstract, key words, Introduction (without heading), Materials and Methods, Results and/ or Discussion, Tables, Figures and References) are presented in the Journal, should be followed strictly.

The manuscript should be typed in double – space throughout on durable paper on one side of the page only. On the first page of manuscript itself, author for correspondence must be indicated with his/her e-mail and postal address, fax and telephone numbers. For colour – toned photographs, the authors(s) will have to remit Rs. 3500 (Rupees three thousand five hundred only) and for black and white Rs. 100 (Rupees hundred only) for each photograph.

The manuscript in triplicate for publication in the Journal of Ornamental Horticulture and any other correspondence in that connection, may please be address to the EDITOR, Indian Society of Ornamental Horticulture, Division of Floriculture and Landscaping, Indian Agriculture Research Institute, New Delhi – 110 012, Telephone No.011-25842708 (Residence).

The articles forwarded to the Editor for publication are understood to be offered to the Journal of Ornamental Horticulture, exclusively. It is also presumed that the authors have obtained the approval of their Department, Faculty or Institute in case wherever such approval is essential. The Society or Editorial Board takes no responsibility for the fact or opinion expressed in this Journal, which rests entirely with the authors thereof. No reprints of will be supplied to the authors. The manuscript found unsuitable for publication will be returned back to the corresponding author.

# CHAPTER - 85

# Journal of Plant Growth Regulation

## Aims and Scope

*Journal of Plant Growth Regulation* is an international journal publishing original articles on all aspects of plant growth and development. We welcome manuscripts reporting question-based research using hormonal, physiological, environmental, genetical, bio-physical, developmental or molecular approaches to the study of plant growth regulation. Beginning in 2000, each issue of the journal will be thematic, containing papers solicited by a guest editor on various aspects of growth and development, in addition to contributed papers on any topic. Papers dealing only with the optimization of cell cultures without asking a scientific question and without advancing our knowledge in plant growth and development will not be considered for publication. Now ranked as #16 of 136 journals in Plant Sciences, JPGR has increased its 2003 impact factor to 2.778.

## Instructions to Authors

### Electronic Submissions

We are pleased to announce that we have moved to an online system of manuscript tracking called Manuscript Central. Authors are encouraged to submit their articles online. This will allow even quicker and more efficient procession of your manuscript.

Please log directly onto the site http://mc.manuscriptcentral.com/ jpgr and upload your manuscripts following the instructions given on the screen.

## System Requirements

- Netscape 4.xor MS Internet Explorer 4.x/5.x
- Adobe Acrobat browser plug-in
- Electronic files of their article text
- Electronic files of their article graphics (scanned or exported)

## Author Accounts

Authors entering the journal's manuscript Central site can either create a new account or sure an existing one. When you have an existing account, use it for all your submissions and you can track their status on the same page.

## Getting Started

Once you have logged into your account, Manuscript Central will lead you through the submission process in a step-by-step orderly process. If you cannot finish your submission in one visit, you can save a draft and re-enter the process at the same point for that manuscript.

While submitting your electronic manuscript, you will be required to enter data about your manuscript in the system. These include title, subtitle, author names and affiliations, and so forth. Support for special characters is available. At any point during this process, there are Help buttons available to see common questions and a support link to ask a specific question via email.

## Uploading Files

Electronic files can be uploaded as PDF, PostScript, or RTF. PDF and PostScript files should already contain the graphics within the file (PostScript files are converted by the system into PDF so that Editors and reviewers may share them.

RTF (Rich Text Format) is a common export property of most popular word processors. Check your word processor to see if it can export or "SaveAs" your file in RTF format. MS Word and WordPerfect both contain this function. After uploading the RTF for text, you will be prompted for uploading graphics. Common graphics files such as GIF, JPEG, EPS,TIFF and  many others are supported. After uploading the parts of the article in this manner, the system will convert the files to PDF. You will see the result of the conversion with the Acrobat plug-in in your browser. Keep copies of your word-processing and graphics files. You may want to revise the manuscript during the review process and you will need the original files if your manuscript is accepted. At any point during this process, there are Help buttons available to see common questions and a support link to ask a specific question via email.

Your will also be notified by email that your submission was successful.

## Graphics Quality

If you are submitting electronic graphics that you have scanned, be prepared to send the hard copy originals upon request. While the electronic files you have created are satisfactory for the review process, they may not be of sufficient quality for printing. This also holds true for files created in low-resolution graphics environments such as MS Powerpoint, etc.

## Keeping Track

After submission, you may return periodically and monitor the progress of your submission through the review process.

JPGR Editorial Office

1002 Stonebriar Drive

Verona, WI 53593, USA

Fax : 608.848.4592, Email; jpgr@tds net

## Preparation of Manuscripts

Please follow these instructions closely when preparing a manuscript. Careful preparation of the manuscript will facilitate copy editing and typesetting, and can expedite publication.

## Form of the Manuscript

Paper must be clearly and concisely written in correct English; this is the responsibility of the authors. The entire manuscript should be double-spaced and lines numbered. Every page should be numbered. The elements of the paper should be presented in the following sequence:

TITLE, with a shortened version for page headings that does not exceed 40 characters.

Complete AUTHOR NAME for each author, plus their mailing address and institutional affiliation. Include phone/FAX numbers and an e-mail address for the corresponding author.

ABSTRCT of up to 250 words that highlights the objectives, results, and conclusions of the paper.

KEYWORDS (6 to 10), to identity the subjects under which the article may be indexed.

TEXT of the manuscript. Subheadings should be used as appropriate.

ACKNOWLEDGMENTS: Individuals who were of direct help to the authors and any sources of financial support should be acknowledged in a brief statement.

REFERENCES

APPENDICES (optional) Each appendix must have a title.

FIGURES should be separate pages (not embedded in the text).

FIGURE LEGENDS  should be on a separate page. Every figure must be cited in the text.

TABLES should be on separate pages. Tables should have a clear and rational structure. All tables should be numbered consecutively with Arabic numerals. Every table must be cited in the text. Tables should be designed to fit  a space on larger than the full-page width of 16.9 cm. Footnotes to tables should be indicated by lower-case superscript letters, beginning with superscript a for each new table.

TABLE LEGENDS  should be on the same page as the table to which they correspond. Provide enough information in legends so that each table is understandable without reference to the text.

## New Category : Short Communications

Short communications will consist of not more than 10 double spaced type-written pages excluding figures and tables. The number of figures and tables should not exceed 4 per manuscript and the reference list will be limited to 20 entries. Follow general Journal of Plant Growth Regulation style guidelines but combine Results and Discussion into one section.

We will consider papers in this category dealing with the isolation, identification and structure of plant growth  regulators, provided that the information is new for a particular growth regulator in a given plant species. The requirements for publication are otherwise as for regular papers.

## Additional Guidelines

*Style Manual :* Guidelines for references, symbols, abbreviations, units of measurement, etc. may be found in the Council of Biology Editors 194 Scientific style and format: the CBEZ manual for authors: editors and publishers. 6th ed. New York: Cambridge University Press, 825 pp

*Footnotes :* These should not be used; information should be integrated into the text.

*Metric System :* The metric system should be used throughout. If required, equivalent values in other systems may placed in parentheses immediately after the metric value.

*Mathematical Symobls :*  Marginal notes to the copy editor should be used to explain mathematical symbols used in the text.

*Illustrations :* Illustrations should be of high quality (i.e. professionally drawn or generated by graphics software). Hard copies of figures must be submitted in duplicate with the final manuscript.

Each figure must labeled on its back, indicating figure number, author name, and an indication of the top of the figure where necessary. Figures should be submitted at their final size. Submit figures as .eps  or .tif files, in addition to hard copies.

*Color Illustraions :* Color may be used without charge for the electronic edition of the journal but will appear in the printed version of the journal at the author's expense: $575 per illustration. If color is essential to the article and funding is unavailable, please contact the Editorial Office at the time of submission.

*References :* Only essential references should be used, and only references cited in the article should appear in the Reference List. When citing works by more than two authors in the text, only the first author should be named, followed by "and others". Responsibility for the accuracy of the references rests with the author. "In press" citations must include the name of the journal that has accepted the paper. For specific guidelines on reference style, see Preparing a 'Reference List'.

***Proofs and Reprints :*** Proofs and a reprint order form are sent to the Corresponding Author unless the Editorial Office is advised otherwise. Authors will be notified via email when proofs are available online (with instructions on where to return proofs). At that time, authors will be able to download a Reprint Order Form and order reprints at cost.

***Copyright :*** Submission of a manuscript implies: that the work described has not been published before (except in the form of an abstract or as part of a published lecture, review, or thesis); that it is not under consideration for publication, elsewhere; that its publication has been approved by all co-authors; that, if and when the manuscript is accepted for publication, the authors agree to transfer the copyright to the publisher; that the manuscript will not be published elsewhere without the consent of the copyright holder; that written permission of the copyright holder is obtained by the authors for material used from other copyrighted sources; and that any costs associated with obtaining this permission are the responsibility of the authors.

## Guidelines for Preparing a Reference List

Double space the list. Check the text citations against the Literature Cited list to make sure there are no gaps or inconsistencies. Names of journals should be abbreviated according to the Bibliographic Guide for Editors & Authors (The American Chemical Society, Washington  D.C.) Do not include abstracts or unpublished material in the Reference List. Instead, cite them in the text as Personal Observations, Personal Communications, or Unpublished Data/ Unpublished Manuscript. Identify authors of unpublished work.

## Use the following formats for Reference List style:

- Journal Article

  Vanderhoef LN, Dute RR. 1981. Auxin regulated wall loosening and sustained growth in elongation. Plant Physiol 67:146-149

- Aricle in Edited Book

  Lang  A. 1980. Inhibition of flowering in long-day plants. In: Skoog F, editor. Plant growth substances 1979. Berlin Heidelberg New York: Springer-Verlag. P 310-322.

- Book

  Lepold AC, Kriedemann PE. 1975, Plant growth and development. 2nd ed. New York: McGraw-Hill, xxxp.

## Guidelines for Electronically Produced Illustrations for Print

**General**

Send illustrations separately from the text (i.e. files should not be integrated with the text files). Always send printouts of all illustrations.

### Vector (Line) Graphics

Vector graphics exported from a drawing program should be stored in EPS format. Suitable drawing program. Adobe Illustrator. For simple line art the following drawing programs are also acceptable: Corel Draw, Freehand, Canvas.

- No rules narrower than .25 pt.
- No gray screens paler than 15% or darker than 60%
- Screens meant to be differentiated from one another must differ by at least 15%

### Readsheet/Presentation Graphics

- Most presentation programs (Excel, PowerPoint, Freelance) produce data that cannot be stored in an EPS format. Therefore graphics produced by these programs cannot be used for print.

## Half Tone Illustrations

- Black & White and color illustrations should be saved in TIFF format.
- Illustrations should be created using Adobe Photoshop whenever possible.

## Scans*

- Scanned reproductions of black and white photographs should be provided as 300 ppi TIFF files.
- Scanned color illustrations should be provided as TIFF files scanned at a minimum of 300 ppi with a 24-bit color depth.
- Line art should be provided as TIFF files at 600 ppi.

*We do prefer having the original art as our printers have drum scanners which allow for better reproduction of critical medical halftones.

## Graphics From Videos

- Separate files should be prepared for frames from a video that are to be printed in the journal. When preparing these files you should follow the same rules as listed under Halftone Illustrations.

## Guidelines for Electronically Produced Illustrations for ONLINE

### Video

- Quicktime (.mov) is the preferred format but .rm, .avi, .mpg, etc. are acceptable.
- No video file should be larger than 2MB. To decrease the size of your file, consider changing one or more of the following variables: frame speed, number of colors/greys, viewing size (in pixels), or compression. Video is subject to Editorial review and approval.

# Chapter - 86

# Journal of Plant Nutrition

## Instruction for Authors

*Aims and Scopes :* This authoritative journal serves as a comprehensive, convenient source of new and important findings exploring the influence of currently known essential and nonessential elements on plant physiology and growth – offering prompt publication of outstanding original research and review papers in this vital area of plant and soil science. Includes special symposium issues that focus on essential nutrients, heavy metals, and trace elements ! Refereed by an internationally renowned editorial board ensuring the high level of scholarship, the Journal of Plant Nutrition provides insightful coverage of nutritional topics, such as hydroponics, nutrient requirements for greenhouse crops, container production, media analysis of pine bark, peat and artificial media, floriculture production, vegetable crop production, fruit crop production, ornamental production, tropical crops, foliage plants, agronomic crops, forestry, and much more!

*Submission of Manuscripts :* Please submit your original manuscript, along with two copies, to Dr. Harry A. Mills, 183 Paradise Boulevard, Suite 104, Athens, Georgia 30607 USA. Authors are required to submit manuscript, on disk along with their hard copies. The disk should be prepared using MS Word or WordPerfect and should be clearly labeled with the authors' names, file name, and software program. Each manuscript must be accompanied by a statement that it has not been published elsewhere and that it has not been submitted simultaneously for publication elsewhere. Authors are responsible for obtaining permission to reproduce copyrighted material from other sources and are required sign an agreement for the transfer of copyright to the publisher. All accepted manuscripts, artwork, and photographs become the property of the publisher.

All parts of the manuscript should be typewritten, double-spaced, with margins of at least one inch no all sides. Number manuscript pages consecutively throughout the paper. Authors should also supply a shortened version of the title suitable for the running head, not exceeding 50 character spaces. Each article should be summarized in an abstract of not more than 150 words. Avoid abbreviations, diagrams, and reference to the text within the abstract. Authors must give from three to ten  key words that identify the most important subjects overed by the paper.

***Affiliation*** *:* On the title page include full names of authors, academic and/or other professional affiliations and the complete mailing address of the author to whom proofs and correspondence should be sent. Do not submit position titles.

***References*** *:* Include only references to books, articles, and bulletins actually cited in the text. Examples of the journal style are listed here. Cite in the text by author and date (Smith, 1983). Prepare reference list in accordance with the Chicago Manual of Style.

## Examples

***Book Chapter*** *:* Hsu, P.H. 1987. Aluminum hydroxides and oxyhydroxides. In *Minerals in soil environments*, eds, J.B. Nixon and S. Weed, 99-143. Madison, Wisconsin:SSSA.

***Journal Article*** *:* Tian, G.,and G.O. Kolawole.2004. Comparison of various plant residues as phosphate rock amendment on Savanna sols of West Africa. *Journal of Plant Nutrition* 27: 571-583.

***Book*** *:* New. T.R. 1991. *Insects as predators.* Kensington, Australia: New South Wales University Press.

***Illustrations*** *:* Illustrations submitted (line drawings, halftones, photos, photomicrographs, etc.) should be clean originals or digital files. Digital files are recommended for highest quality reproduction and should follow these guidelines:

- 300 dpi or higher
- sized to fit on journal page
- EPS, TIFF, or PSD format only
- submitted as separate files, not embedded in text files

Color illustrations will be considered for publication; however, the author will be required to bear the full cost involved in their printing and publication. The charge for the first page with color is $900.00. The next three pages with color are $450.00 each. A custom quote will be provided for color art totaling more than 4 journal pages. Good-quality color prints or files should be provided in their final size. The publisher has the right to refuse publication of color prints deemed unacceptable.

***Tables and Figures*** *:* Tables and figures should not be embedded in the text, but should be included as separate sheets or files. A short descriptive title should appear suitably identified below. All units must be included. Figures should be completely labeled, taking into account necessary size reduction. Captions should be typed, double-spaced, on a separate sheet. All original figures should be clearly marked in pencil in the reverse side with the number, author's name, and top edge indicated.

***Proofs*** *:* All proofs must be corrected and returned to the publisher within 48 hours of receipt. If the manuscripts are not returned within the allotted time, the editor will proofs read the article and it will be printed per his instruction. Only correction of typographical errors is permitted at the proof stage.

***Offprints*** *:* The corresponding author of each article will receive one complimentary copy of the issue in which the article appears. Additional copies and offprint may be ordered form Taylor and Francis by using the order form included with page proofs.

# CHAPTER - 87

# Journal of Plantation Crops

## Guidelines for Authors

### General

Full length research papers and short scientific notes based on original research in any topic on plantation crops viz. coconut, arecanut, oil palm, cashew, spices, tea, coffee, rubber, cocoa, or plantation crop-based farming systems are accepted for publication. It should not be submitted simultaneously or published in any other technical or scientific journal. Review articles are received by invitation only. They will summarize and analyze the existing state of knowledge on a particular topic on plantation crops. Original papers including Tables, illustrations and references should not exceed 4000 words. Short communications should not exceed 1300 words. All the authors must be a member of the Indian Society for Plantation Crops. The manuscripts and correspondence concerning editorial matters should be addressed to the Editor, Journal of Plantation Crops, Central Plantation Crops Research Institute, Kasaragod – 671 124, Kerala, India.

### Manuscripts

Manuscripts must conform to the journal format and must be typed in English one side of DIN A4 sheets of bond paper (21x28cm) with a margin of 4 cm on the left hand side and top, and 3 cm each on the right hand side and bottom. The entire manuscript must be double-spaced including Tables, legends, references and foot notes. Three copies of the manuscript should be submitted. The manuscript should be arranged in the following order :

1. Title page,
2. Abstract,
3. Key words,
4. Introduction
5. Materials and Methods,
6. Results and Discussion,
7. Acknowledgements
8. References
9. Tables,
10. Legends for figures,
11. Figures

In addition, please submit the manuscript in 3 ½ diskettes, prepared in MS DOS/Windows-compatible format, clearly indicating the file name(s).

*Title Page :* Provide a separate title page with the following items:

*Title* of the paper should be informative and concise.

*By-line* should contain the name(s) and initials of the author(s) and footnote symbols to indicate corresponding author and addresses.

*From-line* should contain the name and address of the institution where the research work was carried out.

*Keywords* used in the article should be given.

*Running title* A short title not exceeding 50 characters should be provided for the running head lines.

*Title page foot notes*. To indicate the corresponding address (es) of author(s).

**Abstract :** The abstract should contain a brief but informative and accurate summary of the contents and conclusions of the paper. It should be intelligible without reference to the full paper. It should not exceed 250 words.

**Introduction :** The introduction should explain the aim of the paper. A brief historical or critical review may be included, but it should be confined to the immediate subject to the paper only.

**Materials and Methods :** Should be brief, but informative enough for reproduction of the work.

**Results and Discussion :** Should end with definite conclusions drawn from the results obtained.

**Acknowledgement :** This should be brief and specific.

**References :** The list of references should include only publications cited in the text. References to unpublished data, private communications and documents with limited circulation should be avoided wherever possible. Otherwise, they may be given in the text only. An article should not be referred to as "in Press" unless it has been accepted for publication. The name of the journal in which such an article has been accepted should be given in the reference cited. The references should be cited in alphabetical order under the first author's name followed by co-authors. The following examples may be adopted:

*Articles from journals.* Name(s) and initial(s) of author(s); Year of publication, further distinguished by the addition of small letters a,b,c, etc., where there are citations to more than one paper published by the same authors(s) in one year. Title of the paper, Name of the journal as per the "World list of periodicals;" Volume Number (in bold) and Inclusive pages.

Wigley, T.M. Briffa, K.R. and Jones, P.D. 1984. Predicting plant productivity and water resources.Nature (Lord.) **312**: 102 - 103

Articles from symposia volumes and similar collective publications. Name(s) and initial(s) of author(s), Year of Publication, Title of the article, Inclusive pages of the article followed by "In": title of volume, name(s) of editor(s) preceded by "Ed(s)"; names of publishers, place of publication.

Eg : Baldesdent, J. and Maroitti, A. 1996. Measurement of soil organic matter turnover using $^{13}$C natural abundance. In: *Mass spectroscopy of soil.* (Bds) Bouten, S. and Yamasaki, S.; Marcel Dekker, New York, pp.83-111.

**Books :** Name(s) and initial(s) of the author(s); Year of publication; Complete title of the book in italics; Total number of pages; Edition (if applicable); Name of publisher and Place of publication.

Giller, K.E. and Wilson, K.J. 1991. Nitrogen fixation in tropical cropping systems. CAB International Wallingford, UK, 313 p.

**Tables :** Tables should be typed on separate sheets, numbered consecutively in Roman numerals and carry appropriate titles. The approximate preferred position of the Tables may be indicated on the typescript. Presenting Tables that are too large to print across the page should be avoided.

**Illustrations :** Figures should be consecutively numbered in the order of their first citation in the text. Send sharp, glossy, black-and white/colour photographic prints, not larger than 8 × 10 inches. High quality laser printouts and graphs are acceptable. Letters, numbers, symbols should be clear and even throughout and of sufficient size so that each item will still be legible when reduced for publication. Titles and detailed explanations should be given separately in the legends for figures, not on the figures themselves. Each figure should have a label on its back indicating the number of the figures and author's name.

**Units :** All data in the paper should be presented in metric units using standard abbreviations only.

**Proofs :** Proofs will normally be sent to the author. They should be returned to the Editor without delay. Alterations in proof other than printing errors should be avoided. Excessive corrections are chargeable to the author.

# CHAPTER - 88

# Journal of Pottassium Research

## Imformation for Contributors

*Journal of Potassium Research* will accept full length research papers or short communication in English only.

**Book reviews** of titles related to the subject matter of the Journal will also appear as and when such occasions arise.

The full length article not exceeding **3000 words and** short communications within 750 words should comprise original research investigations, reporting original experimental data/methods or new analyses of already existing data. **All contributions should primarily concern with the studies on potassium in agriculture.**

**The papers submitted for publication in the** *Journal of Potassium Research* must not carry material already published in same form nor should it be offered for publication elsewhere.

The **write up** (both for full papers and short communications) should be arranged in sections such as, Title; Author(s), Abstract, Introduction, Materials and Methods, Results and Discussion, Acknowledgments (if any), References etc. The author(s) will send a short title of approximately 35 characters for a running head on a separate sheet and an Abstract of **150 words** with *key words* (arranged alphabetically) not exceeding 10. The introduction should mention why work has been done and not to be a mere review of past work. The Materials and Methods should inform the reader about the appropriate choice of procedure etc. and help one to verify the results. The Results *must not be repeated* in both tables and figures and the Discussion should relate to the significance of observations made.

The references should include the names of all authors, year, full title of the article, full name of the journal (no abbreviation is allowed), volume number, issue and pages. In case of books/monographs etc. the name of publishers and place and year must also be given. The number of references should be **restricted as far as possible.**

Tables and figures are expensive and should be kept to the minimum. Tables are to be typed on separate sheets with complete titles. The illustrations should be in black Indian ink on superior grade *white art card or tracing cloth* with uniform thickness of lines and all numbers and letters must be **written with suitable stencils** (not free-hand). At least one set of original art-work must be enclosed.The legends should be clearly drawn and included for each figures. Figure numbers and title are to be given *separately* and also written below in *pencil.* Photographs (black and white only), wherever necessary, must be *sharp having adequate contrast* and to be on glossy paper with title of paper, plate number and legend written clearly on the reverse side in *pencil.*

The **original** manuscripts, *neatly typed* on good quality *quarto* size paper with double line spacing and a margin of 3 cm on all sides, along with one carbon copy, and original drawings (of figures and diagrams) should be sent to the Editor, *Journal of Potassium Research,* Potash Research Institute of India, Sector 19, Dundahera, Gurgaon-122 001 (Haryana). A copy of the manuscript in MS-WORD on a 3.5" Floppy or CD should also be sent.

# CHAPTER - 89

# Journal of Spices and Aromatic Crops

## Instructions to Authors

All manuscripts should be addressed to:

Chief Editor

Journal of Spices & Aromatic Crops (JOSAC)

C/o National Research Center for Spices

P.O. Box. 1701, Marikunnu P.O.

Calicut – 673 012, Kerala, India

No page charges for publication in JOSAC

All papers except invited ones are subject to peer review.

JOSAC welcomes general papers, review papers, original research papers, short research communications and book reviews for publication.

General papers describing original research should not exceed 12 pages of printed text, including tables, figures and references (one page of printed text = approx 600 words).

Review papers not exceeding 25 pages of printed text including tables, figures and references will be usually published within six months following acceptance.

Papers already published or in press elsewhere will not be accepted.

Manuscripts should be written in standard English and submitted in triplicate (the original manuscript plus two photocopies, each including all tables, figures and references). The author should retain a complete copy of the manuscript.

Two copies of the paper should be sent without the author(s) names(s), affiliation(s) and acknowledgment. Manuscripts should be typed clearly double-spaced throughout on one side of A4 paper with margins of 3-5 cm. All papers (including the tables, figures, legends and references) should be numbered consecutively.

For guidance, please consult: Council of Biology Style Manual, available from the Council of Biology Editors Inc., 9650 Rockville Pike, Bethesda, MD 20814, USA.

The manuscript should be organized in the following order.

Title page (Page 1)

- The title should be brief but informative.
- The author's full name (if more than one, use '&' before the last name and indicate to whom correspondence should be addressed).
- Affiliation(s)/ Address(es) should be completed.
- Provide a short running title for the paper.

Key words/Abstract/Abbreviations (Page 2)

- Key words ( a maximum of 6, in alphabetical order, suitable for indexing).
- Abstract (brief and informative, not to exceed 250 words). No abbreviations should be used in the abstract.
- Abbreviations (arranged alphabetically, only those which are not familiar and /or commonly used).

**Main text**

- The text should be developed under the following headings: Introduction, materials and methods, Results/Discussion, and Conclusions.
- The relative importance of subheadings when used should be clear. The approximate location of figures and tables should be indicated in the margin.
- New paragraphs should be indicated by clear indentation.
- The use of footnotes should be avoided.

**After the main text**

- Acknowledgment (also grants, support, etc. if any) should follow the text and precede the references.
- Notes, if any, should be numbered consecutively in the text with superscript numerals and listed in numerical order after the Acknowledgments.

**References**

- Literature references should be listed alphabetically, typed double spaced, and in the text refereed to by author name and year of publication enclosed in parentheses, e.g. (Patel 1987, John sons & White 1989).
- Citation of personal communications and unpublished data should be avoided.
- Abbreviate titles of periodicals according to the style of the Bibliography Guide for Editors and Authors (Biosis, Chemical Abstract Service and Engineering Index, Inc., 1974).

- References should contain: author(s) name(s) followed by author(S) initials, year, title of article (only first word and proper nouns capitalized), journal (not underlined), volume number, and inclusive page numbers. Books must include the location and name of the publisher.

## Examples

### Periodicals

Davidonis G H & Hamilton R H 1983 Plant regeneration from callus tissue of *Gossypium hirsutum* L. Plant Sci.Lett. 32 : 89-93.

Books (Edited by some one other than the author of the article)

Rhodes MJC 1985 Immobilized plant cell cultures. In: Wiseman A (Ed.) Topics in Enzyme and Fermentation Biotechnology 10 (pp. 51-57). Ellis Horwood Limited, Chichester.

### Monograph

Bewley J D & Black M 1982 Physiology and Biochemistry of Seeds. Springer Verlag, New York.

## Tables

- Each table should be typed on a separate page.
- Tables should be numbered with Arabic numerals, followed by the title. Horizontal rules allowed; vertical rules should not be used. Table footnotes should be marked with superscript numbers.
- If required, tables will be edited to permit more compact typesetting.

## Figures

- All figures should be cited in the text. Line drawings must be in black ink on tracing film and should not contain shading.
- Extremely small type should be avoided as figures are often reduced in size.
- Photographs should be supplied as black and white high contrast glossy prints. They may be arranged into plates (by taking into account the page size) if photographs are more than two.
- Figures as well as legends should be identified by Arabic numbers and headed 'Fig. 1' etc.
- Where multi-part figures are used, each part should be clearly identified in the legend with (lower case) letters.
- The top of the figure should be indicated on the back and each figure should be identified by lightly writing title and figure number on the back.
- Draw bar scales directly on the figures to indicate magnification.

## Abbreviations and units

- SI units should be used, e.g. mg, g, kg, m, cm, mm, ppm, cpm, iCi (micro Curie), 1 (litre), ml, s (second), min (minute), h (hour), mol, $m^3$ kg per ha or kg ha$^{-1}$ (the minus index form is always to be used in tables).

- Use mg 1$^{.}$1, not mg/1 Growth regulator concentrations should be quoted as iM or mg 1$^{-1}$ (plant growth regulator is the preferred general term and not 'plant hormones').

- If a non-standard  abbreviation is to be used extensively, it should be defined in full on page 2 and follow the abstract.

## Proofs

The final proof only will be sent to author for corrections, that too only when it is absolutely essential, which should be returned immediately. The corrections should be limited to errors in type-setting.

# CHAPTER - 90

# Journal of the Indian Society for Cotton Improvement

## Information to Contributors

The Journal of the Indian Society for Cotton improvement is issued in April, August and December of every year. Material for publication in the Journal and Book for review should be sent to the Secretary. Editorial Board. Journal of the Indian Society for Cotton Improvement, c/o. Central Institute for Research on Cotton Technology, Adenwala Road, Matunga, Mumbai – 400 019.

Contribution may be original papers, critical reviews or research notes which have not been published earlier or have not been simultaneously offered for publication elsewhere without the consent of the Editor of this Journal.

Research Papers and Review Articles may deal with any aspect of cotton research, development or utilization, such as breeding and genetics, bio-technology, agronomy, soil science, physiology, entomology, pathology, mechanical and chemical processing, utilization of by-products, marking, etc. Normally, each paper/article should not exceed 5000 words. Brief reports or research notes should normally not exceed 800 words.

The manuscript should be submitted in triplicate, typed in double space one side of the paper with 4 cm margin. Illustrations should preferably laser prints. Photographs should be on glossy paper.

Results represented in diagrammatic form should not also be given in the form of Tables. Column headings should be kept to a minimum. If a brief heading is not possible, a number or letter may be used and an explanatory note given at the foot of the table.

The use of Greek characters should be kept to a minimum. Formulae should be typed wherever possible; otherwise, these should be neatly and unambiguously handwritten with marginal notes to avoid possible confusion.

An abstract must be provided for each paper. The abstract will precede the main text of the paper and draw attention to its salient points.

References are to be indicated by author's name with year. These should be arranged in alphabetical order at the end of the paper.

*For Journal :*    The name/s of the author/s, the name of the journal, volume and page numbers.

*For book :*    Name/s of the author/s. the title of the book, name and location of the publishers and page numbers.

"Personal communications" and "un-published work" do not go into references and should be incorporated in the text itself.

Acknowledgements should be brief.

The authors should send the revised manuscript of the paper after incorporating the changes/alterations suggested by the referee along with a floppy (MS word).

Membership of ISCI, preferably Life Membership is a prerequisite for publishing an article in the Journal for all the authors.

All articles are subject to the approval of the Editorial Board of the Journal.

Two copies of the appropriate issue of the Journal will be supplied free of charge on request to the first /sole author of each paper. In the case of dual or multiple authors, the copies will be supplied only to the first-named author for distribution as required. Additional copies are available for sale at a special price.

Reprints will be supplied in lots of 50 if intimation is given at the time of submission of  article for publication. Based on the length of the article, cost of reprints will be intimated when the paper is accepted for publication. The amount should be paid within the specified time; otherwise, reprints will not be taken at the time of printing the Journal.

Copyright in respect of material published in the Journal vests with the ISCI. Requests for permission to reproduce matter/figures, etc. from the printed articles should be addressed to The Secretary, Editorial Board, Indian Society for Cotton Improvement. However, authors are free to reproduce part or whole of the papers, including data and figures, in their teaching notes, books etc.

Business correspondence relating to remittances, subscriptions, advertisements etc. should be addressed to the Hon. Secretary, Indian Society for Cotton Improvement, C/o. Technology, Adenwala Road, Matunga, Mumbai-400 019.

# CHAPTER - 91

# Journal of the Indian Society of Agricultural Statistics

## Guidelines to Authors for Submission of Papers

Manuscripts must be original contributions and should be submitted exclusively to the Secretary, Indian Society of Agricultural Statistics, **IASRI** campus, Library Avenue, New Delhi-110 012 (INDIA).

The objective of the Journal is to provide a forum for dissemination and exchange of findings of research on Agricultural Statistics and Computer Application in Agriculture. Purely descriptive material is not appropriate for such a journal. Papers dealing with (i) new developments in research and methods of analysis, or (ii) which apply existing empirical research methods and techniques to new problems or situations, or (iii) which attempt to test new hypotheses, theoretical formulations or modifications of existing statistical techniques to explain biological phenomena especially in Indian context, or (iv) papers based on research done by the authors bringing out new facts on data presented in an analytical frame will be preferred. Younger statisticians are advised to seek guidance from their seniors in the preparation of the paper for the Journal. A one-page abstract indicating the author's own assessment of the importance and relevance of the findings reported in his/her paper in the context of recent researches is also required in the beginning of the paper.

In view of the exorbitant increase in the cost of printing, it has become necessary to restrict the length of the papers accepted for publication to 10-12 (double-space) typed pages (of A4 size) including tables and appendices (with 1.5 inch margins on all sides). Papers exceeding this limit may be returned to the authors without processing. The authors may however send along with their papers computer diskettes as this may be of help in processing their papers speedily.

Three copies typed in double space on one side of A4 paper preferably in MS WORD 6.0 in the font size of 10 in Times New Roman are required.

To protect the anonymity of authors while referring the papers for expert opinion on their merits, authors are advised to avoid disclosing their identity in the text and to attach a separate page showing the name(s) and affiliation(s) of the author(s) along with any footnotes containing bibliographical information or acknowledgements.

While sending papers the authors should state that the material has not been published elsewhere or is not being published or being considered for publication elsewhere. It is editorial policy not to consider for publication many papers from the same author during a year.

The paper with a theoretical content should have as far as possible a numerical example illustrating the ideas and its potential application in agriculture and allied research.

The title of the paper should be in bold title case followed by name or names of authors along with the name of the institution of work of the authors.

A short summary of about 100 words must be included at the beginning of the manuscript, together with 5 or 6 keywords or phrases to describe the content of the paper.  Short title of the paper should be given.

The text should be arranged under sequentially numbered headings; spellings should follow Chambers Dictionary.

The author to whom proofs are to be sent and the full postal address, and an E-Mail address should be clearly stated on the title page.  No new material may be introduced at the proof-reading stage.

Illustrations should be in a form suitable for direct reproduction, drawn in Indian ink on drawing paper or generated from a high-resolution printer, and all illustrations should preferably require the same degree of reduction.  Captions should be typed in a separate list.  All illustrations (and tables) must be clearly numbered and referred to in the text.

At the end of the paper, the references should be listed in alphabetical order of surnames and should be standardised as follows:

In the text, the authors' surnames only should be given followed by the year of publication quoted in parenthesis in the references.

The details in the reference section should be in alphabetical order of (i) surnames followed by initials, (ii) year of publication in parenthesis, (iii) Title of the paper/book, (iv) shortened name of journal where the paper is published in italics following standard convention of shortening.  In case of books, the name of book should be in title case and italics followed by name of publisher of the book, (v) volume number of the journal in bold letters, and (vi) indication of pages.

Examples for a paper in a journal, a book and a paper presented at a conference are:

Srivastava, V.K.and Dwivedi, T.D. (1985).  Improving estimation of population mean: Some Remarks.*J.ind.Soc.Agril.Statist.,*37,154-157.

Anderson, T.W.(19580. An introduction to Multivariate Analysis.  John Wiley and Sons, New York.

Thompson, R. (1977).  Estimation of quantitative genetic parameters. *Proc.Int.Conf.Quant.Genet.*, 639-658, Iowa State Univ.Press, Ames, USA.

Mathematical expressions should be typewritten.  Each equation should be typed on a separate line and numbered consecutively and punctuated in the usual way Matrix, vector (bold) and script quantities should be identified in the margin where they first occur.  The development of mathematical expressions, if necessary, should be presented in appendices with only the relevant equations stated in the main text.  The order of brackets in nested expressions is [{( )}].

The observance of the above guidelines will help in quick processing of the paper submitted for publication.

The paper will acknowledged and refereed.  Referee's report will be sent but the manuscript of the papers will not be sent back.

For further information or advice, please contact the Secretary at the Society's address:

Phone : 091-011-25842861

Fax : 091-011-25841564,25842861

E-Mail : isas@iasri.res.in

# CHAPTER - 92

# Journal of the Indian Society of Costal Agricultural Research

**Information to Authors**

It is requested that in future all articles submitted for publication in 'Journal of the Indian Society of Coastal Agricultural Research' or for presentation in its Seminars must be typed in MS-Word 97 or later versions with Times New Roman Script in 12 pt. Font size, in double space on A-4 size paper, with at least 1.5 inch margin on the left side and one inch margin on all other sides.

Two hard copies of the manuscript along with a CD or 1.44 MB (3.5") floppy disk must be sent for consideration. Further the same may also be sent by e-mail as an attachment (Word document) to iscar@rediffmail.com, cssri@wb.nic.in, arijit_sen@vsnl.net.

The author(s) are also requested to sent their e-mail address if any, for quick correspondence.

# CHAPTER - 93

# Journal of the Indian Society of Soil Science

## Hints for the preparation and submission of papers

Full-length articles, short communications, and book reviews, and review articles are published in the Journal. *From June 2004 onwards, rupees one hundred per printed page have been introduced for the articles (exempting being invited lectures, reports, obituary) published in the Journal.* Review articles and book reviews are only by invitation. Full-length articles and short communications should report results of original investigations in Soil Science of kindred branches of science. The author or at least one of the authors (in case of joint authorship) should be member of the Indian Society of Soil Science and not in arrears of subscriptions.

Once it is decided to contribute an article or a short communication for publication in the Journal, it is advisable for the contributors to consult a current issue of the Journal of the Indian Society of Soil Science for style of presentation and other minute details. The intending contributors are also requested to follow rigorously the 'Instructions for the Preparation of Manuscript for the Journal of the Indian Society of Soil Science' as printed in the Journal (Journal of the Indian Society of Soil Science, Volume 50, No. 4, pp 513-518, 2002).

No abbreviations of periodicals are acceptable. The list of references should be typed as follows:

Black, C.A. (1968) *Soil-Plant Relationships,* Second Edition, John Wiley and Sons, New York. pp New Delhi. Pp 40-45.

Kanwar, J.S. and Raychaudhuri, S.P. (1971) *Review of Soil Research in India,* Indian Society of Soil Science, New Delhi.pp 30-36

Mukherjee, J.N. (1953) The need for delineating the basic soil and climatic regions of importance to the plant industry. *Journal of the Indian Society of Soil Science* 1, 1-6.

Khan, S.K. Mohanty, S.K. and Chalam, A.B. (1986) Integrated management of organic manure and fertilizer nitrogen for rice. *Journal of the Indian Society of Soil Science* 34, 505-509.

Thind, H.S., Bhajan Singh and Gill, M.S. (1984) Relative efficiency of nitrogenous fertilizers for rice-wheat rotation. In: Nitrogen in Soils, Crops and Fertilizers. *Bulletin of the Indian Society of Soil Science* 13, 181-184.

Bijay-Singh and Yadvinder-Singh (1997) Green manuring and biological N fixation: North Indian perspective. In: *Plant Nutrient Needs, Supply, Efficiency and Policy Issues: 2000-2025* (J.S. Kanwar and J.C. Katyal, Eds.) pp 29-44. National Academy of Agricultural Sciences, New Delhi, India.

Figures include diagrams and photographs. Laser print outs of line diagrams are acceptable while dotmatrix  print outs will be rejected. Alternatively, each illustration can be drawn on white art card or tracing cloth/paper, using proper stencil. The lines should be bold and of uniform thickness. The numbers and letterings must be stenciled; free-hand drawing will not be accepted. Size of illustrations as well as numbers, and letterings should be sufficiently large to stand suitable reduction in size. Overall size of the illustrations should be such that on reduction, the size will be the width of single or double column of the printed page of the Journal. Legends, if any, should be included within the illustration. Each illustration should have a number followed by a caption typed/typeset well below the illustration. Title of the article and name(s) of the author(s) should be written sufficiently below the caption. The photographs (black and white) should have a glossy finish with sharp contrast between the light and the dark areas. *Colour photographs/figures are not normally accepted.* **One set of the original figures must be submitted along with the manuscript, while the second set can be photocopy.**

Careful observation of hints as indicated above will facilitate quick handling of the manuscript and printing of the paper.

**One soft copy (preferably through email) + two hard copies or typed manuscript (one top copy and a carbon copy), complete in all respects should be sent to :**

*Secretary*

*Indian Society of Soil Science*

*Division of Soil Science and Agricultural Chemistry*

*Indian Agricultural Research Institute*

*New Delhi – 110 012 (India)*

*Phone:* 0091-11-25841991; *Fax:* 0091-11-25841529;

*Email:* isss@vsnl.com; *Website:* www.isss-india.org

# CHAPTER - 94

# Journal of the South Carolina Academy of Science

## Guidelines and Information for Authors

- The SCAS Journal is an electronic journal. All submissions for journal consideration must be in electronic format. Email submissions (in Microsoft Word format) as an attachment to the editor-in-chief. ->

- *Please note :* files larger than 3.5 MB may be automatically rejected by the USC email system. If your file is large (>3 MB) and/or you have difficulty emailing me, please call 864-503-5685 for an alternate email address to use for manuscript submission

- Do not send submissions directly to section editors - they will not be reviewed.

- Research articles, review papers, and notes (short articles and reports) are welcome.

- There are no geographic restrictions for articles submitted. Manuscripts are not restricted to research conducted in South Carolina.

- There are no page charges. One author must be a member of the South Carolina Academy of Sciences. Please note that access to the full text of articles is restricted to members of the South Carolina Academy of Science.

- The estimated time for a first review of submitted manuscripts is 8-12 weeks. If you have not received notification regarding the status of your manuscript by 8 weeks post-submission, please contact the editor-in-chief for an update. Lengthy manuscripts, submissions near holidays, and submissions during the summer may require additional review time.

- At least one copy of the manuscript, with line numbers added for reference, will be returned to the author. Reviewer's comments will accompany the manuscript. Manuscripts and comments may be returned by conventional mail or email (discretion of the editor). It is recommended that authors read the Guidelines for Section Editors and Reviewers to better understand the Journal's review process and requirements.

- All submissions, with the exception of meeting abstracts, extended abstracts, and announcements must go through the peer-review process.

- THE LEGAL STUFF - Manuscripts published by the Journal of the South Carolina Academy of Science (also referred to as the Journal, or SCAS Journal) are copyrighted by the South Carolina Academy of Sciences. Authors have permission to replicate their own articles for professional purposes (reprints) and contents from the Journal may be distributed in accordance with "fair use" copyright laws. Indexing and abstract services have standing permission to include abstracts or summaries of articles published in the Journal. Distribution or publication of material from the Journal in excess of "fair use", particularly, but not limited to, the re-distribution or re-posting of contents in an electronic format, is prohibited without written permission from either the SCAS President, Journal editor-in-chief, or the SCAS Journal Advisory Board.

**Before emailing your submission, please insure that your submission meets the following requirements:**

- At least one author must be a member of the Academy. If the author is not a member, please join prior to submission of the manuscript to insure review of the manuscript.

- The manuscript must address one of the subject areas of interest to the Journal: Agricultural Sciences, Anthropology/Archaeology, Biochemistry, Bioengineering, Biology, Chemistry, Computer Science, Economics, Engineering/Engineering Technology, General Science, Geography, Geology/Paleontology, History, Mathematics/Statistics, Medical Science, Nursing Science, Philosophy, Physics/Astronomy, Psychology, Science Education, Social Work, or Sociology.-> Please specify in your email the category most appropriate to your manuscript and any sub-specialty your manuscript addresses (i.e. Biology, herpetology).

- Check your manuscript prior to submission with current anti-virus software.

- Submissions are to be in Microsoft Word 2003 or earlier Microsoft Word format, 12 point font size. The typeface required is Times or Times New Roman.

- Language for all submissions is English.

- If the manuscript file size is larger than 3MB, email : DKFERRIS@USCUOSTATE.EDU

- PRIOR to submission for an alternate email address that will accept large file sizes.

- The manuscript must have a title page with:
    - Manuscript title

Author(s) - please underline the names of authors who are members of the Academy Institution(s) and address of all authors E-mail address of author for correspondence (VERY IMPORTANT) Keywords for subject index

- In addition to the title page, the manuscript must have the following sections with headings (as applicable):

  - **Abstract :** Abstract must be the only text on the second page. The abstract should be self explanatory and without citations. (This section is required)

    **Introduction :** Introduction should include a description of the background and aims of the work and what has been accomplished to date.

    **Materials and Methods :** Include full descriptions of all experimental procedures.

    **Results :** Results should be clearly stated and supported by figures, tables, or graphical representations of the findings.

    **Discussion :** The discussion should address the importance of the major findings of the work, expanding upon the results. A discussion of the implications of the research to the general area of investigation is particularly useful.

    **Acknowledgements :** Should be brief (if any).

## References

References must follow the Council of Biology Editors (CBE) name-year style. See the latest copy of the Journal for format. Within the article, if a citation has more than two authors, use the first author followed by "et al." All authors must be listed and Journal names should *not* be abbreviated in the "References" section. Complete coverage of this reference style is found in Scientific Style and Format: The CBE Manual for Authors, Editors and Publishers, 6th edition. The internet links below also provide examples of the CBE style in use:

- http://www.monroecc.edu/depts/library/cbe.htm (particularly helpful!)
- http://www.lib.ohio-state.edu/guides/cbegd.html
- http://acadprojwww.wlu.edu/vol4/BlackmerH/public_html/xliberty/biology/style.html
- http://lib.colstate.edu/tutorials/style/pcbe.shtml
- http://www.bedfordstmartins.com/online/cite8.html

**Figures and Tables :** Figures and tables should be included (embedded) in the text at a point soon after the reference to the figure or table is made. Figures and tables should be numbered sequentially as they appear in the article. When embedding figures, photos, or scans, please verify that all graphics are "stand-alone" and do not require any external reference or external program to display properly.

**If you have any additional questions, please contact David K. Ferris at DKFERRIS@USCUOSTATE.EDU**

<table>
<tr><td>

The web site

http://www.uscupstate.edu/~dkferris/ scas.html is generously hosted by the University of South Carolina Upstate.

Information on this page is presented by SCAS. The contents on this page and anywhere within this site have not been reviewed or approved by the University of South Carolina Upstate

</td><td>

Dr. David Ferris
USC Upstate
Division of Natural Sciences & Engineering

DKFERRIS@USCUOSTATE.EDU

Last page update : 10/11/04

</td></tr>
</table>

# CHAPTER - 95

# Journal of Tropical Agriculture

## Hints for the Preparation and Submission of Papers

Full-length original research articles and research notes are published in the journal. All the authors of the article (except invited articles) should be the subscribers of the journal at least during the year in which their article is published. The authors are advised to consult a current issue of the journal for style and format followed in the journal. The papers are accepted for publication on the understanding that the matter contained in the article has neither been published nor is being considered for publication elsewhere. The concurrence of all the authors and the institution that sponsored the research programme is necessary and a certificate to this effect has to be furnished (please visit the website www.kau.edu/journalmain.htm).

Two copies of the article entered in MS Word should be sent to the Editor (paper size:A4, line space: double, ample margin all sides). Full-length articles (not exceeding six printed pages) will have the title , names of authors, name and address of the sponsoring institution, abstract, key words, and sections entitled introduction, materials and methods, results and discussion, acknowledgement and references. Current address of the author can be indicated as a footnote. In the case of research notes (not exceeding three printed pages), there will be no section headings. But paragraphs corresponding to introduction, materials and methods, results and discussion and acknowledgement will be presented in a condensed form.

Title of the paper must be brief and should contain words useful for indexing. The abstract not exceeding 200 words should indicate the objective of the study, experimental details and significant results obtained. Four to six key words useful for indexing and information retrieval any be given in alphabetical order. All measurements should be given in SI units. Scientific names should be given along with the common names and authority of the scientific name should be quoted at the first instance. Abbreviations for the title of the journal should be in accordance with the *World List of Scientific Periodicals,* Butterworths, London. Tables may be entered in separate files using the table format of MS Word with required number of rows and columns (without incorporating into the text). Use MS Excel for the preparation of charts. The title of the chart need not be given while preparing the Excel charts; instead, it may be typed in a separate sheet of paper. Illustrations (drawings), if any, may be made in Indian ink. Magnification should be indicated in the drawing itself. Figure title and labels need not be given in the original drawing; but, may be typed/written in a xerox copy of the drawing and attached. Free-hand drawing and lettering are not acceptable. Two copies of black and white photographs (glossy print), if any, may be supplied. Colour photographs are not accepted. Magnification should be indicated in the photograph itself.

Normally, the article will be returned to the author for incorporating the suggestions made by the referee. One copy of the print out and the floppy disk of the revised article should be sent to the Editor along with the referee-corrected manuscript. The authors are therefore advised to preserve the floppy containing the text of the article (MS Word), tables (MS Word) and charts (MS Excel) given for publication. The instructions for the preparation of the revised article will be supplied at the time of revision of the article. The intending author should pay the subscriptions of all the authors of the article while furnishing the revised article.

Changes in the current address of the intending author should be promptly intimated. All correspondence should be addressed to the Managing Editor, Journal of Tropical Agriculture. College of Forestry, KAU-PO, Thrissur 680 656, Kerala, India (Phone 0487-2370050, Fax 0487-2371040, website ww.kau.edu and e-mail: bmkumar53@yahoo.co.uk).

# CHAPTER - 96

# Karnataka Journal of Agricultural Science

## Instructions to Authors

### General

Research papers for publication must contribute substantially to the advancement of knowledge in any branch of Agricultural Sciences including Soil, Plant, Animal, Engineering and Home Sciences **and should be routed through proper channel. Review articles on any aspect and papers presented in conferences, seminars, symposia, workshops etc. will not be accepted for publication**. The Editorial Committee has right to accept or reject a paper and the committee does not shoulder any responsibility for the opinion(s) expressed in the paper by the author(s). Once a paper is accepted for publication, it should not be published elsewhere either in the same or abridged form or in any other language, without the written permission of the Editor. They should also specify that they are not sending the paper to any other Journal simultaneously. **The manuscripts are published on cost basis (For every printed page Rs. 100 and \$5 will be charged for Indian and Foreign author, respectively). Therefore, the authors are requested to co-operate in bearing the printing cost if the manuscript is accepted for publication. All correspondence should be addressed to The Editor, Karnataka Journal of Agricultural Science, University of Agricultural Sciences, Dharwad – 580 005.**

## Manuscript and its arrangement

Manuscript, **not more than the and 3-4 pages inclusive of tables, diagrams and references for Research Paper and Research Note, respectively** should be typed in double line space on A4 size bond paper on one side of the page only, with at least 5 cm margin on the left and 2 cm margin on the right. The original and the first copy of the typed paper should be sent. The language should be simple and care should be taken to check up the spellings, punctuations etc. All tables should be serially numbered, should not be too lengthy and should have a heading stating concisely the contents. Each table should be typed on separate sheet and not with the running matter.

The contents of Research Paper should be organized as **Title, Abstract, Introduction, material and Methods, Results and Discussion and References**. Acknowledgement, if any, may be included in the last paragraph of the text before the **References** part. The contents for Research Note should be organized without mentioning the subheadings as above but, with the Title and References.

*Cover page :* The cover page of the manuscript should carry the Title of the research paper, Name(s) of the Author(s) and their affiliation. **The authorship and its order furnished at the time of registration of the manuscript is final.**

If the paper form a part of M.Sc./Ph.D. thesis, a foot-note should indicate the same. Further, a Certificate by the Head of the Department to the effect that the authors were associated with the research work in the capacity of Chairman/ Member of the Advisory Committee must accompany the paper. In such cases, the first name should necessarily be of the student concerned and **should be routed through the Director of Instruction (PGS).**

*Note :* No manuscript should carry Author(s) Name(s) and addresses any where on the body of the manuscript other than on cover page.

*Title :* This should be informative but concise. While typing the title of the paper / Note only first letter of each word must be in capitals. The titles must be typed just before the commencement of abstract.

*Abstract  :* The abstract must be brief and informative and should not be more than 300 words.

*Introduction :* The introduction should be brief and state the objective of the experiment. The review of literature should be pertinent to the problem.

*Material and  Methods :* This should be precise and whenever the methods of other authors and followed, it would suffice if reference to their paper is made instead of repeating the procedure, should include experimental design, treatments and techniques employed.

*Results and Discussion :* This should govern the presentation and interpretation of experimental data only and each of the experiments should be properly titled. Common names of plant species, micro-organisms, insects etc. should be supported with authentic, latest Latin names, which should be underlined in the typescript. When such names are first mentioned, the full generic name and the species name with the authority abbreviated and the authority be deleted.

All headings must be typed in lower case from the left hand margin. The sub-headings must be typed in lower order capitals and must start from the left hand margin and underlined. The para under a sub-heading must start from a line below the sub-heading

*Tables :* Every tables should be on a separate sheet and be clear with proper reference in the text. The units of the data should be defined. Appropriate statistical tests should be applied to the data presented. Footnote should be seldom used.  Lengthy tables should be avoided.

*Illustrations :* The illustrations and text-figures should be clear and capable of reproduction in print.  They should be drawn neatly on tracing paper in Indian ink with serial numbers in clear, uniformly large letters.  Photo prints should be on glazed paper with proper contrast and at least of half plate size pasted on thick mount.  If there are many photo prints, they are to be grouped compactly. Whenever photo prints are included, the magnification should be indicated. **Photostat copies of any illustrations are not accepted and in all cases, originals should be sent.** Legends to figures and photos should be typed on the foot of the figure/photo.

Abbreviations should be used sparingly if advantageous to the reader. All new or unusual abbreviations should be defined when they are used for the first time in the paper. Ordinarily, the sentences should not begin with abbreviations or numbers.

*Reference :* This should be typed in double line space along with the body of the paper but starting on a fresh page. These should be listed in the alphabetical and chronological order. In the text, the references should be cited as Hayman(1970) or (Hayman, 1970), Sen and Bhowal (1961), Tosh *et al.*(1978) or (Tosh *et al.*, 1978), when there are more than two authors.

**While listing the References, the following examples should be followed.**

**While citing an article from :**

## Journal
ABBOT, W.S., 1925, A method of computing the effectiveness of an insecticide. *Journal of Economic Entomology*, 18:265-267.

**Papers presented in Symposium / Seminar / Workshop**
BHASKARAN, P., NARAYANASWAMY, BALASUBRAMANIAM, M. AND RAGHUNATHAN, V., 1976, Field evaluation of insecticide – fungicide spray combinations against rice pests. Paper presented at the *All India Symposium of modern Concepts in plant protection,* Udaipur, 26-28 March, 1976.

**Proceedings of Seminar / Symposia / Workshop (published)**
BOYCE, H.R., 1961, Insecticidal activity of manels formulations aginst greenhouse whitefly. *Proceedings* of Entomological Society of Ontario, 92: 196-200.

**Edited Book**

BURGES, H.D 1981, Progress  in the microbial control of pests. In *Microbial Control of Pests and Plant Diseases,* 1970-80. Ed Burges, H.D.,Academic Press, London, pp.1-5.

**Bulletin**

GRAY, P., 1914, The compatibility of insecticides and fungicides. *Monthly bulletin of California,* July 1914.

# CHAPTER - 97

# Kisan World

**For the Kind Attention of our Contributors**

With globalization of trade and commerce, World Trade Organisation (WTO) becomes operational from 1-1-2005. The economy of the country, particularly the agricultural economy has to face a number of challenges. In countries such as USA, Australia, Brazil, etc. each farmer on an average holds over 1000 hectares cultivable land against an average holding of 1.41 hectares in our country. With least overheads in those countries the cost of production will be low. The products from such countries will have unrestricted  access to our market. Our farmers, with all the constraints, have to empower themselves to meet such challenges. Countries such as Israel, Japan, Singapore etc. have prospered even under severe constraints. We have to adopt the winning strategies of such countries to improve the earnings and prosperity of the nation. In such a situation Kisan World strives to improve the lot, in every sphere of economic activity.

Articles on scientific practical approaches to improve the yield, production quality and productivity in sectors viz. agriculture, horticulture, animal husbandry, agro industries, small scale industries, food processing of prosperity of the people and value addition to the economy, are invited.

The articles should be clearly typewritten in double space on one side of white paper. Photographs of authors/crops or situations are also welcome with articles. We would appreciate if the article does not exceed two pages in the printed form.

Success stories with photographs in all such areas to inspire and motivate the readers/subscribers to adopt such novel approaches to improve their knowledge and prosperity are welcome.

Articles/Success stories may also be sent through E-Mail Chennai @ sakthi sugars.com

# CHAPTER - 98

# Legume Research

## Instructions to Authors

1. Three copies of the article one copy computer typed in double space on one side of the paper giving authors name and address where work was conducted and two copies typed on both the sides of the paper without giving names and address (to be sent to referee) should be submitted to the Managing Editor, Agricultural Research Communication Centre, 1130, Sadar Bazar, Near Post office, KARNAL-132 001, Haryana (INDIA).

2. The latest issue of the journal may be consulted for arrangement of the text and style of the references.

3. The registration number of article and some more information will be communicated to author on receipt of the article.

4. The journal should be subscribed in your Institute Library.

5. One of the authors has to give a certificate on a plain paper that the article or its data is neither printed nor sent/shall be sent to any other journal and it is seen by all the authors.

6. The article will be evaluated by the referee/experts and the editor. The author has to revise the article in the light of referee's comments/ suggestions.

7. Two copies of the revised article typed on one side of each paper & prepared as per the style and format of the journal and fed in the CD should be submitted for final approval.

8. On receipt of the revised article and CD an approval letter will be issued to author for remitting a nominal part of printing charges @ Rs. 50/- or US $ 5/- per page. On receipt of the amount a final acceptance letter will be issued to the author and the article will be sent to press for printing.

9. A copy of the journal in which the article is published will be sent to author as soon as the journal is ready for distribution/circulation.

# CHAPTER - 99

# Molecular Breeding New Strategis in Plant Improvement

## Online Manuscript Submission

Springer now offers authors, editors and reviewers of *Molecular Breeding* a fully web-enabled online manuscript submission and review system. To keep the review time as short as possible (no postal delays!), we request authors to submit manuscripts online to the journal's editorial office. Our online manuscript submission and review system offers authors the option to track the progress of the review process of manuscripts in real time. Manuscripts should be submitted to: http://molb.edmgr.com

The online manuscript submission and review system for *Molecular Breeding* offers easy and straightforward log-in and submission procedures. This system supports various file formats, e.g. Word, WordPerfect, RTF, TXT and LaTex for text and TIFF, GID, JPEG, EPS, PPT and Postscript for figures.

Detailed Author Instructions are available on the journal website (http://www.springeronline.com/jounral/11032). No page charges or costs for handling manuscripts submitted for publication are levied on the authors. Fifty offprints will be supplied free of charge: additional offprints can be ordered.

## Aims and Scope

*Molecular Breeding* is an international journal publishing papers on applications of plant molecular biology, i.e. research most likely leading to practical applications. The practical applications might relate to the Developing as well as the industrialized World and have demonstrable benefits for the seed industry, farmers, processing industry, the environment and the consumer.

All papers published should contribute to the understanding and progress of modern plant breeding, encompassing the scientific disciplines of molecular biology, biochemistry, genetics, physiology, pathology, plant breeding, and ecology among others.

*Molecular Breeding* welcomes the following categories of papers: full papers, short communications, papers describing novel methods and review papers.

All submission will be subject to peer review ensuring the highest possible scientific quality standards.

## Molecular Breeding core areas

*Molecular Breeding* will consider manuscripts describing contemporary methods of molecular genetics and genomic analysis, structural and functional genomics in crops, proteomics and metabolic profiling, abiotic stress and field evaluation of transgenic crops containing particular traits. Manuscripts on market assisted breeding are also of major interest, in particular novel approaches and new results of marker assisted breeding, QTL cloning, integration of conventional and marker assisted breeding, etc.

Manuscripts submitted to *Molecular Breeding* will focus on applications but we will also accept fundamental science papers as long as they are of direct relevance to crop plants and not model systems. *Molecular Breeding* also welcomes relevant articles addressing intellectual property issues, regulation and public attitudes to plant biotechnology and also significant technology advances in applied plant molecular biology and (trans) gene expression technology.

# CHAPTER - 100

# Paddy and Water Environment

## Manuscript Preparation

### General Remarks

To help you prepare your manuscript, Springer offers a template that can be used with Winword 7 (Windows 95), Winword 6 and Word for Macintosh. For details see next chapter (Access to the online template), otherwise authors should submit their manuscript on double line space including line numbers.

Papers up to 10 printed pages are published free of charge. There will be a charge of • 50 or US $ 45, plus 16% VAT, for each printed page exceeding this limit. 3 manuscript pages equal approximately one printed page when you use the template. The space required for the figures and tables should be calculated on the basis of their final printed size.

All manuscripts are subject to copy editing.

A manuscript submitted for publication must be previously unpublished research works written in English.

Authors who are not native speakers should have their manuscripts proofread by a native English-speaker prior to submission.

## Types of Manuscripts

### Articles

Articles cover full reports of research work that must be written following the guidelines described below with the minimum length required for a precise description and clear interpretation of theoretical or experimental work. Over 12 printed pages (36 template pages) will not be accepted.

## Technical Reports

Technical reports must present original, practical information, preliminary or partial results of research, engineering engineering applications in the field covered by Paddy and Water Environment. Over 12 printed pages (36 template pages) will not be accepted.

## Reviews

Reviews dealing with all aspects of paddy- farming related scientific and technological interest in agricultural engineering will be accepted, and authoritative and critical reviews of the current state of knowledge are preferred. There is no prescribed layout for reviews, but the tables and manner of citations should conform to the guidelines for "Articles". Over 12 printed pages will not be accepted.

## Short communications

Short Communications are concise but complete explanation of limited investigation on the critical issues in the field of Paddy and Water Environment. Over 4 printed pages will not be accepted.

## Title page

- The name(s) of the author(s)
- A concise and informative title
- The affiliation(s) and address(es) of the author(s)
- The e-mail address, telephone and fax numbers of the communicating author

## Abstract

Each paper must be preceded by an abstract presenting the most important results and conclusions, within 150-250 words.

## Keywords

Three to seven keywords taken from the text (excluding the title) should be supplied next to the Abstract for indexing purposes.

## References

The list of References should only include works that are cited in the text and that have been published or accepted for publication. Personal communications should only be mentioned in the text.

Citations in the text should be identified by using the author-date system such as (Japan Public Health Association, 1995), and the list of references at the end of the paper should be both alphabetized under the first author's surname. References by the same author or team of authors should be listed in chronological order. If available, the Digital Object Identifier (DOI) of the cited literature should be added at the bend of the reference in  question

## Examples are

### (a)  Journal articles

Breve MA, Skagga RW, Parsons JE, Gillams JW (1997) A nitrogen modelfor artificially drained soils. Transactions of the ASAE 40, pp. 1067–1075.

Molden D, Ishikawa M, Khan NM, Samad MA (2000) Conservation potential for  Egypt's Nile waters. Agricultural Water Management 46(2), pp. 137–148, DOI 10.10007/s002148900025.

**(b)  International Congress or Proceedings (symposia, workshop, Conference):**

Levine B, Sack G, Cirelli G, Jeffrey P, Brissaud F (2000) Role of water reuse in enhancement of integrated water management in Europe and Mediterranean countries. In Proc. 1st World Water Congress of the International Water Association, Vol. 8. Paris, France, pp. 33–40

Nedler K, Khan NM, Tanaka Y, Sato Y (2001) Environmental land degradation analysis using satellite data and GIS techniques. 22nd Asian Conference on Remote Sensing (ACRS), Singapore, July 12–18, pp. 133–138

**(c)  Books and/or multi-author publications**

Brock TD (1988) Biology of Microorganisms. 5th edn. Prentice Hall, Englewood Cliffs, USA, pp. 42–59

Khan NM, Yaoka T, Nedler K (2002) Crop Response under Stressed Condition. In Tanji KK, Nedler K (ed) Salinity management in arid regions, 4th edn. Springer, Berlin Heidelberg New York, pp. 167–180

United States Environmental Protection Agency (1983) Methods for Chemical Analysis of Water and Wastes. EPA-600, Cincinnati, USA, pp. 35–52

## Illustrations and Tables

All figures (photographs, graphs or diagrams) and tables should be cited in the text, and each numbered consecutively throughout. Figure parts should be identified by lower-case roman letters. The placement of figures and tables should be indicated in the left margin. For submission of figures in electronic form see below

## Line drawings

Please submit goodquality prints. The inscriptions should be clearly legible.

## Half-tone illustrations (black and white and color)

Please submit well-contrasted photographic prints with the topside indicated on the back. Magnification should be indicated by scale bars.

## Size of figures

The figures should either match the width of the column (8.6 cm) or be 17.8 cm or 23.6 cm wide. The maximum length is 23.6 cm.

## Figure legends

Figure legends must be brief, self-sufficient explanations of the illustrations. The legends should be placed at the end of the text indicating the figure number.

## Tables

Tables should have a title and a legend explaining any abbreviation used in that table. Footnotes to tables should be indicated by superscript lower-case letters (or asterisks for significance values and other statistical data).

## Color illustrations

For color illustrations the authors will be expected to make a contribution (• 950, US $ 1150, plus VAT) towards the extra costs, irrespective of the number of color figures.

## Access to the online template

Preparing your manuscript

### Text

The template is available :

    via ftp :

        Address: ftp.springer.de

        User ID: ftp

        Password: your own e-mail address

        Directory: /pub/Word/journals

        File names: sv-journ.zip or sv-journ.doc and sv-journ.dot

    via browser:

        ftp://ftp.springer.de/pub/Word/journals

        File names: sv-journ.zip or sv-journ.doc and sv-journ.dot

        The zip file should be sent unencoded.

## Layout guidelines

- Use a normal, plain font (e.g., 12-point Times Roman) for text
  Other style options :

  – for textual emphasis use italic types.

  – for special purposes, such as for mathematical vectors, use boldface type.
- Use the automatic page numbering function to number the pages.
- Do not use field functions.
- For indents use tab stops or other commands, not the space bar.
- Use the table functions of your word processing program, not spreadsheets, to make tables.
- Use the equation editor of your word processing program or MathType for equations.
- Place any figure legends or tables at the end of the article.
- Submit all figures as separate files and do not place them within the text.

## Data formats
Save your file in RTF (Rich Text Format) or Microsoft Word compatible formats.

## Illustrations
The file name (one file for each figure) should include the figure number. Figure legends should be included in the text and not in the figure file.

## Scan resolution
Scanned line drawings should be digitized with a minimum resolution of 800 dpi relative to the final figure size. For digital halftones, 300 dpi is usually sufficient.

## Vector graphics
Fonts used in the vector graphics must be included. Please do not draw with hairlines. The minimum line width is 0.2 mm (i.e., 0.567 pt) relative to the final size.

## General information on data delivery
Please send a zip file (text and illustrations in separate files) to

> The Chief Managing Editor
>
> Dr. Soon-Jin HWANG
>
> either on email
>
> Dr. Soon-Jin HWANG
>
> either on email
>
> (sjhwang@konkuk.ac.kr and pawees@ksae.re.kr)

or

on any of the following media:

- On a diskette [you may use .tar, .zip, .gzip (.gz), .sit, and compress (.Z)]
- On a ZIP cartridge
- On a CD-ROM

Please always supply the following information with your data:

- journal title
- operating system
- word processing program
- drawing program
- image processing program
- compression program

The file name should be memorable (e.g. author's name), have no more than 8 characters, and include no accents or special symbols. Use only the extensions that the program assigns automatically.

## Proofreading

Authors should make their proof corrections on a printout of the pdf file supplied, checking that the text is complete and that all figures and tables are included. After online publication, further changes can only be made in the form of an Erratum, which will be hyperlinked to the article.

The author is entitled to formal corrections only. Substantial changes in content, e.g. new results, corrected values, title and authorship are not allowed without the approval of the responsible editor.

In such a case please contact

the Editor-in-Chief

(ynakano@ggr.kyushu-u.ac.jp

and cc: sjhwang@konkuk.ac.kr)

before returning the proofs to the publisher.

## Offprints

50 offprints of each contribution are supplied free of charge. If you wish to order additional offprints you must return the order form with the corrected page proofs.

# CHAPTER - 101

# Pest Management and Economic Zoology

*Pest Management and Economic Zoology,*  a half yearly journal, publishes original research finding and critical research reviews related to applied ecology, behaviour, physiology, social engineering and management of economically important pests, beneficial animals and wildlife. Each author who is  not a member of the Society will be charged @ Rs. 100/- per article. Manuscript (MS), submitted in duplicate, should exclusively for the journal, MS typed on good quality paper in double space throughout should not normally exceed 10 typed pages (3000 words) including table and figures. TITLE should be precise, specific and informative followed by name of the author(s), institution where the study was carried out and PIN Code. Change of address should be given in a foot note. A RUNNING TITLE should be provided. MS should be divided into Abstract, Introduction, Materials & Methods, Results and Discussion, Acknowledgments, if any and References. ABSTRACT should be concise and should convey the important findings. INTRODUCTION should be brief and pertinent to the problem, relevant and recent REVIEW OF LITERATURE should be incorporated. MATERIALS AND METHODS should  include details of the materials used, place and period of study, techniques followed and statistical methods used. RESULTS AND DISCUSSION  should be written together. Results should contain pertinent data in analysed form in tables typed separately along with title. Weights and measures should be written in SI (metric) units. ILLUSTRATION, if any, should be of good quality and drawn on glossy paper in Indian ink. Same data should not be presented in tables and figures simultaneously. Discussion should be restricted to merits and limitations of research work and compared with the relevant reports by other workers. SHORT COMMUNICATION NOT exceeding 4-5 typed pages, should include an abstract followed by text in running paragraph, acknowledgements, if any and references. ACKNOWLEDGEMENT should be brief. Arrange REFERENCES in alphabetical order as given below :

Bhalla O P. Verma A K and Dhaliwal H S 1982 Foraging activity of insect pollinators  visiting stone fruits *Journal of  Entomological Research* 7 : 91-4.

Croft B A 1978. Potentials for research and implementation of integrated pest management in deciduous tree fruits pp 101-115, *Pest Control Strategies,* E H Smith and D Pimentel (eds). Academic Press, London 334p.

Southwood T R E 1976. *Ecological Methods with particular reference to study of insect populations* Mathuen and Co Ltd., London, 391p.

Wolfersberger M G 1990. Specificity and mode of action of *Bacillus thuringiensis* insecticidal crystal proteins toxic to lepidopteran larvae : recent insights from studies utilizing midgut brush border membrane vesicles. *Proceedings and Abstracts of the Fifth International Colloquium on Invertebreta Pathology and Microbial Control 20-24* August 1990, Adelaide, Australia, pp 278-282.

Manuscript for publication should be sent to Chief Editor, Pest Management and Economic Zoology, Department of Entomology and Apiculture, Dr Y S Parmar University of Horticulture and Forestry, Solan, H P 173 230, India. Submission of manuscript is understood to imply that article is unpublished and has not been submitted elsewhere for publication.

**Purchase of 25 reprints (at nominal cost) is mandatory**

# CHAPTER - 102

# Physiological and Molecular Plant Pathology

## Guide for Authors

### Submission checklist

It is hoped that this list will be useful during the final checking of an article prior to sending it to the journal's editor for review. Please consult this Guide for Authors for further details of any item.

Ensure that the following items are present:

- One author designated as corresponding author:
  - E-mail address
  - Full postal address
  - Telephone and fax numbers
- All necessary files have been uploaded
- Keywords
- All figure captions
- All tables (including title, description, footnotes)

### Further considerations

- Manuscript has been "spellchecked"
- References are in the correct format for this journal
- All references mentioned in the Reference list are cited in the text, and vice versa
- Permission has been obtained for use of copyrighted material from other sources (including the Web)
- Colour figures are clearly marked as being intended for colour reproduction or to be reproduced in black-and-white

**For any further information please contact the Author Support Department at authorsupport@elsevier.com**

## Aim and Scope

*Physiological and Molecular Plant Pathology* Publishes original research papers, reviews, and commentaries on all aspects of the molecular biology, biochemistry, physiology, cell biology, genetics, cytology, ultrastructure, and evolution of plant-microbe interactions involving pathogenic or mutualistic organisms. Work of a trivial nature, even though it may have been competently performed, will not be published. Papers on new techniques, or work based on studies in pure culture, must have a direct bearing on the host-microbe interaction to be acceptable.

## Submission of articles

### General

It is essential to give a fax number and e-mail address when submitting a manuscript. Articles must be written in good English.

Submission of an article implies that the work described has not been published previously (except in the form of an abstract or as part of a published lecture or academic thesis), that it is not under consideration for publication elsewhere, that its publication is approved by all authors and tacitly or explicitly by the responsible authorities where the work was carried out, and that, if accepted, it will not be published elsewhere in the same form, in English or in any other language, without the written consent of the Publisher.

Upon acceptance of an article, authors will be asked to transfer copyright (for more information on copyright see http://authors.elsevier.com  to ensure the widest possible dissemination of information. A form facilitating transfer of copyright will be sent to you when your paper is sent to production.

If excerpts from other copyrighted works are included, the author(s) must obtain written permission from the copyright a owners and credit the source(s) in the article. Elsevier has preprinted forms for use by authors in these case: contact Elsevier Global Rights Department, P.O. Box 800, Oxford, OX5 1DX, UK; phone: +44(0) 1865 843830, fax: +44 (0) 1865 85333, e-mail: permissions@elsevier.com. Requests may also be completed online via the Elsevier homepage (http://www.elsevier.com/locate/permissions).

**US National Institutes of Health (NIH) voluntary posting ("Public Access") policy** Elsevier facilitates author response to the NIH voluntary posting request (referred to as the NIH "Public Access Policy". See http://www.nih.gov/about/publicaccess/index.htm) by posting the peer-reviewed author's manuscript directly to PubMed Central on request from the author, 12 months after formal publication. Upon notification from Elsevier of acceptance, we will ask you to confirm via e-mail (by e-mailing us at NIHauthorrequest@elsevier.com) that your work has received NIH policy request, along with your NIH award number to facilitate processing. Upon such confirmation, Elsevier will submit to PubMed Central on your behalf a version of your manuscript that will include peer-review comments, for posting 12 months after formal publication. This will ensure that you will have responded fully to the NIH request policy. There will be no need for you to post your manuscript directly with PubMed Central and any such posting is prohibited.

**Authors' rights**

As an author, you retain rights for large number of author uses, including use by your employing institute or company. These rights are retained and permitted without the need to obtain specific permission from Elsevier. To see what rights the author retains, see the Copyright Information in http://authors.elsevier.com.

Should authors be requested by the editor to revise the text, the revised version should be submitted within 6 weeks. After this period, the article will be regarded as a new submission.

## Submissions Process

Submission to this journal proceeds totally on-line. Use the following guidelines to prepare your article. Via the "Author Gateway" page of this journal (http://www.authors.elsevier.com) you will be guided stepwise through the creation and uploading of the various files. The system automatically converts source file to a single Adobe Acrobat PDF version of the article, which is used in the peer-review process. Please note that even though manuscript source files are converted to PDF at submission for the review process, these source files are needed for further processing after acceptance. All correspondence, including notification of the Editor's decision and requests for revision, takes place by e-mail and via the Author's homepage, removing the need for a hard-copy paper trail.

The above represents a very brief outline of this form of submission. It can be advantageous to print this "Guide for Authors" section from the site for reference in the subsequent stages of article preparation.

If online submission is not possible, manuscripts may be submitted by sending the source files on disk together with a matching hard copy (both text and figures and tables) by registered mail to the editor. (Please note that this is not the preferred way of submission and could cause a delay in publication of the article.)

Professor Ray Hammerschmidt

*Physiological and Molecular Plant Pathology*

Elsevier Editorial Services Office

Bampfylde Street, Exeter,

Devon, EX1 2AH, UK

Telephone: +44 (0) 1392 251558

Fax: +44 (0) 1392 425370

E-mail: pmpp@elsevie.com

Please submit, with the manuscript, the names and addresses (including fax number and e-mail address) of 3 potential referees.

## Electronic format requirements

### General points

Word or WordPerfect is preferred. Always keep a backup copy of the electronic file for reference and safety. Save your files using the default extension of the program used.

## Word processor documents

It is important that the file be saved in the native format of the word processor used. The text should be in single-column format. Keep the layout of the text as simple as possible. Most  formatting codes will be removed and replaced on processing the article. In particular, do not use the word processor's options to justify text or to hyphenate words. However, do use bold face, italics, subscripts, superscripts etc. Do not embed 'graphically designed' equations or tables, but prepare these using the word processor's facility. When preparing tables, if you are using a table grid, use only one grid for each individual table and not a grid for each tow. If no grid is used, tables, not spaces, to align columns. The electronic text should be prepared in a way very Author Gateway's Quiguide: http://author.elsevier.com). Do not import the figures into the text file but, instead, indicate their approximate locations directly in the electronic text and on the manuscript. See also the section on Preparation of electronic illustrations.

## Preparation of Text

Please write your text in good English (American or British usage is accepted, but not a mixture of these). Italics are to be used for expressions of Latin origin, for example, *in vivo, et al. per se*. Use decimal points (not commas); use a space for thousands ( 10 000 and above).

*English language help service :* Upon request, Elsevier will direct authors to an agent who can check and improve the English of their paper (*before submission*). Please contact authorsupport@ Elsevier.com for further information.

Use double spacing and wide (3 cm) margins and it helps to have each line numbered in the left hand margin. (Avoid full justification, i.e. do not use a constant right-hand margin.) Ensure that each new paragraph is clearly indicated. Present tables and figure legends on separate pages at the end of the manuscript. If possible, consult a recent issue of the journal or refer to the sample issue online at http://sciencedirect.com/science/hjournal/08855765 to become familiar with layout and conventions.

The following information should be provided on the title page (in the order given).

*Title*. Titles should be concise and informative. They are often use in information-retrieval systems. Avoid abbreviations and formulae where possible.

*Author names and affiliations.* Where the family name may be ambiguous (e.g. a double name), please indicate this clearly. Present the authors' affiliation addresses (where the actual work was done) below the names. Indicate all affiliations with a lower-case superscript letter immediately after the author's name and in front of the appropriate address. Provide the full postal address of each affiliation, including the country name, and, if available, the e-mail address of each author. If an author has moved since the work described in the article was done, or was visiting at the time, a 'Present address' (or 'permanent address') may be indicated as a footnote to that author's name. The address at which the author actually did the work must be retained as the main, affiliation address. Superscript Arabic numerals are used for such footnotes.

*Corresponding author.* Clearly indicate who is to handle correspondence at all stages of pre- and post-publication. **Ensure that telephone and fax numbers (with country and area code) are provided in addition to the e-mail address and the complete postal address.**

*Abstract.* A concise and factual abstract is required (maximum length 100 words). An abstract is often presented separate from the article, so it must be able to stand-alone. References should therefore be avoided, but if essential, they must be cited in full, without reference to the reference list. Non-standard or uncommon abbreviations should be avoided, but if essential they must be defined at their first mention in the abstract itself.

*Keywords.* Immediately after the abstract, provide a list of keywords, avoiding general and plural terms and multiple concepts (avoid, for example, 'and' , 'of'). The keywords should include the complete scientific names of all plants and micro-organisms used and should also cover the topic investigated and special techniques used. Only abbreviations firmly established in the field may be eligible. These keywords will be used for indexing purposes.

*Abbreviations.* Define abbreviations that are not standard in this field in a footnote to be placed on the first page of the article. Ensure consistency of abbreviations throughout the article.

The following abbreviations may be used without definition except in the title and abstract:

| | |
|---|---|
| absorbance (e.g. absorbance at 310) | $A(A_{310})$ |
| adenosine, mono-, di-, triphosphate | AMP, ADP, ATP |
| approximately | approx. (*not* c.or ca.) |
| becquerel | Bq (1 Ci = $3.7 \times 10^{10}$ Bq) |
| bovine serum albumin | BSA |
| centigrade | use Celsius;(°C) |
| centimeter ($10^{-2}$) $\times$ m | Cm |
| coenzyme A and acetyl derivatives | CoA and Acetyl CoA |

| | |
|---|---|
| colony forming units | cfu |
| concentration | concn (in tables only) |
| counts per minute | ct min |
| cultivar | cv. |
| dalton | Da |
| deci ($10^{-1}$ ×) | d; e.g. dm |
| degree absolute (Kelvin) | $^{\circ}K = {}^{\circ}C + 273$ |
| deoxyribonucleic acid, deoxyribonuclease complementary DNA | DNA, Dnase, cDNA |
| disintegrations per minute | d min$^{-1}$ |
| dry weight | d. wt |
| Einstein(s) | E |
| electron microscope (transmission and scanning) | TEM, SEM |
| ethylene diaminetetraacetate | EDTA |
| experiment | Expt (in tables only) |
| femto ($10^{-15}$ x ) | f; e.g. fg |
| fresh weight | fresh wt |
| gas chromatography-mass spectrometry | GC-MS |
| gas liquid chromatography | GCL |
| grams(s) | g |
| hectare | ha |
| high performance liquid chromatography | HPLC |
| hour | h |
| joule (kg m$^2$ s$^{-2}$) | J, I calorie = 4.18 J |
| kilo ($10^3$ × ) | k; e.g. kg, km |
| least significant difference | LSD |
| litre | I, do not abbreviate when confusion could arise with number one |
| mass spectrometry | MS |
| mega ($10^6$ × ) | M |
| metre | m |
| Michaelis constant | Km |
| Micro ( $10^{-6}$ × ) | ì; e.g. ìg |
| micromolar | ìM |

| | |
|---|---|
| milli ($10^{-3}\times$) | m; e.g. mm, mg |
| millmolar | mM |
| milliequivalents | meq |
| minute | min |
| molar (mol $1^{-1}$) | M |
| mole ( a gram molecule) | mol |
| molecular weight | mol. Wt |
| nano ($10^{-9}\times$) | n; e.g. nm |
| Newton | N |
| nicotinamide adenine dinucleotide and reduced form | NAD, NADH |
| nicotinamide adenine dinucleotide phosphate and reduced form | NADP, NADPH |
| number | No. (in tables only) |
| pascal (unit of pressure) | Pa; 100 kPa = I bar = 0.987 atmospheres |
| per | use minus index, e.g. mg $1^{-1}$ except when units is a culture vessel or organism |
| pico ($10^{-12}\times$) | p; e.g. pg |
| precipitate | ppt (in tables only) |
| probability (statistical) | P; use P = 0.05, etc. |
| radiant | use W m-2 (energy flux density) or photon flux density- ìmol m$^{-2}$ s$^{-1}$ or µE m$^{-2}$ s$^{-1}$ |
| relative humidity | RH |
| retardation factor | $R_F$ |
| ribonucleic acid | RNA, messenger RNA = mRNA ect. |
| second | s |
| sodium dodecylsulphate | SDS |
| species | sp,; plural spp. |
| standard deviation of sample | SD |
| standard error of mean | SE |
| temperature | temp. (in tables only) |
| thin-layer chromatography | TLC |
| ultraviolet light | u.v. |

| volume(s) | vol. (in tables only) |
| volume/volume(concentration) | v/v |
| water potential | $\psi$ |
| watt (IJ s$^{-1}$ ) | wt  W$^w$ |
| weight | wt (in tables only) |
| weight/ volume (concentration) | w/v |

## Structure of the Article

*Subdivision of the article.* Divide your article into clearly defined and numbered sections. Subsections should be numbered 1.1 (then 1.1.1, 1.1.2, ) 1.1, etc. (the abstract is not included in section numbering also for internal cross-referencing: do not just refer to 'the text.' Any subsection may be given a brief heading. Each heading should appear  on its own separate line.

*Introduction.* State the objectives of the work and provide an adequate background, avoiding a detailed literature survey or a summary of the results.

*Materials and methods.* Provide sufficient detail to allow the work to be reproduced. Methods already published should be indicated by a reference; only relevant modifications should be described.

## Results

*Discussion.* This should explore the significance of the results of the work, not repeat them. A combined Results and Discussion section is often appropriate. Avoid extensive citations and discussion of published literature.

*Acknowledgements.* Place acknowledgements, including information on grants received, before the references, in a separate section, and not as a footnote on the title page.

*References.* See *separate section, below.*

*Figure legends, tables, figures, scheme.* Present these, in this order, at the end of the article. They are described in more detail below. High-resolution graphics files must always be provided separate from the main text file (*see Preparation of illustrations*).

*Footnotes.* Footnotes should be used sparingly. Number them consecutively throughout the article, using superscript Arabic numbers. Many word processors build footnotes into the text, and this feature may be used. Should this not be the case, indicate the position of footnotes in the text and present the footnotes themselves on a separate sheet at the end of the article. Do not include footnotes in the Reference list.

*Tables.* Number tables consecutively in accordance with their appearance in the text. Place footnotes to tables below the table body and indicate them with superscript lowercase letters. Avoid vertical rules. Be sparing in the use of tables and ensure that the data presented in tables do not duplicate results described elsewhere in the article.

*Nomenclature and units.*   Follow internationally accepted rules and conventions: use the international system of units (SI).  If other quantities are mentioned, give their equivalent in SI.

The complete Latin name (genus, species, authority, together with cultivar, strain or culture number where appropriate) should be cited for every organism at first mention.  Thereafter the generic name may be abbreviated to the initial except where this could cause confusion.  No further abbreviation is permitted.

DNA *sequences and GenBank Accession numbers.*  Many Elsevier journals cite "gene accession numbers" in their running text and footnotes. Gene accession numbers refer to genes or DNA sequences about which further information can be found in the databases at the National Center for Biotechnical Information (NICBI)at the National Library of Medicine. Elsevier authors within to enable other scientists to use the accession numbers cited in their papers via links to these sources, should  type this information in the following manner:

*For each and every*  accession number cited in an article, authors should type the accession number in bold, underlined text.  Letters in the accession number should always be capitalized. (See Example below). This combination of letters and format will enable Elsevier's typesetters to recognize the relevant texts as accession numbers and add the required link to GenBank's sequences.

## Example
"GenBank accession nos. AI631510, AI631511, AI632198, and BE223228,) a B-cell tumor from achronic lymphatic leudemia (GenBank accession no. BE675048), and a T-cell lymphoma (GenBank accession no. AA361117)".

Authors are encouraged to check accession numbers used very carefully. An error in a letter or number can result in a dead link.  In the final version of the printed article, the accession number text will not appear bold or underlined. In the final version of the electronic copy, the accession number text will be linked to the appropriate source in the NCBI databases enabling readers to go directly to that source from the article.

*Preparation of  supplementary data.* Elsevier now accepts electronic supplementary material to supplementary material to support and enhance your scientific research. Supplementary files offer the author additional possibilities to publish supporting applications, movies, animation sequences, high-resolution images, background datasets, sound clips and more. Supplementary files supplied will be published online alongside the electronic version of your article in Elsevier web products, including ScienceDirect: http://www.sciencedirect.com. In order to ensure that your submitted material is directly usable, please ensure that data is provided in one of our recommended file formats. Authors should submit the material in electronic format together with the article and supply a concise and descriptive caption for each file. For more detailed instructions please visit our Author Gateway at http://authors.elsevier.com.

## References

Responsibility for the accuracy of bibliographic citations lies entirely with the authors.

*Citations in the text:* Please ensure that every reference cited in the text is also present in the reference list (and vice versa). Any references cited in the abstract must be given in full. Unpublished results and personal communications should not be in the reference list, but may be mentioned in the text. Citation of a reference as 'in press' implies that the item has been accepted for publication and a copy of the title page of the relevant article must be submitted.

*Citing and listing of web references.* As a minimum, the full URL should be given. Any further information, if known (author names, dates, reference to a source publication, etc.), should also be given. Web references can be listed separately (e.g., after the reference list) under a different heading if desired, or can be included in the reference list.

**Text :** Indicate references by number(s) in square brackets in line with the text. The actual authors can be referred to , but the reference number(s) must always be given.

**List :** Number the references (numbers in square brackets) in the list in the order in which they appear in the text.

## Examples

Reference to a journal publication:

1. Van der Geer J, Hanraads JAJ, Lupton RA. The art of writing a scientific article. J Sci Commun 2000;163:51-9.

## Reference to a book

2. Strunk Jr W, White EB. The elements of style. 3$^{rd}$ ed. New York: Macmillan; 1979.

## Reference to a chapter in an edited book

3. Mettam GR, Adams LB. How to prepare an electronic version of your article. In: Jones BS, Smith RZ, editors. Introduction to the electronic age, New York: E-Publishing Inc; 1999, p. 281-304.

*Note* shortened form for last page number. E.g. 51-9, and that for more than 6 authors the first 6 should be listed followed by 'et al'. For further details you are referred to "Uniform Requirements for manuscripts submitted to Biomedical Journals" (J Am Med Assoc 1997;277-934) (see also http://www.nejm.org/general/text/requirements/htm)

## Preparation of Illustrations

### Preparation of electronic Illustrations

Submitting your artwork in an electronic format helps us to produce your work to the best possible standards, ensuring accuracy, clarity and a high level of details.

#### *General points*

- Make sure you use uniform lettering and sizing of your original artwork.
- Save text in illustrations as "graphics" or enclose the font.
- Only use the following fonts in your illustrations: Arial, Courier, Helvetica, Times, Symbol.
- Number the illustrations according to their sequence in the text.
- Use a logical naming convention for your artwork files, and supply a separate listing of the files and the software used.
- Provide all illustrations as separate files.
- Provide captions to illustrations
- Produce images near to the desired size of the printed version.

A detailed guide on electronic artwork is available on our website: http://authors.elsevier.com/artwork **You are urged to visit this site; some excerpts from the detailed information are given here.**

### Formats

Regardless of the application used, when your electronic artwork is finalized, please "save as" or convert the images to one of the following formats (Note the resolution requirements of line drawings, halftones, and line/halftone combinations given below)

*EPS :* Vector drawings. Embed the font or save the text as "graphics".

*TIFF :* Colour or greyscale photographs (halftones): always use a minimum of 300 dpi

*TIFF :* Bitmapped line drawings: use a minimum of 1000 dpi

*TIFF :* Combination bitmapped line/half-tone (color or greyscale): a minimum of 500 dip is required.

DOC, XLS or PPT: If your electronic artwork is created in any of these Microsoft Office applications please supply "as is".

### Please do not

- Supply embedded graphic in your word processor (spreadsheet, presentation) document;

- Supply file that tare optimized for screen use (like GIF, BMP, PICT, WPG); the resolution is too low;
- Supply files that are too low in resolution;
- Submit graphic that are disproportionately large for the content.

## Captions

Ensure that each illustration has a caption. Supply captions on a separate sheet, not attached to the figure. A caption should comprise a brief title (not on the figure itself) and a description of the illustration. Keep text in the illustrations themselves to a minimum but explain all symbols and abbreviations used.

## Colour illustrations

Submit colour illustrations as original photographs, high quality computer prints or transparencies, close to the size expected in publication, or as 35 mm slides. Please make sure that artwork files are in an acceptable format (TIFF EPS or MS Office files) and with the correct resolution. Polaroid colour prints are not suitable. The publisher does not charge for figures which must be printed in colour to present the information. For further information on the preparation of electronic artwork, please see http://authors.elsevier.com/artwork.

***Please note :*** Because of technical complications which can arise converting colour figures to ' grey scale' (for the printed version should your not opt for colour in print) please submit in addition usable black and white prints corresponding to all the colour illustrations.

## Proofs

When Elsevier receives your manuscript it is considered to be in its final form.

Elsevier will send by e-mail page proofs in PDF format to the corresponding author, to be checked for typesetting/editing. No changes in, or additions to, the accepted (and subsequently edited) manuscript will be allowed at this stage. Proofreading is solely your responsibility.

A form with queries from the copyeditor may accompany your proofs. Please answer all queries and make any corrections or additions required.

Elsevier will do everything possible to get your article corrected and published a quickly and accurately as possible. In order to do this we need your help. When you receive the (PDF) proof of your article for correction, it is important to ensure that all of your corrections are sent back to us in one communication. Subsequent corrections will not be possible, so please ensure your first sending is complete. Note that this does not mean you have any less time to make your corrections, just that only one set of corrections will be accepted.

Elsevier reserves the right to proceed with publication if corrections are not communicated. Return corrections within 2 days of receipt of the proofs. Should there be no corrections, please confirm this.

## Offprints

Twenty-five offprints will be provided free of charge. Additional reprints may be purchased at the proof stage.

# CHAPTER - 103

# Physiological Entomology

## Notice to Contributors

*Physiological Entomology* is a journal designed primarily to serve the interests of experimentalists who work on the behaviour of insects and other arthropods. It thus has a bias towards physiological and experimental approaches to understanding behaviour, but it also retains the Royal Entomological Society's interest in the general physiology of arthropods. The subjects it covers include: *experimental analysis of behaviour: behavioral physiology and biochemistry; neurobiology and sensory physiology; general physiology; endocrinology; circadian rhythms and photoperiodism.* Relevant review papers are welcome.

Papers should be in clear concise English. They should not exceed 6000 words of text (10 printed pages) though longer papers may be accepted if they are of exceptional merit. Papers submitted must not have been published or accepted for publication by any other journal .

Fifty offprints of each paper are provided free (100 to Fellows and Members of the Society). More copies may be ordered at current prices when proofs are returned.

The name and full postal address and e-mail of the author to whom readers should address correspondence and offprint request should be given on the first page; this will appear as a footnote in the journal.

Recent issues should be examined for details of acceptable style and format. Manuscripts must be typed, on one side of the paper, at least double spaced with wide margins, preferably on A4 paper. Tables must be on separate sheets, should be self-explanatory, and should not be ruled up. Figure legends should be grouped together on a separate sheet. In the full-text online edition of the journal figure legends may be truncated in abbreviated links to the full screen version. Therefore the first 100 characters of any legend should inform the reader of key aspects of the figure. Three copies of text and figures are required.

Figures should be of publishable quality and be submitted about twice their printed size, labeled with stenciled or preprinted lettering (large enough to allow reduction), and numbered serially. Photographs, in the form of glossy prints, are acceptable for half-tone reproduction when they are a real contribution to the text; authors may submit them mounted in lay-outs to fill the printed page if they wish. Colour plates are printed by special arrangement only.

When data are expressed as means, whether in figures or tables, some indication of their distribution about the mean should be given. The same data should not be given in both tables and figures.

References should conform to the 'named-and-date' system: *titles of periodicals must be given in full* in the list of references at the end of the paper.

The title of the paper should be informative but preferably not exceed twenty words. A short title (for page headlines) of not more than forty letters should also be supplied. Taxonomic affiliation and authority for the principle species involved in the paper should be given in the Abstract and at the first mention in the text, not in the title. The paper should include a self-contained abstract of less than 250 words, to follow underneath the title. Authors should provide a maximum of ten key words.

**Electronic artwork**. We would like to receive your artwork in electronic form. Please save vector graphics (e.g. line artwork) in Encapsulated Postscript Format (EPS). and bitmap files (e.g. half-tones) in Tagged Image File Format (TIFF). Detailed information is available at: http://www.balckwellpublishing.com/authors/digill.asp

**Disks**. The journal requires submission of accepted manuscripts on disk which should be IMB-compatible. An accurate hardcopy must accompany each disk, together with details of the type of hardware used, the software employed and the disk system, if known. Do not justify. Particular attention should be taken to adhere exactly to the journal style in all respects. Disks will not be returned to authors.

**Proofs.** The corresponding author will receive an e-mail alert containing a link to a web site. A working e-mail address must therefore be provided for the corresponding author. The proof can be downloaded as a PDF (portable document format) file from this site. Acrobat Reader will be required, which can be downloaded (free of charge) from: http://www.adobe.com/products/acrobat/readstep2.html.

This will enable the file to be opened, read on screen and printed out in order for any corrections to be added. Further instruction will be sent with the proofs. Hard copy proofs will be posted if no e-mail address is available. Excessive changes made by the author in the proofs, excluding typesetting errors, will be charged separately.

**Paper**. Blackwell Publishing's policy is to use permanent paper from mills that operate a sustainable forestry policy and which has been manufactured from pulp that is processed using acid-free and elementary chlorine-free practices. Furthermore, Blackwell Publishing ensures that the text paper and cover board used in all our journals has met acceptable environmental accreditation standards.

**Disclaimer.** The Publisher, the Royal Entomological Society and Editors cannot be held responsible for errors or any consequences arising from the use of information contained in this journal: the views and opinions expressed do not necessarily reflect those of the Publisher, the Royal Entomological Society and Editors, neither does the publication of advertisements constitute any endorsement by the Publisher, the Royal Entomological Society and Editors of the products advertised.

**Cover caption**. *Background:* Scanning electron micrograph of compound eye of *Drosophila melanogaster* showing omatidia and hairs© 2002 Robert Harding. *Inset:* Locust antennal sensilla © Mami Yamamoto-Kihara.

# CHAPTER - 104

# Plant Biotechnology Journal

## Information for Contributors

Plant biotechnology journal will ONLY accept manuscripts submitted in electronic form. Any articles submitted in paper format will be returned to the authors for re-submission. Plant Biotechnology Journal encourages online submission of manuscripts at http://plantbiotechjournal.manuscrptcentral.com/. Submission online enables the quickest possible review and allows online manuscript tracking. Manuscript submission online can be in Word document (.doc), Rich Text Format (.rtf), Portable Document Format (.pdf) or PostScript (.ps). Figures can be embedded in the native word processor file or may be uploaded separately in one of the following formats: GIF (.gif), IPEG (.jpg), TIFF (.tif), ESP(.eps). Full upload instructions and support are available from the submission site via the 'Get Help Now' button. You can also email for online submission support. Please note that you should submit your covering letter or comments to the Editor-in-Chief, when prompted online. Please do not duplicate your submission by submitting online and by post. Please note that following acceptance of your manuscript for publication in Plant Biotechnology Journal, you will be required to supply your manuscript and figures on disk (for disk instructions see Disk Submission below).

Authors should provide a list of five possible reviewers for their manuscript. Each submission will be considered by a member of the *Senior Editorial Team* to determine whether it lies within the scope of the Journal. Those that do will be assigned to a handling editor (one of the Senior Team) and will be sent out for full review. Those that do not will be returned to the authors quickly so that resubmission elsewhere will not be unduly delayed. Full instructions for submission of revised manuscripts can be found in on *http:// www.blackwellpublishing.com/journals/pbi.* If accepted, papers become the copyright of the Journal. No page charges will be levied. All queries should be directed to the Editorial Office. For the address, see inside front cover.

Papers are accepted on the understanding that no substantial part has been, or will be, published elsewhere. Publication of a paper in *Plant Biotechnology Journal* implies that authors will distribute freely to researchers, for non-commercial purposes, all plant cultivars, cell lines, DNA, antibodies and other similar materials that were used in the experiments reported. Seeds of mutants described must be deposited at the appropriate stock centre and accession numbers provided.

## Presentation of manuscripts

Authors are advised to keep a copy of all material as the Editors do not accept responsibility of damage to or loss of papers submitted. Only manuscripts in English will be published.

The title page should page include: the full title of the paper; the full names of all the authors; the name(s) and address (es) of the institution(s) at which the work was carried out (the present addresses of the authors, if different from the above, should appear in a footnote); the name, address, telephone and fax numbers and e-mail address of the author to whom all correspondence and proofs should be sent; the e-mail addresses all the authors if possible; a suggested running title of not more than 50 characters, including spaces; six key words to aid indexing; accession numbers for the EMBL sequence database and seed stock if required; word count.

Generally, all papers should be divided into the following sections and appear in the order: (1) *Summary*, not exceeding 250 words, (2) *Introduction,* (3) *Results,* (4) *Discussion,* (5) *Experimental procedures*, (6) *Acknowledgements,* (7) *References,* (8) *Tables,* (9) *Figure legends,* (10) *Figures.* The results and Discussion sections may be combined and may contain subheadings. Experimental procedures should be sufficiently detailed to enable the experiments to be reproduced. Trade names should be capitalized and the manufacturers name and address given.

## Submission of revised manuscripts

Revised manuscripts must be in their final form when submitted. The proofs received later are for correction of typographical errors only. They should not be used for final changes to articles; such changes must be made to the manuscript before it goes to the publishers. Major alterations to the text at proof stage will be charged to the author and may delay publication. manuscripts should be  checked very carefully for correct designation of genes and proteins. It is important to differentiate between genes and proteins. Gene names and loci should be  italic, proteins should be  upright. It is essential that authors return one signed copy of the Copyright form to the Editorial Office with their revised manuscripts. A completed colour work agreement form must also be returned with the revised manuscript if it contains colour. This form is also available as a pdf file from our website. Failure to return the forms will delay the publication of manuscripts if they are accepted.

## Supplementary Material

Supplementary material, such as data sets or additional figures or tables, that will not be published in the print edition of the journal, but which will be viewable via the online, can be submitted. It should be clearly stated at the time of submission that the Supplementary Material is intended to be made available through the online edition. Supplementary Material will be made available in electronic form free of charge either through the online pages or on request from Blackwell Publishing. Alternatively, if the size or format of the Supplementary Material is such that it cannot be accommodated on the journal website, the author aggress to make the Supplementary Material available free of charge on a permanent website, to which links will be set up from the Journal website.

## Disk submission of manuscripts

When supplying your manuscript on disk, please follow these instructions and return the disk, *File Description Form* and hard copy of your manuscript to the Editorial Office.

- Final version of the hard copy and the floppy disk must be the same. It is essential that the disk contains any last-minute changes.

- Be consistent. Use the same presentation for all headings, etc. that are to appear the same in the finished printing.

- Tables might not be set from disk, and it is essential that an adequate hard copy is supplied.

- Advise us about any matter that you think could cause a problem.

- *A File Description Form* is available to print out or can be obtained from the Editorial Office.

- Do not use the carriage return (enter) at the end of lines within a paragraph

- Disks will not be returned to authors.

*Units and abbreviations.* System International (SI) units should be used as given in Units Symbols and Abbreviations, published by the Royal Society of Medicine Services Ltd. London. Abbreviations should be given in full the first time they are used.

*Scientific names.* Complete scientific names should be given when organisms are first mentioned. The generic name any subsequently be abbreviated to the initial.

*References.* References should be cited in the text by author and date, e.g. Shah and Klessing (1999). Joint authors should be referred to by et al. if there are more than two e.g. Sambrook *et al.* (1989). The full titles of papers, chapters and books should be given, the abbreviated names of journals, with the first and last page numbers.

## Examples

Lacomme, C. and Santa Cruz, S. (1999) Bax-induced cell death in tobacco is similar to the hypersensitive response. *Proc. Natl Acad, Sci, USA,* **96**, 7956-7961.

Sambrook, J.,Fritsch, E.F. and Maniatis, T. (1989) *Molecular Cloning: A laboratory Manual, 2nd edn.* Cold Spring Harbor, NY: Cold Spring Harbor Labaratory Press.

Shah, J. and Klessig, D.F. (1999) Salicylic acid: signal perception and transduction. In: Biochemistry and Molecular Biology of Plant Hormones (Hooykaas, P.P.J., Hall, M.A. and Libbenga, K.R.,eds), pp. 513-541. New York: Elsevier Science.

Work that has not been accepted for publication and personal communications should not appear in the reference list, but may be referred to in the text. It is the author's responsibility to obtain permission from colleagues to include their work as a personal communication.

*Figures.* Two sets of all figures and photographs should be supplied and submitted as either laser copies or high-quality glossy prints as appropriate in the size that they will appear in the Journal. Figure sections should be designated with lower case letters. Magnification bars should be given on electron and light microscope photographs. All figures should be identified with the figure number and authors names in soft pencil or on labels on the back, and the top edge should be indicated.

Figure legends should be typed on sheets separate to those containing the figures and contain sufficient information to be understood without reference to the text, but should not contain methods. Each should begin with a short title for the figure. All symbols and abbreviations used in the figures should be explained.

The Journal welcomes colour photographs. It is the policy of Plant Biotechnology Journal for authors to pay the full cost for the reproduction of their colour artwork. Therefore, please note that if there is colour artwork in your manuscript when it is accepted for publication. Blackwell Publishing require you to complete and return a colour work agreement form before your paper can be published. This form can be as a PDF from the Internet. The web address for the form is: http://www.blackwellpublishing.com/pdf/ SN_X_CoW.pdf. Once completed, please return the form to the Production Editor (address inside front cover). Under exceptional circumstances, authors may request waiver of these charges. This must be requested at the time of submission of the manuscript, and authors must justify to the Editors that inclusion of colour figures is essential for interpretation of the results presented. Note that review articles may contain colour artwork, free of charge.

*Tables.* Tables should be on separate sheets. They should have a brief descriptive title and be self-explanatory. No vertical rules should be used. Units should  appears in parentheses in the column headings, not in the body of the table. Repeated words or numerals on successive lines should be written in full.

## Electronic artwork

We would like to receive your artwork in electronic form. If you are unable to submit in the required format, please send good quality originals as detailed above. Please save vector graphics (e.g. line artwork) in Encapsulated Postscript Format (.eps), and bitmap files (e.g. half-tones) in Tagged Image File Format (.tif). TIFF files should be supplied at a minimum resolution of 300 ppi (pixels per inch) at the final size at which they are to appear in the journal. Colour files should be in CMYK format. Detailed information on our digital illustration standards is available at: http://www.backwellpublishing.com/authors/digill.asp

## Proofs

Proofs will be sent via e-mail as an Acrobat PDF (portable document format).file. The corresponding author should therefore supply their email address when they submit their manuscript. Acrobat Reader will be required in order to read this file. This software can be downloaded (free of charge) from the Adobe website. Further instructions will be sent with the proof. Proofs will be posted if no e-mail address is available; in your absence, please arrange for a colleague to access your e-mail to retrieve the proofs. Major alterations to the text will be charged to the author and may delay publication.

*Offprints.* A PDF offprint of each article will be supplied free. Additional offprints in units of 100 may be purchased if ordered on the form supplied with the proofs.

*Registration of sequences.* New nucleotide sequence data reported in papers published in the *Plant Biotechnology Journal* must be submitted and deposited in the DDBI/EMBL/GenBank databases and accession numbers obtained.

## Cover photograph

Photographs of high quality suitable of the cover of the *Plant Biotechnology Journal* are welcomed. They should be sent to the Editorial Office and be accompanied by a brief descriptive summary. It is  preferred, but not essential, that these should be related to submitted papers.

## Disclaimer

The Publisher, the SEB, AAB and Editors cannot be held responsible for errors or any consequences arising from the use of information contained in this journal; the views and opinions expressed do not necessarily reflect those of the Publisher, SEB, AAB and Editors, neither does the publication of advertisements constitute any endorsement by the Publisher, SEB, AAB and Editors of the products advertised.

# CHAPTER - 105

# Pest Management in Horticultural Ecosystems

## Guidelines for Authors

*Pest Management in Horticultural Ecosystems,* an international journal, covers the field of pest (insects, mites, nematodes, slugs, snails, birds, mammals) management in its broadest sense. It includes bioecology, population modeling, biocontrol, host plant resistance, use of plant product, semiochemicals, insecticide resistance, insect biotechnology and IPM in fruits, vegetables, ornamental, medicinal and aromatic crops, plantation crops, tuber crops, spices, etc. The journal is published twice in a year, in June and December.

Two copies of the manuscript and an electronic form (3 ½" floppy, MS Word) should be submitted to the Chief Editor. The manuscript should preferably pertain to the research work carried out during the last five years. Author(s) must certify that the manuscript has not been submitted elsewhere. All papers will be refereed. Short communications on significant research findings or new records are welcome. Only invited review papers will be considered. Decision of the Chief Editor is final. Authors are permitted to photocopy their articles for non-commercial and scientific purposes.

*Title :* Title of the article should be bold and in capital. Scientific names should be in running and italicized. Author name(s) should be in capital and bold, with full address in running.

*Abstract :* The abstract should not exceed 200 words. It should be suitable for indexing and publication in abstracting journals.

*Text  :* The text should be typed in double space on one side of bond paper (21 × 29 cm) with 3 cm margin on all sides and in clear and concise English. The title should be brief, specific and describe the scope of the paper. Below the title, author(s) name(s), and address of the first author should be given. Address (es) of the other author(s) should be given as footnotes and indicated by consecutive superscript numbers. The paper should be divided into subheadings such as Abstract, Key Words, Introduction, Materials and Methods, Results and Discussion, Acknowledgements, and References. Subheadings should be on the left margin. It is desirable that authors get their papers edited for English grammar prior to submission.  The length of the paper should not exceed 2000 words.

*Nomenclature :* Common names of organisms should be qualified at the first mention by full Latin name and authority, followed by the order and family in parenthesis.  Pesticides should be referred by their accepted common names. The trade name may be used with the chemical formula given in parenthesis.

*Units and Abbreviations :* All units should be in metric (SI) system.

*Tables :* Each table should be on a separate sheet at the end of the paper, numbered in the order in which it appears in the text.  Table should carry a short title and be numbered in Arabic numerals.  All data reported must be subjected to appropriate statistical analysis.  Standard abbreviations for column headings should be followed.

*Illustrations :* Illustrations should be relevant to the article and referred to in the text.  Line drawings should be laser printed, of 10 x 15 cm dimension with legends and captions.

*References :* References should be cited in the text in the form: (Granett and Dunbard, 1975).  The term *et al.,* should be used when there are more than two authors.  The letters a, b, c... should be used following the year.  To distinguish between two or more papers by the same author(s) in one year.  References at the end of the text should be given in the following form:

Morris, R.F.1954. Single factor analysis in population dynamics. *Ecology* 40: 580-589.

Takabayashi, J., Diske, M.and Posthumus, M.A. 1991.  Introduction of indirect defence against spider mites in uninfested Lima bean leaves. *Phytochemistry* 30 : 459-1462.

Litsinger, J.A. and Moody, K.1986. Integrated pest management in multiple cropping systems.  In *Multiple Cropping* Ed. Solly, M., American Society of Agronomy, Wisconsin, pp.293-316.

Singh, D.P.1986. *Breeding for resistance to diseases and insect pests.* Springer Verlag.  New York.

# CHAPTER - 106

# Productivity

## Guidelines for Contributors

### Scope and Coverage

*PRODUCTIVITY* is the principal journal of the National Productivity Council of India. The Journal aims at disseminating information on concepts of and data on productivity and its growth in India and elsewhere. It also aims at disseminating knowledge on techniques and methods of productivity improvement through effective management of all types of resources. Thus, contributions for a lager spectrum of disciplines are accepted for publication. Only those manuscripts that present the results of original research and analysis are acceptable to the Journal. The managerial/policy implications of the study should be highlighted separately towards the end of the paper.

## Format of Manuscripts

Contributions should be of about 5,000 words length. Tables, illustrations, chart, figures, exhibits etc. should be serially numbered and typed in separate pages and should  not be mixed with the main text. The text should be addressed to the Editor, **PRODUCTIVITY**, National Productivity Council, Utpadakta Bhawan, Lodi Road, New Delhi-110 003.

## About the References

Only those references which are actually utilized in the text should be included in the reference list. In the text, references should be cited with the surname of the author(s) alongwith the year of  publication and the page number, all in brackets. If there are more than one reference by the same author during any year, the year may be subscripted with 'a' or 'b'. For instance, reference may be given at the end of the sentence as : (Szendrovits, 1988a, p. 337). Sources of data need to be given below each table unless otherwise mentioned in the text. Reference list  should be alphabetically arranged. Each reference should carry the surname of the author, followed by other names, the title of the paper in quotes, the name of the journal underlined, volume and issue numbers and the year of publication. In the event of a book, the title should be followed by the publisher's name and year of publication. In the event of a report from an organisation, the name of the organisation may be cited in the place of the author.

## Accompanying Material

The manuscripts should be accompanied by:

1. An abstract of the paper not exceeding 100 words.
2. A declaration that the paper is original and has not been submitted elsewhere for publication.
3. A note about the author(s) not exceeding 50 words.

# CHAPTER - 107

# Rural Sociology

## Submission Guidelines (Revised Spring 2006)

Submission to *Rural Sociology* assumes adherence to the anonymous peer review process. Final judgment is made by the editor, who relies on reviewers' recommendations.

Authors submitting manuscripts for possible publication in *Rural sociology* must abide by the following guidelines.

Your submission should include:

- A cover letter

- Two hard copies of your manuscript
  - o *Note :* international authors are not required to submit any hard copies.

- An electronic copy of your manuscript in MSWord or WordPerfect format
  - o You may send this on a diskette, CD, or by email. Please append any tables or graphs to the end of the manuscript so that you send only one file.

- $25 USD
  - o Payable by check, money order, or credit card. Please email us to arrange payment by credit card.

*Rural Sociology* publishes in standard American English and therefore prefers to receive manuscripts written in this style. However, we recognize and value the research of rural sociologists whose primary language is not English, and we understand that it requires considerable effort to translate a manuscript into another language.

As a service to international authors and as a way of broadening the scope of our publication, *Rural Sociology* will now take manuscripts written in Spanish for an "initial review." In the initial review, one or more reviewers fluent in Spanish will evaluate your manuscript to determine its suitability for the standard

*Rural Sociology* peer review process. Acceptance in the initial review stage does NOT guarantee that the manuscript will ultimately be published – it simply means that the reviewers believe that the content of your manuscript and its quality is promising enough to be submitted to *Rural Sociology*. If your manuscript is accepted in the initial review, we will recommend that you have it translated into standard American English and then submit it normally.

If you wish to submit a Spanish-language manuscript for initial review, send an email (in English) to rslsu@lsu.edu with an electronic version of your paper attached.

## Formatting Your Manuscript

To maximize readability and electronic compatibility, use a standard variable-width font such as Times New Roman/Times (recommended) or Arial/Helvetica. Double-space all content, including abstracts, footnotes, appendices, and references. Use twelve point type, six-inch lines, and US Letter page size.

The first page of your manuscript should include:

- Title
- Author(s)' names
- Corresponding author's name, address, phone number, and email
- Acknowledgments and/or funding source (if any)

The first page will be removed before the manuscript is sent out for review.

The second page of your manuscript should include ONLY the title and abstract.

The title should be repeated once again on the third page before the text of the manuscript begins.

## A Note on Anonymity and Citing Yourself

You must not refer to yourself in a way that reveals your identity after the first page of your manuscript. To do so compromises the standard of the anonymous peer review process. We recognize that authors sometimes must cite their own work and we find no problem with this, however, you must cite yourself in the correct manner. For example:

*Unacceptable :* "As I have stated previously, rural areas remain important for sociological study despite the continuing urbanization of society (Doe 1996)."

*Acceptable :* "As I have stated previously, rural areas remain important for sociological study despite the continuing urbanization of society (XXXXXX 1996)."

*Recommended:* "Rural areas remain important for sociological study despite the continuing urbanization of society **(see Doe 1996).**"

## Manuscript Style Guide

Rural Sociology uses the American Sociological Association (ASA) manuscript style. (See American Sociological Association. 1997. American Sociological Association Style Guide. 2nd ed. Washington, DC: American Sociological Association.)

Use footnotes, only to explain material that cannot be justified for inclusion in text or tables. Number the notes consecutively, beginning with footnote 1. Print double spaced.

Place each table and each figure on a separate sheet. Indicate in the text where each table should be inserted (e.g., "Table 1 about here"). Follow the American Sociological Association (ASA) style for tables.

Use headings to organize the article. Three headings are generally adequate.

1. **First-Level Head** : center justify, bold, use upper and lower case.

   **Example**

   Data and Methods

2. **Second-Level Head** : left justify, bold, use upper and lower case.

   Example

   Resource Management

3. **Third-Level Head** : indent, italicize capitalize first letter of first word only, follow with a period, include as part of text. Example:

   *Independent variables.*  The independent variables include…

## Reference Citations in Text

Cite all references in the text, where appropriate, by the author's last name, publication year, and (where applicable) page number(s). Footnotes are NOT to be used for citations.

Depending on sentence construction, the citation will appear as follows: Bowen and Finegan (1999). If a page number is use, it follows the publication year and is set off by a colon: Kennedy and Silverman (1985:276).

Enclose a series of citations within parentheses, separated by semicolons. Place multiple citations in alphabetical order: (Clemente and Kleiman 1997; Kennedy and Silverman 1995; Lee 1992).

Make subsequent citations of a source exactly as cited the first time. If an author has two citations in the same year, distinguish them by attaching a or b to the year in both the text and the references: (Ploch 1995a, 1995b).

For works by two authors, cite both last names. For three authors, cite all three last names in the first citation in the text: (Carr, Smith, and Jones 1992a:

366); thereafter use "et al." in the citation: (Carr et al. 1992a: 366). If a work has more than three authors, use "et al." in the first citation and in all subsequent citations.

## Reference Citation Style

**Article in journal :** Deseran, F.A. and D.Keithly.  1994.  "Teenagers in the U.S.Labor Force : Local Labor Markets, Race, and Family." *Rural Sociology* 59:668-92.

**Book :** Lobao, L.M. 1990. *Locality and Inequality.* Albany, NY: SUNY Press.

**Article or chapter in an edited volume :** Zuiches, J.J.1982.  "Residential Preferences."  Pp.247-63 in *Rural Society in the U.S.: Issuues for the 1980s,* edited by D.A. Dillman and D.J.Hobbs.  Boulder, CO: Westview.

**Government document :** Beale, C.L. 1975. *The Revival of Population Growth in Nonmetropolitan America.* ERS-605, U. S. Department of Agriculture. Washington, DC: U.S. Government Printing Office.

**Dissertation :** Dorsi, L. 1999.  "Ecology and Agriculture: A Partial Replication and Update of Maro's Study in Gkirdle Bend, Oregon."  PhD dissertation, Department of Ecology, Atlantis University.

**Unpublished manuscript :** Mundi, G.1998.  "Environmentalism and Youth Activities."  Department of Sociology, St. Pippin's college, Cincinnati. Unpublished manuscript.

**Presented paper :** Anderson, A. 1999.  "Forest Preservation in the Midwest." Presented at the annual meetings of the Rural Sociological Society, August 16, Chicago.

**Newspaper article (print) :** Goldstein, A. 1997.  "Dying Patients' Care Varies Widely by Place, Study Says." *Washington Post,* October 15, p. Al.

**Machine readable data file :** American Institute of Public Opinion.  1976. Gallup Public Opinion Poll #965 [MEDF].  Princeton, NJ: American Institute of Public Opinion [producer].  New Haven, CT: Roper Public Opinion Research Center, Yale University [distributor].

**On line journal article :** Jacobsen, J.W., J.A. Mulick, and A.A. Schwartz. 1995.  "A History of Facilitated Communication: Sciene, Pseudoscience, and Anti-Science." *American Psychologist* 50750-65. Retrieved January 25, w996 (http://www.apa.org/journals/jacobsen.html).

**On-line newspaper article :** Goldstein, A.  1997. "Dying Patients' Care Varies Widely by Place, Study Says." *Washington Post, October 15, p.AJ.* Retrieved October 15, 1997 (http://www.washingtonpost.com/Wplate/1997-10/15/061-101-idx.html).

**On-line abstract :** Swidler, A. and J. Arditi. 1994. "The New Sociology of Knowledge" (abstract). *Annual Review of Sociology* 20:305-29. Retrieved October 15, 1997 (http://www.annurev.prg/series/sociology/Vo120/so20abst.htm).

# CHAPTER - 108

# Sankhya

## Instructions for Authors

Sankhya is published in two Series. Series A publishes papers in Probability and Mathematical Statistics and appears in February, June and October Series B publishes papers in Statistical Methods and Applied Statistics reading Quantitative Economics and appears in April, August Statistics member. Series B is mainly intended for analysis of original data of broad interest, novel applications of methodology and development of new methods and techniques of immediate practical use. We have no backlog of accepted papers. Sankhya papers are abstracted/reviewed in Statistical Theory and Method Abstracts, mathematical Reviews and Zentralblatt fur Mathematik. We do not supply reprints. Authors of a published paper receive one copy of the particular issue. The corresponding author also receives the LATEX file of his/her paper.

Manuscripts may be submitted in any one of the ways: (a) Three copies to any Editor/Co-Editor and one copy for reference to the Managing Editor at Sankhya office. (b) Four copies to any managing Editor at the Sankhya office. (c) Any form of TEX file or Postscript file by e-mail or on a 3.5" DOS floppy along with one hard copy to the Managing Editor.

Manuscripts must be typed on one side, double-spaced with wide margins. *Please consult the latest issue for style.* Primary and secondary AMS subject classification numbers and some keywords and phrases should be mentioned at the bottom of the first page. A short title should be given for use as running title. Present institutional affiliation for each author should be given after each author's name. A self-contained summary not exceeding 120 words is required. Diagrams should be in a form suitable for direct reproduction. Formulae should be numbered consecutively with numbers in brackets ( ) appearing on the right margin. Tables should have short explanatory headings and be numbered as Table 1, Table 8, etc. References appearing in the text should be as Roy (1956), Morgan *et al.* (1972) etc. References at the end should be listed alphabetically. Permanent institutional address of each author should be given at the end. E-mail address if available should be included.

Our address is : *Sankhya, Indian Statistical Institute, 203 B.T. Road Calcutta – 700035, India.* Fax: 91 33 577 6680.

Email : sankhya@isical.ac.in

# CHAPTER - 109

# Sarvekshana

The object of this journal is to present all the results of the National Sample Survey (NSS) as and when they are ready for publication. Analytical articles based on the results previously published are also included in this journal. Sarvekshana welcomes contributions based on the NSS data and those relating to its methodology.

- Authors are requested to send three typewritten copies their papers/articles to the Chief Executive Officer, National Sample Survey Organization. Sardar Patel Bhavan, Parliament Street, New Delhi – 110001.

- Typescript should be double-spaced. Margins should be 1.5 inch wide on the left and 1 inch on the right.

- Mathematical formulae, symbols etc. should be neatly written, if typing is not possible. The graphs, charts and diagrams, if any, should be complete with all details and be such that direct reproduction can be obtained.

- Reference to other papers, books etc. should be given in full *e.g.* name of author, title of publication, name of journal, date of publication etc.

- A summary of the paper not exceeding 150 words should also be sent.

- The paper sent for this journal should not have been published elsewhere.

- Contributors will receive 25 reprints of their papers (15 copies for each of the joint authors) free.

- Subscription to Sarvekshana should be sent to the Controller of Publications, Department of Publication, Civil Lines, Delhi-110 054.

# CHAPTER - 110

# Scientia Horticulturae

## Guide for Authors

### Aims and scope

Scientia horticulturae is an international journal publishing research related to horticultural crops. Articles in the journal deal with open or protected production of vegetables, fruits, edible fungi and ornamentals under temperate, subtropical and tropical conditions. Papers in related areas (biochemistry, micropropagation, soil science, plant breeding, plant physiology, phytopathology, etc.) are considered, if they contain information of direct significance to horticulture. Papers on the technical aspects of horticulture (engineering, crop processing, storage, transport etc.) are accepted for publication only if they relate directly to the living product. In the case of plantation crops, those yielding a product that may be used fresh (e.g. tropical vegetables, citrus, bananas, and other fruits) will be considered, while those requiring processing (e.g. rubber, tobacco, tea, and quinine) will not.

## Types of contributions

1. Original full papers(Regular Papers)

2. Review articles

3. Short Communications

   3.1 Report of preliminary results of important research (pilot investigation; e.g. no duplications or with other restrictions)

   3.2 Report on research with restricted originality; modification of earlier reported research with no or limited explanatory aspects.

   3.3 Newly developed methodology or modification of existing methodology, possibly description of first test.

   3.4 Results of the application of an earlier published research methodology on other crops or under different conditions (fact finding or recipes), that are nevertheless of interest to an international leadership.

4. Book Reviews will be included in the journal on a range of relevant books which are not more than 2 years old.

Original papers should report the results of original research. The materials should not have been previously published elsewhere, except in a preliminary form.

Reviews should cover a part of the subject active current interest. They may be submitted or invited.

Short Communications should be as completely documented, both by reference to the literature and description of the experimental procedures employed, as a regular paper. They should not occupy more than 4 printed pages (about 8 manuscript pages, including figures, etc.)

For consultation or suggestions please contact

Dr. John Bower

Department of Horticultural Science

University of KwaZulu-Natal

P.Bag X01

Scottsville

Pietermartizburg 3209

South Africa

DR. Robert Geneve

Department of Horticulture and Landscape Architecture

College of Agriculture

N 318 Ag Science Bldg North

Lexinton

K Y 40546 – 0019

USA

## Online Submission of manuscripts

Submission of an article implies that the work described has not been published previously (except in the form of an abstract or as part of a published lecture or academic thesis), that it is not under consideration for publication elsewhere, that its publication is approved by all Authors and tacitly or explicitly by the responsible authorities where the work was carried out, and that, if accepted, it will not be published elsewhere in the same form, in English or in any other language, without the written consent of the Publisher.

Upon acceptance of an article. Authors will be asked to transfer copyright (for more information on copyright see http://authors.elsevier.com). This transfer will ensure the widest possible dissemination of information. A letter will be sent to the corresponding Author confirming receipt of the manuscript. A form facilitating transfer of copyright will be provided.

## Guide for Authors

**US national Institutes of Health (NIH) voluntary posting ("Public Access") policy** Elsevier facilitates author posting in connection with the voluntary posting request of the NIH (referred to as the NIH "Public Access Policy" , see http://www.nih.gov/about publicaccess/index.htm) by posting the peer-reviewed author's manuscript directly to PubMed Central on request from the author, after formal publication. Upon notification from Elsevier of acceptance, we will ask you to confirm via e-mail (by e-mailing us at NIHauthorrequest@elsevier.com) that your work has received NIH funding (with the NIH award number, as well as the name and e-mail address of the Prime Investigator) and that you intend to respond to the NIH request. Upon such confirmation, Elsevier will submit to PubMed Central version of your manuscript that will include peer-review comments, for posting 12 months after the formal publication date. This will ensure that you will have responded fully to the NIH request policy. There will be no need for you to post your manuscript directly to PubMed Central, and any such posting is prohibited. Individual modifications to this general policy may apply to some Elsevier journals and its society publishing partners.

## Authors' rights

As an author you (or your employer or institution) may do the following:

- make copies (print or electronic) of the article for your own personal use, including for your own classroom teaching use
- make copies and distribute such copies (including through e-mail) of the article to research colleagues, for the personal use by such colleagues (but not commercially or systematically, e.g. via an e-mail list or list server)
- post a pre-print version of the article on Internet website including electronic pre-print servers, and to retain indefinitely such version on such servers or sites
- post a revised personal version of the final text of the article (to reflect changes made in the peer review and editing process) on your personal or institutional website or server, with a link to the journal homepage (on Elsevier.com)
- present that article at a meeting or conference and to distribute copies of the article to the delegates attending such a meeting
- for your employer, if the article is a 'work for hire' , made within the scope of your employment, your employer may use all or part of the information in the article for other intra-company use (e.g. training)
- retain patent and trademark rights and rights to any processes or procedure described in the article
- include the article in full or in part in a thesis or dissertation (provided that this is not to be published commercially)

- Use the article or any part thereof in a printed compilation of your works, such as collected writings or lecture notes (subsequent to publication of your article in the journal)

- prepare other derivative works, to extend the article into book-length form, or to otherwise re-use portions or excerpts in other works, with full acknowledgement of it s original publication in the journal.

If excerpts from other copyrighted works are included, the Author(s) must obtain written permission from the copyright owners and credit the source(s) in the article. Elsevier has preprinted forms for use by Authors in these cases: contact Elsevier Rights Department, Oxford, UK: phone: (+44) 1865 853333, e-mail: permissions@elsevier.com. Requests may also be completed on-line via the Elsevier homepage (http://www.elsevier.com/locate/permissions).

## On-line submission to the journal prior to acceptance

Submission to this journal proceeds totally online. Use the following guidelines to prepare your article. Via the "Author Gateway" page of this journal (http://authors.elsevier.com) you will be guided stepwise through the creation and uploading of the various files. The system automatically converts source files to a single Adobe Acrobat PDF version of the article, which is used in the peer-review process. Please note that even though manuscript source files are converted to PDF at submission for the review process, these source files are needed for further processing after acceptance. All correspondence, including notification of the Editor's decision and requests for revision, takes place by e-mail and via the Author's homepage, removing the need for a hard-copy paper trail.

The above represents a very brief outline of this form of submission. It can be advantageous to "Guide for Authors" section from the site for reference in the subsequent stages of article preparation.

## Electronic format requirements of accepted articles

### General points

We accept most wordprocessing formats, but WordPerfect or LaTex is reference. Always keep a backup copy of the electronic file for reference and safety.

## Wordprocessor documents

*It is important that the file be saved in the native format of the wordprocessor used. The text should be in single-column format. Keep the layout of the text as simple as possible. Most formatting codes will be removed and replaced on processing the article. In particular, do not use the wordprocessor's options to justify text or to hyphenate words. However, do use boldface, italics, subscripts, superscripts etc. Do not embed 'graphically designed' equations or tables, but prepare these using the wordprocessor's facility. When preparing tables, if you are using a table gird, use only one grid for each individual table and not a grid for each row. If no grid is used, use tabs, not spaces, to align columns. The electronic text should be prepared in a way very similar to that of conventional manuscripts (see also the Author Gateway's Quickguide:* http://authors.elsevier.com). *Do not import the figures into the text file but, instead, indicate their approximate locations directly in the electronic text and on the manuscript. See also the section on Preparation of electronic illustrations.*

To avoid unnecessary errors you are strongly advised to use the 'spellchecker' function of your wordprocessor.

## Preparation of  manuscripts

1. Articles must be written in good English. Authors whose native language is not English are strongly advised to have their manuscripts checked by an English-speaking colleague prior to submission. Language Polishing: For authors who require information about language editing and copyediting services pre- and post-submission  please visit http://www.elsevier.com/wps/find/authorshome.authors/langu  polishing  or  contact authorsupport@elesevier for more information. Please note Elsevier neither endorses nor takes responsibility for any products, goods or services     Elsevier.com/wps/find/termsconditions.cws_home/termscondtions.

2. Manuscripts should be prepared with numbered lines, with wide margins and double spacing throughout, i.e. also for abstract, footnotes and references. **Every page of the manuscript, including the title page, references, tables, etc. should be numbered. Authors are requested to submit, with their manuscripts, the names and contact details of four potential referees.** However, in the text no reference should be made to page numbers; if necessary one may refer to sections.

3. Manuscripts in general should be organized in the following order:
   - Title (should be clear, descriptive and not too  long)
   - Name(s) of author(s)
   - Complete postal address(es) of affiliations
   - Full telephone, fax no. and e-mail address of the corresponding author must be included
   - Present address(es) of author(s) if applicable
   - Complete correspondence address to which the proofs should be sent
   - Abstract
   - Key words (indexing terms), normally 3-6 items
   - Introduction
   - Material studied, area descriptions, methods, techniques
   - Results

- Discussion
- Conclusion
- Acknowledgments and any additional information concerning research grants, etc.
- References
- Tables
- Figure captions

4. In typing the manuscript, titles and subtitles should not be run within the text. They should be typed on a separate line, without indentation. Use lower-case lowertype.

5. Elsevier reserves the privilege of returning to the author for revision accepted manuscripts and illustrations which are not in the proper form given in this guide.

## Abstracts

The abstract should be clear, descriptive and not longer than 400 words.

## Tables

1. Authors should take notice of the limitations set by the size and lay-out of the journal. Large tables should be avoided. Reversing columns and rows often reduce the dimensions of a table.

2. If many data are to be presented, an attempt should be made to divide them over two or more tables.

3. Tables should be numbered according to their sequence in the text. The text should include references to all tables.

4. Each table should be typewritten on a separate page of the manuscript. Tables should never be included in the text.

5. Each table should have a brief and self-explanatory title.

6. Column headings should be brief, but sufficiently explanatory. Standard abbreviations of units of measurement should be added between parentheses.

7. Vertical lines should not be used to separate columns. Leave some extra space between the columns instead.

8. Any explanation essential to the understanding of the table should be given as a footnote at the bottom of the table.

## Preparation of electronic illustrations

Submitting your artwork in an electronic format helps us to produce your work to the best possible standards, ensuring accuracy, clarity and a high level of detail.

### *General points*

- Always supply high-quality printouts of your artwork, in case conversion of the electronic artwork is problematic.
- Make sure you use uniform lettering and sizing of your original artwork.
- Save text in illustrations as "graphics" or enclose the font.
- Only use the following fonts in your illustrations: Arial, Courier, Helvetica, Times, Symbol.
- Number the illustrations according to their sequence in the text.
- Use a logical naming convention for your artwork files, and supply a separate listing of the software used.
- Provide all illustrations as separate files.
- Provide captions to illustrations separately.
- Produce images near to the desired size of the printed version.

A detailed guide on electronic artwork is available on our website:http://authors.Elsevier.com/artwork

You are urged to visit this site: some excerpts from the detailed information are given here.

## Formats

Regardless of the application used, when your electronic artwork is finalized, please "save as" or convert the images to one of the following formats (Note the resolution requirements for line drawings, halftones, and line/halftone combinations given below.):

*EPS :* Vector drawings. Embed the font or save the text as "graphics"

*TIFF :* Colour or greyscale photographs (halftones): always use a minimum of 300 dpi.

*TIFF:* Bitmapped line drawings: use a minimum of 1000 dpi.

*TIFF :* Combinations bitmapped line/half-tone (colour or greyscale): a minimum of 500 dpi is required.

DOC, XLS or PPT: If your electronic artwork is created in any of these Microsoft Office applications please supply "as is".

**Please do not :**

- Supply embedded graphics in your wordprocessor (spreadsheet, presentation) document;
- Supply files that are optimized for screen use (like GIF, BMP, PICT, MWPG;) the resolution is too low;
- Supply files that are too low in resolution;
- Submit graphics that are disproportionately large for the content.

## Colour illustrations

If, together with your accepted article, you submit usable colour figures then Elsevier will ensure, at no additional charge, that these figures will appear in colour on the web (e.g. Science Direct and other sites) regardless of whether or not these illustrations are reproduced in colour in the printed version. For colour reproduction in print, you will receive information regarding the costs from Elsevier after receipt of your accepted article. For further information on the preparation of electronic artwork, please see http://authors.elsevier.com/artwork.

***Please note :*** Because of technical complications which can arise by converting colour figures to 'grey scale' (for the printed version should you not opt for colour in print) please submit in addition usable black and white prints corresponding to all the colour illustrations.

Mark the appropriate Position of a figure in the article.

## Supplementary files

Supplementary files offer the author additional possibilities to publish supporting applications, movies, animation sequences, high-resolution images, background datasets, sound clips and more. Supplementary files supplied will be published on line alongside the electronic version of your article in Elsevier web products, including ScienceDirect:http://www.sciencedirect.com. In order to ensure that your submitted material is directly usable, please ensure that data is provided in one of our recommended file formats. Authors should submit the material in electronic format together with the article and supply a concise and descriptive caption for each file. For more detailed instructions please visit the journal's home page and click on the left hand side link to the Guide of Authors.

## References

1. All publications cited in the text should be presented in a list of references following the text of the manuscript. The manuscript should be carefully checked to ensure that the spelling of author's names and dates are exactly the same in the text as in the reference list.

2. In the text refer to the author's name (without initial) and year of publication, followed – if necessary – by a short reference to appropriate pages.

   ***Examples :*** "Since Peterson (1988) has shown that ..." "This is in agreement with results obtained later (Kramer, 1989, pp. 12-16)".

3. If reference is made in the text to a publication written by more than two authors the name of the first author should be used followed by "et al." This indication, however, should never be used in the list of references. In this list names of first author and co-authors should be mentioned.

4. References cited together in the text should be arranged chronologically. The list of references should be arranged alphabetically on authors' names.

And chronologically per author. If an author's name in the list is also mentioned with co-authors the following order should be used: publications of the single author, arranged according to publication dates - publications of the same author with one coauthor – publications of the author with more than one co-author. Publications by the same author(s) in the same year should be listed as 1974a, 1974b, etc.

5.  Use the following system for arranging your references:

(a)    For periodicals

Schmitt, E.R. Feucht, W., 1993. Content of linolenic acid in senescing cherry leaves. Sci.Hort. 55, 273-282.

(b)    *For edited symposia, special issues, etc., published in a periodical*

Rice, K., 1992. Theory and conceptual issues. In: Gall, G.A.E. and Staton, M. (Eds.), Integrating Cjonservation Biology and Agricultural Production. Agriculture Ecosystmes and Environment, 42, 9-26.

(c)    For books

Guagh Jr., H.G., 1992. Statistical Analysis of Regional Yield Trials. Elsevier, Amsterdam.

(d)    *For multi-author books*

Baker, J., 1993. Insects. In: de Hertogh, A. and Le Nard, M. (Eds.), The Physiology of Flower Bulbs, Elsevier, Amsterdam, pp. 101-153.

6.  Abbreviate the titles of periodicals mentioned in the list of references, according to the International *List of Periodical Title Word Abbreviations*.

7.  In the case of publications in any language other than English, the original title is to be retained. However, the titles of publications in non-Latin alphabets should be transliterated, and a notation such as "(in Russian)" or "(in Greek, with English abstract)" should be added.

8.  Work accepted for publication but not yet published should be refered to as "in press".

9.  References concerning unpublished data and "personal communications" should not be cited in the reference list but may be mentioned in the text.

The digital object identifier (DOI) may be used to cite and link to electronic documents. The DOI consists of a unique alpha-numeric character string which is assigned to a document by the publisher upon the initial electronic publications. The assigned DOI never changes. Therefore, it is an ideal medium for citing a document, particularly 'Articles in press' because they have not yet received their full bibliographic information.

The correct format for citing a DOI is shown as follows (example taken from a document in the journal *Physics Letters B)*: doi: 10.1016 j.physletb.2003.10.071

When you see the DOI to create URL hyperlinks to documents on the web, they are guaranteed, never to change.

## Formulae

1.  Subscripts and superscripts should be clear and not too small.

2.  Give the meaning of all symbols immediately after the equation in which they are first used.

3.  For simple fractions use the solidus(/) instead of a horizontal line, e.g. $I_p/2_m$ rather than $I_p/2_m$.

4.  Equations should be numbered serially at the right-hand side in parentheses. In general only equations explicitly referred to the text need be numbered.

5.  The use of fractional powers instead of root signs is recommended. Also powers of e are often more conveniently denoted by exp.

6.  Levels of statistical significance which can be mentioned without further explanation are $*P<0.05$, $** P<0.01$ and $***P0.001$.

7.  In chemical formulae, valence of ions should be given as , e.g. $Ca^{2+}$, and $CO^{2-}_3$, not as $Ca^{++}$ or $CO^-_3$.

8.  Isotope number should precede the symbols, e.g., $^{18}O$.

9.  The repeated writing of chemical formulae in the text is to be avoided where reasonably possible; instead the name of the compound should be given in full. Exceptions may be made in the case of a very long name occurring very frequently or in the case of a compound are being described as the end product of a gravimetric determination (e.g, phosphate as $P_2O_5$).

## Footnotes

1.  Footnotes should  only be used if absolutely essential. In most case it should be possible to incorporate the information in normal text.

2.  If used, they should be numbered in the text, indicated by superscript numbers, and kept as short as possible.

## Nomenclature

1.  Authors and editors are, by general agreement, obliged to accept the rules governing biological nomenclature, as laid down in the *International Code of Botanical Nomenclature,*  the  *International Code of Bacteria,* and the *International Code of Zoological  Nomenclature.*

2.  All biotica (crops, plant, insect, birds, mammals, etc.) should be identified by their scientific names when the English term is first used, with the exception of common domestic animals.

3.  All biocides and other organic compounds must be identified by their Geneva names when first used in the text. Active ingredients of all formulations should be likewise identified.

4. For chemical nomenclature, the conventions of the *International Union of Pure and Applied Chemistry* and the official recommendations of the *IUPAC-IUB Combined Commission on Biochemical Nomenclature* should be followed.

## Proofs

When your manuscript is received by the Publisher it is considered to be in its final form. Proofs are not to be regarded as 'drafts'.

One set of page proofs in PDF format will be sent by e-mail to the corresponding author, to be checked for typesetting/editing. No changes in, or additions to , the accepted (and subsequently edited) manuscript will be allowed at this stage. Proofreading is solely your responsibility.

The Publisher reserves the right to proceed with publication if corrections are not communicated.

Return corrections within 2 days of receipt of the proofs. Should there be no corrections, please confirm this.

Elsevier will do everything possible to get your article corrected and published as quickly and accurately as possible. In order to do this we need your help. When you receive the (PDF) proof of your article for correction, it is important to ensure that all of your corrections are sent back to us in one communication. Subsequent corrections will not be possible, so please ensure your first sending is complete. Note that this does not mean you have any less time to make your corrections, just that only one set of corrections will be accepted.

## Offprints

1. Twenty-five offprints will be supplied free of charge.
2. Additional offprints can be ordered on an offprints order form, which is included with the proofs.
3. UNESCO coupons are acceptable in in payment of extra offprints.

**Information about *Scientia Horticulturae* is available**

**On the World Wide Web at the following address:**

**http://www.elsevier.com/locte/scihorti.**

**Scientia Horticulture has no page charges**

# CHAPTER - 111

# Seed Science and Technology

## Instructions to Contributors

### Introduction

*Seed Science and Technology* publishes original papers and review articles in all areas of seed science and technology, including germination, storage, physiology, processing and pathology.

### Procedure for the submission of papers

Manuscripts shall be written in English, in a language which is grammatically correct, clear and to the point. If you are not a native English speaker, you should consult an expert in order to ensure that the language is correct. Manuscripts should be submitted preferably via E-mail to **sst@ista.**ch or on paper k(two copies and a diskette). The manuscripts should be sent to the Chief Editor, Dr.Anne Bulow-Olsen, Furesf Parkvej 23,DK-2830 Virum, Denmark. The Chief Editor will notify the authors on acceptance and whether or not revisions are needed.

When a paper is finally accepted and any necessary revisions made, authors are requested to submit a final digital version, also preferably via e-mail or alternatively on a 3 ½ inch diskette or a **CD-ROM**. Suitable formats are Word and WordPerfect for PC's. Word processing systems other than those mentioned above are normally acceptable if MSDOS compatible.

When proofs are ready, a **page charge** of Swiss Francs **CHF**25 per printed page (maximum charge Swiss Francs CHF150) must be paid. ISTA members are not subjected to page charges. After payment of the page charges, proofs will be forwarded by E-mail to authors for corrections. These should be restricted to printer's errors. Other substantial alterations can be undertaken only at the author's expense. Corrections must be sent to the editor within three weeks, preferably via E-mail.

Twenty-five **reprints** are provided free. Further copies may be ordered when returning proofs. In the case of joint authorship, proofs, reprint order forms and free reprints will be sent to the corresponding author only.

## Instructions for full-length papers

Papers should be typed double-spaced on one of A4 (210 × 297 mm) paper, with wide margins: all sheets should be numbered consecutively. A shortened version of the title, for use as **running heading**, should be inserted in the top right-hand corner of the first page. The title should be concise but informative, containing key words which describe the subject matter for use in abstracting systems. The **name of the author**(s) and the **full postal address** of the institution where the work was carried out should follow the title. Authors are asked to give an E-mail address where possible. The E-mail address should be given in parentheses after the postal address. A **summary** indicating the scope of the paper in not more than 200 words, should be typed immediately below this. Papers describing experimental work should normally have the subject matter grouped under six main headings: **Introduction, Materials and Methods, Results, Discussion, Acknowledgements, References**. Main headings are printed in heavy type, secondary headings in italics. A scientific ('Latin') name included in an italicized heading appears in Roman type. Abbreviations should be defined the first time they are used, although initial letters, without punctuation, may be used for easily recognized organisations or countries (for example ISTA, USA, FAO).

Capital letters and Roman numerals should be used in both headings and text only when essential (figure 2, not Fig.II or Figure II). Except when occurring in a series of numbers or in conjunction with a specific recognised unit of measurement, numbers from one to nine should be spelt out (8 mm, 5 g . m$^{-2}$, but nine hours, three months). The percentage sign should be used in conjunction with a number (7%), but percentage should be written in full when used as a noun (percentage germination, not per cent germination or % germination). Measurements should be expressed metrically, whenever possible following the Systeme International d' Unites (SI) (published by OFFILIB, 48 rue Gay-Lussac, F-75005 Paris).

Species should be described by their scientific ('Latin') names; at the first mention in the main text the full binomial and authority must be given, but subsequently the genus should be abbreviated to its initial letter and the authority omitted. Authorities are not quoted after Latin names in the title or summary. When a name has been stabilized by ISTA in the ISTA List of Stabilized Plant Names (see http://www.ars-grin.gov/~ sbmljw/ istaintrod.html), synonyms must not be used. For species not included in this list authors are advised to consult http://www.ars-gring.gov/npgs/tax/.

**References** in the text should be cited by author's surname and year of publication in brackets after the statement they support. If there are more than two authors, citations should quote the surname of the first author and the words "*et al.*" All names should be included in the list of references. At the end of the paper, references should be listed alphabetically in the form: Name, Initials (year). Title of journal or other source, volume number, first and last page numbers. The titles of periodicals should be written without abbreviation. Titles of book must also be given in full, together with the publisher's name and the place of publication. The following are correct:

Author, A.B., Author, C.D. and Author, E.F. (1999). Title of article. *Journal Title in Full,* **00**, 123-456.

International seed Testing Association (1999). International Rules for Seed Testing. *Seed Science and Technology,* **27, Supplement**, 333pp.

Author, A.B. and Author, C.D.(1999). Chapter in book, In Book Title, (eds.E.F.Editor and G.H. Editor). vol.3, pp.123-456, Publisher, City.

When a paper is cited which has been published in another language, its title may be translated. The translation should appear between square brackets, as follows:

Author, G.H. and Author, I.J.(1999). Title another language. [Title in English]. *Journal Title in Full, 00,* 12-34.

Tables and figure captions should be typed on separate sheets from the main text and numbered in separate series. **Figures** should be supplied as computer graphics (preferable compatible with Word or WordPerfect). Each figure should be submitted as a separate file. Shadings should be avoided in the figures. Lettering should be kept to a minimum and should be at least 2 mm in the final print. Photographs should only be submitted if they are essential to an understanding of the paper, and may be submitted as scans. In exceptional circumstances based on technical necessity, colour Prints may published and the extra costs will be charged to the authors.

## Special instructions for short reports (Research Notes)

*'Research Notes'* are appropriate for the reporting of research which might not justify a full paper but which nevertheless contains important information of potentially practical importance. For example, a study of novel dormancy-breaking treatments in a series of closely related species might justify a single full length paper describing the principles and experimental basis of the work, followed by a number of 'Research Notes' reporting data for particular species of economic importance.

*'Research Notes'* are also appropriate in the case of early but scientifically rigorous experiments in a novel area where a full length paper is not yet justified but the data warrant immediate publication because of their novelty and importance.

The format for *'Research Notes'* is as for full-length papers except for the following (1) They should be a maximum of two sides of printed text plus a maximum of two figures or photographs. (2) The section headings will: 'Summary'; 'Experimental and Discussion'; 'Acknowledgements'; and 'References'. (*Note:* The 'Experimental and Discussion' section will be included a single narrative text those elements are normally separated in full-length papers, as 'Introduction', 'Materials and Methods', 'Results', and 'Discussion').

## Review articles

Reviews of interest to those working in the seeds sector are welcomed. The content of reviews should have a strong technical base but papers dealing with the economic, infrastructural, or legal aspects of seeds in combination with technical aspects are also published.

## Copyright

Submission of a manuscript implies that the work described has not been published elsewhere, except in the form of poster, an abstract or a thesis, that it is not under consideration for publication elsewhere and that all co-authors have approved the manuscript.

The International Seed Testing Association retains the copyright for the papers published in this journal. Photocopies of papers may be made for personal use. For permission to reproduce other than single copies, apply to Chief Editor.

An abstract of the published papers will appear on the ISTA web site: www.seedtest.org.

# CHAPTER - 112

# Seed Research

## Guidelines to Contributors

**Seed Research** publishes original papers and review articles in all areas of seed science e.g. production, sampling, testing, processing, physiology, entomology, pathology, storage, marketing, etc. Two issues per year are published.

Manuscripts for both the full papers, as well as short notes in English may be sent to the Chief Editor, Seed Research, Division of Seed Science and Technology, Indian Agricultural Research Institute, New Delhi 110012, India. It should be sent in duplicate including illustrations, typed on bond paper in double space with wide margins.

On receipt of an article an acknowledgment giving the registration number of the paper is sent to the author. The number should invariably be quoted while making further enquiry about its publications.

The title should be informative. The name of the authors and of the institution where the work was carried out should follow the title. Authors should give their E-mail address where ever possible. The **E-mail** address should be given in parentheses after postal address. The abstract of not more than 200 words indicating the main findings of the paper should follow.

The paper may have main headings-Abstract, Introduction, Materials and Methods, Results, Discussion and References. However, Results and Discussion may be combined to avoid repetitions. The introductory part should be brief and should emphasise the aim of the experiment.

Reference should be cited in the text by number in brackets []. References should be numbered according to their order appearing in the text. References should be listed in the following format:

1. BHAN, V.M. & R.A. MAURYA (1972). Controlling weeds in wheat. *Indian Farmer's Digest* 5:23-25.

2. ANONYMOUS (1985). International rules for seed testing. *Seed Sci. & Technol.* 13: 307-320.

3. FEKETE, M.A.R. & G.H. VIEWEG (1976). Starch metabolism: synthesis vs. degradation pathways. In: *Plant Carbohydrate Biochemistry*, ed. J.B.Pridham, pp.127-143, Academic Press, London.

4. MAYER, A.M. & A. POLJAKOFF-MAYBER (1989). *The Germination of Seeds*. Pergamon Press. U.K.


Periodicals should be abbreviated according to the Style Manual for Biological Journals, American Institute of Biological Sciences, Washington.

Table should be typed on separate sheets. Photographs should be glossy prints of good quality. They should be about twice the size they will appear in the Journal. Measurements should be expressed metrically. On the top of the first page provide a short running title of less than 40 characters.

When a paper is finally accepted, authors should submit the final corrected version on a computer diskette in MS word or as an E-mail attachment along with two hard copies. The graphic should be drawn using MS Excel.

AUTHORS SHOULD CONSULT A RECENT ISSUE OF THE **Seed Research** BEFORE TYPING.

Proofs will be sent to authors for correction. Although every effort is made by the editors to correct the proof, they assume no responsibility for any error that may remain in the final printing.

Two reprints are supplied free of cost to the first author.,

**E-mail: cd_seedresearch@yahoo.com**

# Soil and Tillage Research

## Guidelines to Contributors

### Aims and Scope

This journal is concerned with the changes in the physical, chemical and biological parameters of the soil environment brought about by soil tillage and field traffic, their effects on both below and above ground environmental quality, crop establishment, root development and plant growth, and the interactions between these various effects. This implies research on: characterization or modelling of tillage and field traffic effects on the soil environment; the selection, adaption of development of tillage systems (including reduced cultivation and direct drilling) suitable for specific conditions of soil, climate, topography, irrigation and drainage, crops and crop rotations, intensities for fertilization, degree of mechanization, etc. and the appropriate use of tillage systems to maintain a balance between acceptable crop production, sustainability and minimum environmental impacts. In this context, papers on the characterization or modelling of tillage effects on: soil physical, chemical and biological properties, processes related to surface and subsurface groundwater quality, soil erosion, carbon and nutrient cycling and crop production, are; most welcome. Papers on soil deformation processes, soil-working tools and traction devices, energy requirements and economic aspects of tillage are also considered. Attention will also be given to the role of tillage in weed, pest and disease control.

## Types of contribution

1. Original research papers (Regular Papers)
2. Review articles
3. Short communications

*Original research papers* should report the results of original research. The material should not have been previously published elsewhere, except in a preliminary form.

*Review articles* should cover subjects falling within the scope of the journal which are of active current interest. They may be submitted or invited.

*A Short communication* is a concise but complete description of a limited investigation, which will not be include in a later paper. Short Communications should be as completely document, both by reference to the literature and description of the experimental procedures employed, as a regular paper. They should not occupy more than 6 printed pages (about 12 manuscript pages, including figures, tables and references).

## Submission of manuscripts

Submission of an article implies that the work described has not been published previously (except in the form of an abstract or as part of a published lecture or academic thesis), that it is not under consideration for publication elsewhere, that its publication is approved by all authors and tacitly or explicitly by the responsible authorities where the work was carried out, and that, if accepted, it will not be published elsewhere in the same form, in English or in any other language, without the written consent of the Publisher. Upon acceptance of an article, authors will be asked to transfer copyright (for more information on copyright see http://authors.elsevier.come). This transfer will ensure the widest possible dissemination of information. A letter will be sent to the corresponding author confirming receipt of the manuscript. A form facilitating transfer of copyright will be provided. If excerpts from other copyrighted works are included, the author(s) must obtain written permission from the copyright owners and credit the source(s) in the article. Elsevier has preprinted forms for use by authors in these cases: contact ELSEVIER, Rights department, P.O. Box 800, Oxford, OX5 IDX, UK; phone: (+44) 1865 843830, fax: (+44) 1865 853333, e-mail: permissions@elsevier.com. Requests may also be completed online via the Elsevier homepage (http://www.elsevier.com/locate/permissions).

Papers for consideration should be submitted to: Elsevier Editorial System

## Electronic manuscripts

Submission to this journal proceeds totally on-line. Use the following guidelines to prepare your article. Via the Author Gateway page of this journal (see http://authors.elsevier.com), you will be guided stepwise through the creation and uploading of the various files. Once the uploading is done, our system automatically generates an electronic (PDF) proof, which is then used for reviewing. It is crucial that all graphical elements be uploaded in separate files. So that the PDF is suitable for reviewing. Authors can upload their article as a LaTex, Microsoft (MS) Word, WordPerfect, PostScript or Adobe Acrobat PDF document. All correspondence, including notification of the Editor's decision and requests for revisions, will be by e-mail.

## Electronic format requirements for accepted articles

We accept most wordprocessing formats, but Word, WordPerfect or LaTeX is preferred. Always keep a backup copy of the electronic file for reference and safety. Save your files using the default extension of the program used.

## Wordprocessor documents

It is important that the file be saved in the native format of the wordprocessor used. The text should be in single-column format. Keep the layout of the text as simple as possible. Most formatting codes will be removed and replaced on processing the article. In particular, do not use the workdprocessor's options to justify text or to hyphenate words. However, do use bold face, italics, subscripts, superscripts etc. Do not embed 'graphically designed' equations or tables, but prepare these using the wordprocessor's facility. When preparing tables, if you are using a table grid, use only one grid for each individual table and not a grid for each row. If no grid is used, use tabs, not spaces, to align columns. The electronic text should be prepared in a way very similar to that of conventional manuscripts (see also the Author Gateway's Quickguide (http: authors.elsevier.come). Do not import the figures into the text file but, instead, indicate their approximate locations directly in the electronic text and on the manuscript. See also the section on Preparation of electronic illustrations.

To avoid unnecessary errors you are strongly advised to use the 'spellchecker' function of your wordprocessor.

## Preparation of manuscripts

1. Manuscripts should be written in English. Authors whose native language is not English are strongly advised to have their manuscripts checked by an English-speaking colleague prior to submission.

   **Language Polishing :** For authors who require information about language editing and copyediting services plre- and post-submission please visit http://www.elsevier.com/wps/find/authorshome.authors/languagepolishing or contact authorsupport@elsevier.com for more information. Please note Elsevier neither endorses nor takes responsibility for any products, goods or services offered by outside vendors through our services or in any advertising. For more information please refer to our Terms & Conditions http://www.elsevier.com/wps/find/termsconditions.cws_home/termsconditions.

2. **Authors are requested to submit, with their manuscripts, the names and addresses and email address of four potential referees.**

3. Manuscripts should be prepared **with numbered lines, with wide margins and double spacing throughout, i.e. also for abstracts, footnotes and references. Every page of the manuscript, including the title page, references, tables, etc. should be numbered.** However, in the text no reference should be made to page numbers; if necessary, one may refer to sections. Underline words that should be in italics, and do not underline any other words. Avoid excessive use of italics to emphasize part of the text.

4. Manuscripts in general should be organized in the following order:
   - Title (should be clear, descriptive and not too long)
   - Name(s) of author(s)
   - Complete postal address (es) of affiliations
   - Full telephone, E-mail and Fax No. of the corresponding author
   - Present address (es) of author(s) if applicable
   - Complete correspondence address and e-mail address to which the proofs should be sent
   - Abstract
   - Keywords (indexing terms), normally 3-6 items
   - Introduction
   - Material studied, area descriptions, methods, techniques
   - Results
   - Discussion
   - Conclusion
   - Acknowledgements and any additional information concerning research grants, etc.
   - Research grants, etc.
   - References
   - Tables
   - Figure captions

5. In typing the manuscript, titles and subtitles should not be run within the text. They should be typed on a separate line, without indentation. Use lower-case letter type.

6. SI units should be used.

7. Elsevier reserves the privilege of returning to the author for revision accepted manuscripts and illustrations which are not in the proper form given in this guide.

## Abstracts

The abstract should be clear, descriptive and not longer than 400 words.

## Tables

1. Authors should take notice of the limitations set by the size and lay-out of the journal. Large tables should be avoided. Reversing columns and rows will often reduce the dimensions of a table.

2. If many data are to be presented, an attempt should be made to divide them over two or more tables.

3. Drawn tables, from which prints need to be made, should not be folded.

4.  Tables should be numbered according to their sequence in the text. The text should include references to all tables.

5.  Each table should be typewritten on a separate page of the manuscript. Tables should never be included in the text.

6.  Each table should have a brief and self-explanatory title.

7.  Column headings should be brief but sufficiently explanatory. Standard abbreviations of units of measurement should be added between parentheses.

8.  Vertical lines should not be used to separate columns. Leave some extra space between the columns instead.

9.  Any explanation essential to the understanding of the table should be given as a footnote at the bottom of the table.

## Electronic Illustrations

Submitting your artwork in an electronic format helps us to produce your work to the best possible standards, ensuring accuracy, clarity and a high level of detail.

1.  Always supply high-quality printouts of your artwork, in case conversion of the electronic artwork is problematic.

2.  Make sure you use uniform lettering and sizing of your original artwork.

3.  Save text in illustrations as "graphics" or enclose the font.

4.  Only use the following fonts in your illustrations: Arial, Courier, Helvetica, Times, Symbol.

5.  Number the illustrations according to their sequence in the text.

6.  Use a logical naming convention for your artwork files, and supply a separate listing of the files and the software used.

7.  Provide all illustrations as separate files. .

8.  Provide captions to illustrations separately.

9.  Produce images near to the desired size of the printed version.

A detailed guide on electronic artwork is available on our website:

http://authors.elsevier.com/artwork

**You are urged to visit this site; some excerpts from the detailed information are given here.**

## Formats

Regardless of the application used, when your electronic artwork is finalised, please "save as" or convert the images to one of the following formats (Note the resolution requirements for line drawings, halftones, and line/halftone combinations given below.):

*EPS :* Vector drawings.  Embed the font or save the text as "graphics".

*TIFF :* Colour or greyscale photographs (halftones): always use a minimum of 300 dpi

*TIFF :* Bitmapped line drawings: use a minimum of 1000 dpi.

*TIFF :* Combinations bitmapped line/half-tone (colour or greyscale): a minimum of 500 dpi is required.

*DOC,XLS(or) :* If your electronic artwork is created in any of these Microsoft Office

*PPT :* applications please supply "as is".

**Please do not :**
- Supply embedded graphics in your wordprocessor (spreadsheet, presentation) document:
- Supply files that are optimised for screen use (like GIF, BMP, PICT,WPG); the resolution is too low;
- Supply files that are too low in resolution:
- Submit graphics that are disproportionately large for the content.

## Non-electronic Illustrations

For illustrations that are unable to be uploaded electronically hard copies will be accepted.

**Please send to :**

The Editorial Office of Soil & Tillage Research

Elsevier Ireland Ltd.

Brookvale Plaza

East Park

Shannon

Co. Clare

Ireland

Provide all illustrations as high-quality printouts, suitable for reproduction (which may include reduction) without retouching.  Number illustrations consecutively in the order in which they are referred to in the text.  Clearly mark all illustrations on the back (or-in case of line drawings-on the lower front side) with the figure number and the author's name and, in cases of ambiguity, the correct orientation.  Mark the appropriate position of a figure in the article.

## Colour illustrations

Submit colour illustrations as original photographs, high-quality computer prints or transparencies, close to the size expected in publication, or as 35mm slides.

Please make sure that artwork files are in an acceptable format (TIFF, EPS or MS Office files) and with the correct resolution. Polaroid colour prints are not suitable. If, together with your accepted article, you submit usable colour figures then Elsevier will ensure, at no additional charge, that these figures will appear in colour on the web (e.g., ScienceDirect and other sites) regardless of whether or not these illustrations are reproduced in colour in the printed version. Please indicate your preference for colour on the web (free of charge) or in print and on the web (free of charge) or in print and on the web (charged) when submitting your article.

For colour reproduction in print, you will receive information regarding the costs from Elsevier after receipt of your accepted article. For further information on the preparation of electronic artwork, please see http://authors.elsevier.com/artwork.

Please note: Because of technical complications which can arise by converting colour figures to grey scale (for the printed version should you opt to not pay for colour in print) please submit in addition usable black and white prints corresponding to all the colour illustrations.

As only one figure caption may be used for both colour and black and white versions of figures, please ensure that the figure captions are meaningful for both versions, if applicable.

## References

1. All publications cited in the text should be presented in a list of references following the text of the manuscript. The manuscript should be carefully checked to ensure that the spelling of author's names and dates are exactly the same in the text as in the reference list.

2. In the text refer to the author's name (without initial)  and year of publication, followed – if necessary – by a short reference to appropriate pages. Examples: "Since Peterson (1988) has shown that…" "This is in agreement with results obtained later (Kramer, 1989, pp. 12-16)".

3. If reference is made in the text to a publication written by more than two authors the name of the first author should be used followed by "et al." This indication, however, should never be used in the list of references. In this list names of first author and co-authors should be mentioned.

4. References cited together in the text should be arranged chronologically. The list of references should be arranged alphabetically on authors' names, and chronologically per author. If an author's name in the list is also mentioned with co-authors the following order should be used: publications of the single author, arranged according to publication dates – publications of the same author with one co-author. Publications by the same author(s) in the same year should be listed as 1994a, 1994b, etc.

5.  Use the following system for arranging your references:

    (a)  *For periodicals*

    Hettiaratchi, D.R.P., 1993. The development of a powered low draught tine cultivator. Soil Tillage Res. 28, 159-177

    (b)  *For edited symposia, special issues, etc., published in a periodical*

    Benites, J.R., Ofori, C.S., 1993. Crop production through conservation – effective tillage in the tropics. In: Lal, R. (Ed.), Soil Tillage for Agricultural Sustainability. Proceedings of the 12th Conference of ISTRO, 8-12 July 1991, Ibadan, Nigeria. Soil Tillage Res. 27, 9-33.

    (c)  *For books*

    Russell, E.W., 1973. Soil Conditions and Plant Growth. 10th ed. Longmans, London.

    (d)  *For multi-author books*

    Kuipers, H., Koolen, A.J., 1989. Interface between implements, tillage and soil structure. In: Larson, W.E.., Blake, G.R., Allmaras, R.R. Voorhees, W.B., Gupta, S.C. (Eds.), Mechanics and Related Processes in Structured Agricultural Soils, NATO ASI Series E, vol. 172. Kluwer Academic Publishers, Dordercht, Netherlands, pp. 105-120.

6.  Abbreviate the titles of periodicals mentioned in the list of references according to the International *List of Periodical Title Word Abbreviations*.

7.  In the case of publications in any language other than English, the original title is to be retained.  However, the titles of publications in non-Roman alphabets should be transliterated, and a notation such as "(in Russian)" or "(in Greek, with English abstract)" should be added.

8.  Work accepted for publication but not yet published should be referred to as "in press".

9.  References concerning unpublished data and "personal communications" should not be cited in the references list but may be mentioned in the text.

## Formulae

1.  Formulae should be typewritten, if possible. Leave ample space around the formulae.

2.  Subscripts and superscripts should be clear.

3.  Geek letters and other non-Roman or handwritten symbols should be explained in the margin where they are first used.  Take special care to show clearly the difference between zero (0) and the letter O, and between one (1) and the letter 1.

4.  Give the meaning of all symbols immediately after the equation in which they are first used.

5. For simple fractions use the solidus (/) instead of a horizontal line.

6. Equations should be numbered serially at the right-hand side in parentheses. In general only equations explicitly referred to in the text need be numbered.

7. The use of fractional powers instead of root signs is recommended. Also powers of e are often, more conveniently denoted by exp.

8. Levels of statistical significance which can be mentioned without further explanation are * $P < 0:05$, **$P < 0:01$ and ***$P < 0:001$

9. In chemical formulae, valence of ions should be given, e.g. as $Ca^{2+}$ and $CO_3^{2}$ rather than as $Ca^{++}$ and $CO_3^{-}$.

10. Isotope numbers should precede the symbol, e.g. $^{18}O$

11. The repeated writing of chemical formulae in the text is to be avoided where reasonably possible; instead, the name of the compound should be given in full. Exceptions may be made in the case of a compound being described as the end product of a gravimetric determination (e.g. phosphate as $P_2O_5$).

## Footnotes

1. Footnotes should only be used if absolutely essential. In most cases it should be possible to incorporate the information in normal text.

2. If used, they should be numbered in the text, indicated by superscript numbers, and kept as short as possible.

## Nomenclature

1. Authors and Editor(s) are, by general agreement, obliged to accept the rules governing biological nomenclature, as laid sown in the *International code of Botanical Nomenclature,* the *International*

2. All biotica (crops, plants, insects, birds, mammals, etc.) should be identified by their scientific names when the English term is first used, with the exception of common domestic animals.

3. All biocides and other organic compounds must be identified by their Geneva names when first used in the text. Active ingredients of all formulations should be likewise identified.

4. For chemical nomenclature, the conventions of the *International Union of Pure and Applied Chemistry* and the official recommendations of the IUPAC-IUB *Combined Commission on Biochemical Nomenclature* should be followed.

## Supplementary data

Elsevier now accepts electronic supplementary material to support and enhance your scientific research. Supplementary files offer the author additional possibilities to publish supporting applications, movies, animation sequences, high-resolution images, background datasets, sound clips and more.

Supplementary files supplied will be published online alongside the electronic version of your article in Elsevier web products, including Science Direct: http://www.sciencedirect.com. In order to ensure that your submitted material is directly usable, please ensure that data is provided in one of our recommended file formats. Authors should submit the material in electronic format together with the article and supply a concise and descriptive caption for each file. For more detailed instructions please visit our Author Gateway at http://authors.elsevier.com.

**US National Institutes of Health (NIH) voluntary posting ("Public Access") policy** Elsevier facilitates author response to the NIH voluntary posting request (referred to as the NIH "Public Access Policy"; see (http://www.nih.gov/about/publicaccess/index.htm) by posting the peer-reviewed author's manuscript directly to PubMed Central on request from the author, 12 months after formal publication. Upon notification from Elsevier of acceptance, we will ask you to confirm via e-mail (by e-mailing us at NIH funding and that your intend to respond to the NIH policy request, along with your NIH award number to facilitate processing. Upon such confirmation, Elsevier will submit to PubMed Central on your behalf a version of your manuscript that will include peer-review comments, for posting 12 months after formal publication. This will ensure that your will have responded fully to the NIH request policy. There will be no need for you to post your manuscript directly with PubMed Central, and any such posting is prohibited.

## Copyright

1. An author, when quoting from someone else's work or when considering reproducing an illustration or table from a book or journal article, should make sure that he is not infringing a copyright.

2. Although in general an author may quote from other published works, he should obtain permission from the holder of the copyright if he wishes to make substantial extracts or to reproduce tables, plates, or other illustrations. If the copyright-holder is not the author of the quoted or reproduced material, it is recommended that the permission of the author should also be sought.

3. Material in unpublished letter and manuscripts is also protected and must not be published unless permission has been obtained.

4. A suitable acknowledgement of any borrowed material must always be made.

## Proofs

When your manuscript is received at the Publisher it is considered to be in its final form. Publisher it is considered to be in its final form. Proofs are not to be regarded as 'drafts'.

One set of page proofs in PDF format will be sent by e-mail to the corresponding author, to be checked for typesetting/editing. No changes in, or additions to, the accepted (and subsequently edited) manuscript will be allowed at this stage. Proofreading is solely your responsibility.

A form with queries from the coy editor may accompany your proofs. Please answer all queries and make any corrections or additions required.

The Publisher reserves the right to proceed with publication if corrections are not communicated. Return corrections within two working days of receipt of the proofs. Should there be corrections, please confirm this.

Elsevier will do everything possible to get your article corrected and published as quickly and accurately as possible. In order to do this we need your help. When you receive the (PDF) proof of your article for correction, it is important to ensure that all of your corrections are sent back to us in one communication. Subsequent corrections will not be possible, so please ensure your first sending is complete. Note that this does not mean you have any less time to make your corrections, just that only one set of corrections will be accepted.

## Offprints

1. Twenty five offprints will be supplied free of charge.
2. One hundred free offprints will be supplied to the first author of a review article.
3. Additional offprints can be ordered on an offprints order form, which is included with the proofs.
4. UNESCO coupons are acceptable in payment of extra offprints.

## Author Service

Authors can also keep a track on the progress of their accepted article, and set up e-mail alerts informing them of changes to their manuscript's status, by using the "Track a Paper" feature of Elsevier's Author Gateway http:// authors.elsevier.com. For privacy, information on each article is password-protected. The author should key in the "Our Reference" code (which is in the letter of acknowledgement sent by the publisher on receipt of the accepted article) and the name of the corresponding author. In case of problems or questions, authors may contact the Author Service Department, E-mail: authorsupport@elsevier.com.

*Soil & Tillage Research* Carries no page charge

# CHAPTER - 114

# Soil Biology and Biochemistry

## Guide For Authors

This journal is a forum for research on soil organisms, their biochemical activities and their influence on the soil environment and plant growth. It publishes original work on quantitative, analytical and experimental aspects of such research. Soil biology and soil biochemistry cover many scientific disciplines but a single journal brings together the results and views of research workers working in a wide variety of research areas. The scope of this journal is wide and embraces accounts of original research on the biology, ecology and biochemical activities of all forms of life that exist in the soil environment. Some of the subjects which have proved to be prominent are the biological transformations of plant nutrients in soil, nitrogen fixation and denitrification, soil-borne phases of plant parasites, the ecological control of soil-borne pathogens, the influence of pesticides on soil organisms, the biochemistry of pesticide and pollution decomposition in soil, microbial aspects of soil pollution, the composition of soil populations, modelling of biological processes in soil systems, the biochemical activities of soil organisms, soil enzymes and the interactions of soil organisms with plants and the effects of tillage on soil organisms and soil biochemistry.

## Types of contribution

1. *Regular papers.* Original full-length research papers which have not been published previously, except in a preliminary form, may be submitted as regular papers.

2. *Short communications.* These should not exceed 1200 words (three printed pages) or their equivalent, excluding references and legends. Submissions should include a short abstract not exceeding 10% of the length of the communication and which summarizes briefly the main findings of the work to be reported. The bulk of the text should be in a continuous form that does not require numbered sections such as Introduction, Materials and methods, Results and Discussion. However, a Cover page, Abstract and a list of Keywords are required at the beginning of the communication and Acknowledgements and References at the end. These components are to be prepared in the same format as used for full-length research

papers. Occasionally authors may use sub-titles of their own choice to highlight sections of the text.

3. *Review articles**. Review articles are welcome but should be topical and not just an overview of the literature. Before submission please contact one of the Chief Editors.

4. *News and Views*. Authors may submit comments and views on any subject covered by the Aims and Scope. The article should be about 1200 words, and submitted to a Chief Editor.

5. *Letters to the Editor.* Letters are published from time to time on matters of topical interest. These should be submitted directly to one of the Editors-in-Chief.

**Review Articles For Soil Biology & Biochemistry*** 1. Readers of Soil Biology & Biochemistry may submit reviews on any topic that falls within the scope of this journal. Authors of Review Articles should aim to provide facts as well as qualified ideas and opinions derived from reliable and relevant publications. Then, from such material develop reasoned arguments and questions for future evaluation and research. Reviewers should provide a list of relevant and appropriate references. They should avoid introducing new facts in the form of unpublished data or personal communications. Thus, the reader will be able to assess the interpretations and evaluate the methodology employed in the publications that are cited in the review.

The Introduction should outline the scope of the review and set the limits to the field it covers. The overall objective of the review may be posed as a question or a series of questions.

The bulk of the review should aim to present or introduce new ideas to the reader, review the literature relevant to these ideas and be specific. The authors of a Review Article might be able to provide alternative and reasoned interpretations or opinions to those advanced in the articles cited in the review.

The review might conclude with a set of hypotheses for future work that could be tested either using available technology or for which current technology could be improved. 2. **Instructions for Authors of Review Articles Review Articles** should consist of the following sections; Cover page, Abstract, Keywords, Introduction, Review per se, Conclusions, Acknowledgements, References and any essential Tables or Figures.

The Introduction, Review and Conclusions should be numbered 1, 2 and 3, respectively. The Review section may be divided into numbered sub-sections, e.g., 2.1., 2.2., etc and sub-sections, e.g., 2.1.1., 2.1.2., etc.

The Reference section must be accurate and complete. It may include references to the more useful databases consulted.

The authors should provide the names and email addresses of no more than three scientists who are experts in the relevant field and who might be willing to provide open reviews of the submitted Review Article. The Editor will also submit the manuscript to experts for blind reviews.

## Submission of Manuscripts

Papers for consideration should be submitted to Elsevier Editorial System which can be accessed at http://authors.elsevier.com/journal/soilbio

Submission of an article implies that the work described has not been published previously (except in the form of an abstract or as part of a published lecture or academic thesis), that it is not under consideration for publication elsewhere, that its publication is approved by all authors and tacitly or explicitly by the responsible authorities where the work was carried out, and that, if accepted, it will not be published elsewhere in the same form, in English or in any other language, without the written consent of the Publisher.

## Electronic manuscripts

Submission to this journal proceeds totally online. Use the following guidelines to prepare your article. Via the homepage of this journal (http://www.elsevier.com/locate/sbb) you will be guided stepwise through the creation and uploading of the various files. The system automatically converts source files to a single Adobe Acrobat PDF version of the article, which is used in the peer-review process. Please note that even though manuscript source files are converted to PDF at submission for the review process, these source files are needed for further processing after acceptance. All correspondence, including notification of the Editor's decision and requests for revision, takes place by e-mail and via the author's homepage, removing the need for a hard-copy paper-trail.

## Electronic format requirements for accepted articles

We accept most wordprocessing formats, but Word, WordPerfect or LaTeX is preferred. Always keep a backup copy of the electronic file for reference and safety. Save your files using the default extension of the program used.

## Wordprocessor documents

It is important that the file be saved in the native format of the wordprocessor used. The text should be in single-column format. Keep the layout of the text as simple as possible. Most formatting codes will be removed and replaced on processing the article. In particular, do not use the wordprocessor's options to justify text or to hyphenate words. However, do use bold face, italics, subscripts, superscripts etc. Do not embed 'graphically designed' equations or tables, but prepare these using the wordprocessor's facility. When preparing tables, if you are using a table grid, use only one grid for each individual table and not a grid for each row. If no grid is used, use tabs, not spaces, to align columns. The electronic text should be prepared in a way very similar to that of conventional manuscripts (see also the Guide to Publishing with Elsevier:

http://www.elsevier.com/wps/find/authorshome.authors/howtosubmitpaper). Do not import the figures into the text file but, instead, indicate their approximate locations directly in the electronic text and on the manuscript. See also the section on Preparation of electronic illustrations.

To avoid unnecessary errors you are strongly advised to use the 'spellchecker' function of your wordprocessor.

## Preparation of manuscripts

1.  The Chief Editors request that papers submitted for publication should be written concisely and clearly. Manuscripts should be written in English. Authors whose native language is not English are strongly advised to have their manuscripts checked by an English-speaking colleague prior to submission. Either the Concise Oxford Dictionary or Webster's New International Dictionary may be used as a standard for English spelling.

    A poorly prepared manuscript often produces a negative reaction in reviewers, irrespective of the quality of the science. The Chief Editors will return to the authors submissions that do not conform to journal style and format or are poorly written.

    ### English language help service:

    Authors who require information about language editing and copyediting services pre- and post-submission please visit http://www.elsevier.com/wps/find/authorshome.authors/languagepolishing or contact authorsupport@elsevier.com for more information. Please note Elsevier neither endorses nor takes responsibility for any products, goods or services offered by outside vendors through our services or in any advertising. For more information please refer to our Terms & Conditions http://www.elsevier.com/wps/find/termsconditions.cws_home/termsconditions

2.  Manuscripts should be prepared **with all lines of text throughout the manuscript numbered consecutively from page to page, with wide margins and double spacing throughout, i.e. also for abstracts, footnotes and references. Every page of the manuscript, including the title page, references, tables, etc. should be numbered.** However, in the text no reference should be made to page numbers; if necessary, one may refer to sections. Underline words that should be in italics, and do not underline any other words. Avoid excessive use of italics to emphasize part of the text.

3.  Authors should provide a separate cover page including:
    Type of contribution
    Date of preparation, number of text pages, number of tables, figures etc.
    Title (should be clear, descriptive and not too long)

Names of authors

Complete postal address(es) or affiliations

Full telephone, Fax No. and E-mail address of the corresponding author

Present addresses of authors if applicable

Complete correspondence address to which the proofs should be sent as a footnote indicated with an asterisk

Special instructions to the printer such as: (a) magnification of photographs, (b) layout of figures, (c) unusual positioning of Figures and Tables in relation to text; (d) if the submitted paper is one of a series of papers to be published in the journal the order in which the papers are to appear should be indicated.

4. Manuscripts should be organised in the following sequence: Cover page (see above)

    Abstract

    Keywords

    Introduction

    Materials and methods

    Results

    Discussion (including Conclusions)

    Acknowledgements and any additional information concerning research grants, etc.

    References

    Tables

    Figure captions

5. In typing the manuscript, titles and subtitles should not be run within the text. They should be typed on a separate line, without indentation. Use lower-case letter type. First and second order headings should be numbered.

6. SI units should be used, but authors may include conversions for unfamiliar units (1 bar=0.1 MPa). Do not include periods. Note the following conventions: e.g. not eg., rev min$^{-1}$ not rpm, mg kg$^{-1}$ or l$^{-1}$ not ppm, 1 bar equals 0.1 MPa, round off units to eliminate unnecessary decimal places, e.g. 124 im not 0.124 mm (note space between number and unit), l not L for litre, kg not Kg, s not sec, min not mins, h not hr, d for day, y not yr, 25 t ha$^{-1}$ not 25 tonnes/ha, 3 mg cm$^{-3}$ not three mg per cubic cm, 23°C and 23% (no spaces), (Keating et al., 1996) not (Keating et al, 1996), al. is an abbreviation of alii (others - Latin). Molar concentrations should appear in small caps.

7. Abbreviations may be used for unwieldy names which occur frequently and such abbreviations must be defined the first time they occur in the text. Conventional abbreviations, e.g. EDTA, ATP, 2,4-D should be used in preference to freshly coined ones.

8. Elsevier reserves the privilege of returning to the author for revision accepted manuscripts and illustrations which are not in the proper form given in this guide. Upon submission, papers will be checked to determine if they conform to the style and format for *Soil Biology & Biochemistry*. Papers that do not comply may be returned to the corresponding author with a check list detailing faults and omissions.

Avoid new or uncommon acronyms. Use single letters (Greek, Roman, italic) for variables with subscripts as appropriate.

## Title

This should be clear, descriptive and brief. Avoid non-specific phrases such as "A study of..." or "The effects of...". Do not give the title a numbered subtitle or series number.

## Abstract

The abstract should be clear, descriptive and not longer than 400 words.

## Keywords

Keywords are index terms or descriptions for information retrieval systems, normally 6 to 10 items. Words selected should reflect the essential topics of the article and may be taken from both the title and the text. Do not select "soil"

## Introduction

This should give the reasons for doing the work. As this is a specialist journal a detailed review of the literature is not necessary. The Introduction should preferably conclude with a final paragraph stating concisely and clearly the Aims and Objectives of the investigation.

## Materials and methods

A full technical description of a method should be given in detail only when the method is new.

## Results

This need only report results of representative experiments illustrated by Tables and Figures. Use well-known statistical tests in preference to obscure ones. Consult a statistician or a statistics text for detailed advice.

## Discussion

This section must not recapitulate results but should relate the authors' experiments to other work and give their conclusions, which may be given in a subsection headed Conclusions.

## Acknowledgements

Do not include grant numbers or institutional journal publication numbers.

## Tables

1. Authors should take notice of the limitations set by the size and lay-out of the journal. Large tables should be avoided. Reversing columns and rows will often reduce the dimensions of a table.

2. If many data are to be presented, an attempt should be made to divide them over two or more tables.

3. Drawn tables, from which blocks need to be made, should not be folded.

4. Tables should be numbered according to their sequence in the text. The text should include references to all tables.

5. Each table should be typewritten on a separate page of the manuscript. Tables should never be included in the text.

6. Tables and their footnotes should be typed using a readable uniform font of the same size as that used in the text. Each text should have a brief and self-explanatory title.

7. Column headings should be brief, but sufficiently explanatory. Standard abbreviations of units of measurement should be added between parentheses.

8. Vertical lines should not be used to separate columns. Leave some extra space between the columns instead.

9. Any explanation essential to the understanding of the table should be given as a footnote at the bottom of the table.

10. Zero results must be represented by 0 and no determination by ND; the dash sign (-) is ambiguous. Report data in such a way that readers can assess the degree of experimental variation and estimate the variability or precision of the findings. Use the standard deviation SD and the mean to summarise data and to show the variability among individuals. Use the standard error of the mean SEM to show the precision of the sample mean. Always state the number of measurements on which means are based. In tables and figures use asterisks to indicate probability values ($P$). In footnotes or text show the degree of significance of $P$, e.g. $P < 0.05*$.

## Electronic Illustrations

Submitting your artwork in an electronic format helps us to produce your work to the best possible standards, ensuring accuracy, clarity and a high level of detail.

1. Always supply high-quality printouts of your artwork, in case conversion of the electronic artwork is problematic.

2. Make sure you use uniform lettering and sizing of your original artwork.

3.  Save text in illustrations as "graphics" or enclose the font.

4.  Only use the following fonts in your illustrations: Arial, Courier, Helvetica, Times, Symbol.

5.  Number the illustrations according to their sequence in the text.

6.  Use a logical naming convention for your artwork files, and supply a separate listing of the files and the software used.

7.  Provide all illustrations as separate files.

8.  Provide captions to illustrations separately.

9.  Produce images near to the desired size of the printed version.

A detailed guide on electronic artwork is available on our website: http://authors.elsevier.com/artwork

**You are urged to visit this site; some excerpts from the detailed information are given here.**

## Formats

Regardless of the application used, when your electronic artwork is finalised, please "save as" or convert the images to one of the following formats (Note the resolution requirements for line drawings, halftones, and line/halftone combinations given below.):

EPS: Vector drawings. Embed the font or save the text as "graphics". TIFF: Colour or greyscale photographs (halftones): always use a minimum of 300 dpi

TIFF: Bitmapped line drawings: use a minimum of 1000 dpi. TIFF: Combinations bitmapped line/half-tone (colour or greyscale): a minimum of 500 dpi is required.

DOC, XLS or PPT: If your electronic artwork is created in any of these Microsoft Office applications please supply "as is".

Please do not:

* Supply embedded graphics in your wordprocessor (spreadsheet, presentation) document;

* Supply files that are optimised for screen use (like GIF, BMP, PICT, WPG); the resolution is too low;

* Supply files that are too low in resolution;

* Submit graphics that are disproportionately large for the content.

## Colour illustrations

If, together with your accepted article, you submit usable colour figures then Elsevier will ensure, at no additional charge, that these figures will appear in colour on the web (e.g., ScienceDirect and other sites) regardless of whether or not these illustrations are reproduced in colour in the printed version.

For colour reproduction in print, you will receive information regarding the costs from Elsevier after receipt of your accepted article. For further information on the preparation of electronic artwork, please see http://authors.elsevier.com/artwork.

Please note: Because of technical complications which can arise by converting colour figures to grey scale (for the printed version should you opt to not pay for colour in print) please submit in addition usable black and white prints corresponding to all the colour illustrations.

As only one figure caption may be used for both colour and black and white versions of figures, please ensure that the figure captions are meaningful for both versions, if applicable.

## References

*Note:* Authors are strongly encouraged to check the accuracy of each reference against its original source.

1.  All publications cited in the text should be presented in a list of references following the text of the manuscript. The manuscript should be carefully checked to ensure that the spelling of author's names and dates are exactly the same in the text as in the reference list.

2.  In the text refer to the author's name (without initial) and year of publication, followed if necessary by a short reference to appropriate pages. Examples: "Since Peterson (1988) has shown that...". "This is in agreement with results obtained later (Kramer,1989, pp. 12-16)".

3.  If reference is made in the text to a publication written by more than two authors the name of the first author should be used followed by "et al.". This indication, however, should never be used in the list of references. In this list names of first author and co-authors should be mentioned.

4.  References cited together in the text should be arranged chronologically. The list of references should be arranged alphabetically on authors' names, and chronologically per author. If an author's name in the list is also mentioned with co-authors the following order should be used: publications of the single author, arranged according to publication dates - publications of the same author with one co-author - publications of the author with more than one co-author. Publications by the same author(s) in the same year should be listed as 1974a, 1974b, etc.

5.  Use the following system for arranging your references, please note the proper position of the punctuation:

a.  *For periodicals*
    Zelles, L., Bai, Q.Y., Beck, T., Beese, F., 1992. Signature fatty acids in phospholipids and lipopolysaccharides as indicators of microbial biomass and community structure in agricultural soils. Soil Biology & Biochemistry 24, 317-323.

b.  *For edited symposia, special issues, etc., published in a periodical*
Rice, K., 1992. Theory and conceptual issues. In: Gall, G.A.E., Staton, M. (Eds.), Integrating Conservation Biology and Agricultural Production. Agriculture, Ecosystems and Environment 42, 9-26.

c.  *For books*
Gaugh, Jr., H.G., 1992. Statistical Analysis of Regional Field Trials. Elsevier, Amsterdam, 278 pp.

d.  *For multi-author books*
DeLacy, I.H., Cooper, M., Lawrence, P.K., 1990. Pattern analysis over years of regional variety trials: relationship among sites. In: Kang, M.S. (Ed.), Genotype by Environment

Interaction and Plant Breeding. Louisiana State University, Baton Rouge, LA, pp. 189-213.

e.  *For website (WWW) entires*
Citations must be confined to peer-reviewed material or official publications, such as annual reports.

CAN/BNQ, 1996. Amenderments organiques-composts. Norme nationale du Canada. CAN/BNQ 0413-200. Conseil canadien des normes. Ottawa, Ont. (on line) http://www.scc.ca/ (11 July, 2004).

6.  In the case of publications in any language other than English, the original title is to be retained. However, the titles of publications in non-Roman alphabets should be transliterated, and a notation such as "(in Russian)" or "(in Greek, with English abstract)" should be added.

7.  Work accepted for publication but not yet published should be referred to as "in press". Authors should provide evidence (such as a copy of the letter of acceptance).

8.  Higher degree dissertations (Ph.D., M.Sc etc.) unpublished data and personal communications should be referred to as 'unpublished observations'in the text and not cited in the References.

## Formulae

1.  Formulae should be typewritten, if possible. Leave ample space around the formulae.

2.  Subscripts and superscripts should be clear.

3.  Greek letters and other non-Roman or handwritten symbols should be explained in the margin where they are first used. Take special care to show clearly the difference between zero (0) and the letter O, and between one (1) and the letter l.

4.  Give the meaning of all symbols immediately after the equation in which they are first used.

5.  For simple fractions use the solidus (/) instead of a horizontal line.

6. Equations should be numbered serially at the right-hand side in parentheses. In general only equations explicitly referred to in the text need be numbered.

7. The use of fractional powers instead of root signs is recommended. Also powers of e are often more conveniently denoted by exp.

8. Levels of statistical significance which can be mentioned without further explanation are $*P < 0.05$, $**P < 0.01$ and $***P < 0.001$.

9. In chemical formulae, valence of ions should be given as, e.g., $Ca^{2+}$, not as $Ca^{++}$.

10. Isotope numbers should precede the symbols, e.g., $^{18}O$.

## Footnotes

Footnotes should only be used to provide addresses of authors or to provide explanations essential to the understanding of Tables.

## Nomenclature

1. Authors and editors are, by general agreement, obliged to accept the rules governing biological nomenclature, as laid down in the International Code of Botanical Nomenclature, the International Code of Nomenclature of Bacteria, and the International Code of Zoological Nomenclature.

2. All organisms should be identified by their scientific names when the English  term is first used,with the exception of common domestic animals. The authority of a species should only be given in the Materials and Methods section.

3. All biocides and other organic compounds must be identified by their Geneva names when first used in the text. Active ingredients of all formulations should be likewise identified.

4. For chemical nomenclature, the conventions of the International Union of Pure and Applied Chemistry and the official recommendations of the IUPAC-IUB Combined Commission on Biochemical Nomenclature should be followed.

*Supplementary data*

Elsevier now accepts electronic supplementary material to support and enhance your scientific research. Supplementary files offer the author additional possibilities to publish supporting applications, movies, animation sequences, high-resolution images, background datasets, sound clips and more. Supplementary files supplied will be published online alongside the electronic version of your article in Elsevier web products, including ScienceDirect: http://www.sciencedirect.com. In order to ensure that your submitted material is directly usable, please ensure that data is provided in one of our recommended file formats. Authors should submit the material in electronic format together with the article and supply a concise and descriptive caption for each file. For more detailed instructions please visit our Author Gateway at http://authors.elsevier.com.

## Copyright

1. An author, when quoting from someone else's work or when considering reproducing an illustration or table from a book or journal article, should make sure that he is not infringing a copyright.

2. Although in general an author may quote from other published works, he should obtain permission from the holder of the copyright if he wishes to make substantial extracts or to reproduce tables, plates, or other illustrations. If the copyright-holder is not the author of the quoted or reproduced material, it is recommended that the permission of the author should also be sought.

3. Material in unpublished letters and manuscripts is also protected and must not be published unless permission has been obtained.

4. A suitable acknowledgement of any borrowed material must always be made.

## Proofs

When your manuscript is received at the Publisher it is considered to be in its final form. Proofs are not to be regarded as 'drafts'.

One set of page proofs in PDF format will be sent by e-mail to the corresponding author (if we do not have an e-mail address then paper proofs will be sent by post). Elsevier now sends PDF proofs which can be annotated; for this you will need to download Adobe Reader version 7 available free from http://www.adobe.com/products/acrobat/readstep2.html. Instructions on how to annotate PDF files will accompany the proofs. The exact system requirements are given at the Adobe site: http://www.adobe.com/products/acrobat/acrrsystemreqs.html#70win.

If you do not wish to use the PDF annotations function, you may list the corrections (including replies to the Query Form) and return to Elsevier in an e-mail. Please list your corrections quoting line number. If, for any reason, this is not possible, then mark the corrections and any other comments (including replies to the Query Form) on a printout of your proof and return by fax, or scan the pages and e-mail, or by post. Please use this proof only for checking the typesetting, editing, completeness and correctness of the text, tables and figures. Significant changes to the article as accepted for publication will only be considered at this stage with permission from the Editor. We will do everything possible to get your article published quickly and accurately. Therefore, it is important to ensure that all of your corrections are sent back to us in one communication: please check carefully before replying, as inclusion of any subsequent corrections cannot be guaranteed. Proofreading is solely your responsibility. Note that Elsevier may proceed with the publication of your article if no response is received.

## Offprints

1. Twenty-five offprints for regular papers will be supplied free of charge.
2. Additional offprints can be ordered on an offprint order form, which is included with the proofs.
3. UNESCO coupons are acceptable in payment of extra offprints.

***Soil Biology & Biochemistry does not have page charges.***

## Author Services

Authors can keep a track on the progress of their accepted article, and set up e-mail alerts informing them of changes to their manuscript's status, by using the "Track a Paper" feature at http://authors.elsevier.com. For privacy, information on each article is password-protected. The author should key in the "Our Reference" code (which is in the letter of acknowledgement sent by the publisher on receipt of the accepted article) and the name of the corresponding author. In case of problems or questions, authors may contact the Author Service Department, E-mail: authorsupport@elsevier.com. Information about Soil Biology & Biochemistry is available on the World Wide Web at the following addresses: http://www.elsevier.com/locate/soilbio

# CHAPTER - 115

# Soil Science Society of America Journal

## Instructions to Authors

### Genral Requirements

Contributions to the Soil Science Society of America Journal (SSSAJ) may be:

1. **REVIEW PAPER** – A review is not simply a collection of papers that are all centered on a common theme. Review papers should provide a synthesis of existing knowledge and give new insights or concepts not previously presented in the literature, or at least not with the same level of detail. The review should identify knowledge gaps for future research. An author should generally be allowed more freedom to provide his/her view on a topic in a review as the papers being reviewed are presumed to already have passed some level of scientific scrutiny by peers. A good review is often one of the most important ways to advance an area of science. The review paper should be targeted, not more than 10 published pages. Ten pages is around 10,000 words, less about 250 words for each table or figure.

2. **ISSUES PAPER** – Soils issues papers include discussion of contemporary soils issues from a combination of scientific, political, legislative, and regulatory perspectives. These papers will often have more of a philosophical bent to them, but must still be based on a foundation of good science. The issue paper should be targeted, not more than 10 published pages.

3. **ORIGINAL RESEARCH ARTICLE** – Original research findings are interpreted to mean the outcome of scholarly inquiry, investigations, modeling, or experimentation having as an objective the revision of existing concepts, the development of new concepts, or the development of new or improved techniques in some phase of soil science. Authors are encouraged to test modeling results with measurements or published data.

4. **NOTE** – Notes focus on studies of limited scope, preliminary data, unique observations, or research techniques and apparartus. The length of a note should be 2 to 3 published pages.

5. Comments and Letters to the Editor– contain (a) critical comments on papers published in one of the Society outlets or elsewhere, (b) editorial comments by Society officers, or (c) personal comments on matters having to do with soil science. Contributions need not have been presented at annual meetings. The SSSAJ also invites submissions for cover illustrations from authors of manuscripts accepted for publication. Refer to SSSA Publication Policy [Soil Sci. Soc. Am. J. 70(1):308-310, 2006] and to the Publications Handbook and Style Manual (ASA-CSSA-SSSA, 2004) at http://www.asa-cssa-sssa.org/style/ for additional information.

The SSSAJ uses a double blind review format. Authors are anonymous to reviewers and reviewers are anonymous to authors. The manuscript title but not the authors' names must appear on the abstract page. If included, the acknowledgment section should appear on the title page rather than preceding the references section (as in published papers), so it can be removed prior to review. From time to time, authors' names are added or deleted from a manuscript between the time of submission and publication. In situations such as this, the ethical and responsible manner of handling this type of change is for the lead author to advise the author being added or deleted of the change and to notify, in writing, the editor and managing editor of the journal.

## Submitting Manuscripts

Manuscripts can be submitted to the SSSAJ Editor through ManuscriptTracker at http://www.manuscripttracker.com/sssaj/. Only electronic submissions are accepted, in the form of a single PDF file containing all text, tables, and figures. To avoid font substitution errors that will delay processing, use of Microsoft Word is strongly recommended in preference to otherword processing software. PDF conversion may be accomplished through ManuscriptTracker. The following information must be provided for each manuscript submitted: corresponding author with e-mail address and other contact information; complete listing of all authors; manuscript title; and a listing of all individuals acknowledged in the manuscript. Additionally, authors are strongly encouraged to specify an appropriate division for review. Instructions on various aspects of ManuscriptTracker are available at http://www.manuscripttracker.com/sssaj/authorhelp.htm.

## Potential Reviewers

Authors are encouraged to provide a list of potentialreviewers through ManuscriptTracker (for electronic submissions). Reviewers must not be subject to a conflict of interest involving the author(s) or manuscript. The SSSAJ editorial board is not obligated to use any reviewer suggested by the author(s).

## Creating the Manuscript

Although manuscript review is done using printed copies or PDF files, Microsoft Word files are required for on-screen editing of all accepted manuscripts, and therefore authors are strongly advised to use this software during manuscript composition. The use of Word Perfect is not recommended for electronic manuscript preparation because font substitution errors are apt to occur upon PDF conversion that will delay manuscript review. Rich-text format (.rtf extension) and TEX files are not acceptable. The file that is sent for typesetting closely resembles a text-only file. Production editors must delete all unnecessary formatting in the manuscript file to prepare it for typesetting. Therefore, authors should avoid using word processing features such as automated bulleting and numbering, footnoting, head and subhead formatting, internal linking, or styles. Avoid using more than one font and font size. Limited use of italics, bold, superscripts, and subscripts is acceptable. All paragraphs (including references) should be double-spaced and line-numbered, with at least 2.5-cm margins.

## Title Page (optional)

The title page should include:

1. A short title not exceeding 12 words. The title should accurately identify and describe the manuscript content.

2. Author-paper documentation. Include author name(s), sponsoring organization(s), and complete address(es). Identify the corresponding author with an asterisk (*). Do not list professional titles. Other information such as funding source(s) may be included here or placed in an acknowledgment, also on the title page. To ensure an unbiased review, the title page will be deleted prior to the review process. The title, but not the byline, should therefore be repeated on the page that contains the abstract.

3. The corresponding author's phone and fax numbers and e-mail address. An e-mail address is essential for manuscript processing with ManuscriptTracker.

4. The acknowledgements section, if any.

## Abstract

An informative, self-explanatory abstract, not exceeding 250 words (150 words for notes), must be supplied on a separate page. It should describe specifically why and how the study was conducted, what the results were, and why they are important. Use quantitative terms. Formatting must be as a single paragraph. References cannot be cited. The title (without author identification) must precede the abstract. A list of any abbreviations used in the text should follow the abstract, alphabetized according to abbreviation.

## Tables

Each table must be submitted on a separate page and must be numbered consecutively. Do not duplicate matter that is presented in charts or graphs. Use the following symbols for footnotes in the order shown: †, ‡ ,§, ¶, #, ††,‡‡, etc. The symbols *, **, and *** are always used to show statistical significance at the 0.05, 0.01, and 0.001 level, respectively, and are not used for other footnotes. Spell out abbreviations on first mention in tables, even if the abbreviation is defined in the text (i.e., a reader should be able to interpret each table without referring to the text).

## Figures

Do not use figures that duplicate matter in tables. Photographs for halftone reproduction should be glossy prints with good dark and light contrast, or digital image files. When creating figures, use font sizes and line weights that will reproduce clearly and accurately when figures are sized to the appropriate column width. The minimum line weight is ½ point (thinner lines will not reproduce well). Screening and/or shaded patterns often do not reproduce well; whenever possible, use black lines on a white background in place of shaded patterns. Color figures are acceptable, but will be subject to a publication surcharge if the manuscript is accepted.

Authors should try to supply photographs and drawings that can be reduced to a one-column width (8.5 cm or 20 picas). Lettering or numbers in the printed figure should not be smaller than the type size in the body of an article as printed in the journal (8-point type) or larger than the size of the main subheads (12-point type). The minimum type size is 6-point type. As an example, a 17- cm-wide figure should have 16-point type, so that when the figure is reduced to a single column, the type is reduced to 8-point type. As with tables, spell out abbreviations on first mention in figure captions, even if they have already been defined in the text.

## References

Note the following in preparing the references section:

1. Format the references with double-spacing and linenumbering.
2. Do not number the references listed.
3. Arrange the list alphabetically by last name of the senior author and then by last name of successive authors.
4. Single-authored articles should precede multipleauthored articles for which the same individual is senior author.
5. Two or more articles by the same author(s) are listed chronologically; two or more in the same year are indicated by the letters a, b, c, etc.
6. All published works cited in the text must be listed as a reference and vice versa.
7. Only literature that is available through libraries can be cited. The reference list can include theses, dissertations, abstracts, or web (URL) listings.
8. Material not available through libraries, such as personal communications or privileged data, should be cited in the text in parenthetical form.

9. Chapter references from books must include, in order, author(s), year, chapter or article title, page range, editor(s), book title, publisher, and city.

10. Symposium proceedings should include editor(s), date and place of symposium, publisher, publisher's location, and page numbers.

## Style Guidelines

All soils discussed in the manuscript should be identified according to the U.S. soil taxonomic system at first mention. The Latin binomial or trinomial and authority must be shown for all plants, insects, pathogens, and animals when first mentioned. Both the accepted common name and the chemical name of pesticides must be provided. SI units must be used throughout the manuscript. Corresponding metric or English units may be added in parentheses at the discretion of the author. If a commercially available product is mentioned, the name and location of the manufacturer should be included in parentheses after first mention.

## Official Sources

1. Spelling: Webster's New Collegiate Dictionary

2. Amendments to the U.S. system of soil taxonomy (Soil Survey Staff, 1975) have been issued in the National Soil Survey Handbook (NRCS, 1982-1996) and in Keys to Soil Taxonomy (Soil Survey Staff, 1996). Updated versions of these and other resources are available at http:// soils.usda.gov/technical/ classification/tax_keys/.

3. Scientific names of plants: A Checklist of Names for 3000 vascular plants of Economic Importance (USDA Agric. Handb. 505, see also the USDA Germplasm Resources Information Network database at http:// www.ars-grin.gov/npgs/searchgrin.html).

4. Chemical names of pesticides: Farm Chemicals Handbook (Meister Publishing, revised yearly)

5. Soil series names: Soil Series of the United States,Including Puerto Rico and the U.S. Virgin Islands(USDA-SCS Misc. Publ. 1483, http:// ortho.ftw.nrcs. usda.gov/osd/osd.html).

6. Fungal nomenclature: Fungi on Plants and Plant Products in the United States (APS Press)

7. Journal abbreviations: Chemical Abstracts Service Source Index (American Chemical Society, revised yearly)

8. The Glossary of Soil Science Terms is available both in hard copy (SSSA, 2001) and on the SSSA Web page (www.soils.org/sssagloss/). It contains definitions of more than 1800 terms, a procedural guide for tillage terminology, an outline of the U.S. soil classification system, and the designations for soil horizons and layers.

## Manuscript Revisions

Authors have three months to make revisions and return their manuscript following receipt of reviewer and associate editor comments. If not returned within three months, the manuscript will be released. To receive further consideration for publication, it must be resubmitted to the Editor as a new manuscript.

## Publication Charges and Manuscript Length

Membership in the Society is not a requirement for publication in the SSSAJ; however, nonmembers will be charged a higher fee than members. The publication fee for volunteered papers is $650 for members and $750 for nonmembers, regardless of length. To qualify for the member rate, at least one author must be an active, emeritus, graduate student, or undergraduate student member of SSSA or ASA on the date the manuscript is accepted for publication. The aforementioned publication fee may be waived for invited review papers, and will not be assessed for Comments and Letters to the Editor or book reviews. If the manuscript is prepared with a word processor using a 12-point proportional font, 1000 words will be approximately equivalent to one printed page of the SSSAJ. For economy of space, some sections are set in small type, including Materials and Methods, Theory, tables, figure captions, and References. Each table and figure will typically occupy ¼ of a printed page. For tabular matter, 10 lines of headings, subheadings, and/or data rows require 1 inch of column space. Tables with up to 60 characters per row (including spaces between characters) can usually be printed in a single column, while tables that exceed this width will require two columns. The depth of a printed figure will be in the same proportion to the width (1 column = 8.5 cm; 2 column = 17.2 cm) as that of the corresponding dimensions in the original drawing. Authors can publish color photos, figures, or maps at their own expense. Please contact the Managing Editor (608-268-4969) for pricing information.

## Manuscript Reviews

Up to three months may be required for the initial review. Thereafter, authors may contact the Editor to obtain information about the progress of the review.

## Accepted Manuscripts

Following notification of manuscript acceptance, both a printed copy or pdf and word processing file of the final accepted manuscript are required. The printed copy or pdf and word processing file must match exactly in all parts of the manuscript. Printed copies and files for tables and figures must also be included. Separate files should be submitted for text, tables, and figures. Send the printed copy and a disk with the manuscript files to:Accepted manuscript Soil Science Society of America Journal 677 South Segoe Road Madison, WI, USA 53711

## Questions?

Send your questions to Rebecca Funck, Managing Editor, SSSAJ (rfunck@agronomy.org).

# CHAPTER - 116

# The Andhra Agricultural Journal

**Guidelines to Contributor**

It is policy of the Editorial Board to publish papers based on original research in the disciplines of Agricultural Sciences including Plant, Animal Food, Home and Social Sciences and Ag. Engineering.

The Journal is open to the members only. All the authors and co-authors must become members for processing the article for publication in the Journal. The Editorial Board reserves the right to accept or reject a paper and the decision of the Editorial Board is final. The research articles / notes accepted for publication in Andhra Agricultural Journal should not have been previously published nor should simultaneously appear in any other journal. The authors should send an undertaking along with the paper stating that **" the research entitled" ........... submitted for publication in the AAJ is a bonafide research work conducted with the involvement of all the authors and the paper has not been previously published or simultaneously submitted for publication in any other scientific or technical journal"**. The author(s) should send two copies (one original) of the article along with the undertaking through the concerned Head of the department or research station.

All the manuscripts of the research paper should be written in English in double space on A4 size paper with 2.5 cm margins.  The paper should contain (1) a short and clear title with initial capital letters for each word, (2) names and addresses or affiliations of the authors (3) a brief and informative abstract of about 200 words followed by all key words. The manuscript should contain four sections *viz.,* **introduction, Materials and Methods, Results and Discussion and  References.**

Introduction should include, a brief of previous literature and objectives. The methodology should clearly specify the year(s) of study, area / laboratory where the experiment was conducted and the procedure(s) followed. Results and Discussion should be clear, crisp and supported by the data.  The illustrations should be clear. Drawings should be with Indian Ink on white drawing / tracing paper with title and legends neatly typed on it. Photostat copies of illustrations are not accepted because of poor reproduction in print.

In the text citations should be presented chronologically and should invariably be listed under References. All weights and measures should be given in Metric system. Each reference should contain the name(s) of the author(s) (bold), the year of publication (bold), the title of the paper, the name of the Journal/periodical/ book in full (italics) without using abbreviations, volume (bold), and page range of the Journal/Publishers of the book as given below. Please confine the total number of references to around 10 to 15. The papers may be confined to less than 10 pages (3500 words) including tables, figures, illustrations ect.

1.  JOURNAL

    **Singh, D.S., Sircar, P., Saxena, P.N.and Dingra, S. 1983.** Comparative toxicity of pyrethroids to Achaea Janata Linn. Indian Journal of Entomology, 45 : 390 – 395.

2.  Thesis

    **Vijayalakshmi, K. 1994.** Tranmission and ecolo;gy of Thrips plami Karny. The vector of peanut bud necrosis virus.Ph.D.Thesis, Acharya N.G. Ranga Agricultural University, Rajendranagar, Hyderabad, A.P.

3.  Text Book

    **Finney, D.J. 1971.** probit analysis. 3$^{rd}$ ed., Cambridge University Press, London, 333 pp.

4.  Part of a Text Book

    **Metcalf, R.L. 1975.** Insecticides in Peat Management, In: Introduction to Insect Pest Management (eds., Matcalf, R.L.I and Luckmann, W.H.). John Wiley and Sons, New York, USA. Pp 235 – 273.

5.  Symposium / Seminar

    **Srinivasa Rao, M., Dharma Reddy, K. and Singh, T.V.K.2002:** Role of bird predation in IPM in Pigeonpea. Proceedings of the National Symposium in Avion Biodiversity, Feb. 7-8,2002. Society for Applied Ornithology, Hyderabad. A.P.

The papers should be prepared by taking maximum care avoiding typographical / other errors. Research articles which are not in line with the above guidelines shall be rejected. The Editorial Board does not take any responsibility for the opinion(s) expressed by the author(s).

# CHAPTER - 117

# The Asian Journal of Horticulture

## Guidelines to the Authors

The Asian Journal of Horticulture(The Asian J.Hort) is an official publication of the Hind Agri-Horticultural Society. It features the original research in all branches of plant and other cognate sciences of sufficient releavance. The journal publishes three types of articles. i.e. **Review/Strategy paper** (exclusively by incitation from the personalities of eminence) Research paper and **Short communication.** The manuscripts should be submitted in triplicate with CD completed in all respect to the **Editor, The Asian Journal of horticulture, Hind Agri-horticultural society, Ashram, 418/4-South civil lines (Numaish camp). MUZAFFARANGAR-251001(U.P). INDIA.** clearnes,brevity and concisencess are essential in form, Style, punctuation, spelling and use of English language. Manuscripts should conform to the S.I.system for numerical data and data should be subjected to appropriate statistical analysis. On receipt of an article at the Editorial Office, an acknowledgment with a number is sent to the corresponding author. This number should be quoted while making any enquiry about its status. **All the authors have to become the annual member of the Journal.**

**Review/Strategy paper:** It should be comprehensive, up-to-date and critical on a recent topic of importance. The maximum page limit is of 16 double spaced typed pages including tables and figures. It should cite latest literatures and identify some gaps for future. It should have a specific Title followed by the Name(s) of the auther(s), Affiliation: Address of the institution(s) where the research was undertaken and corresponding author with address.

**Research Paper:** The paper should describe a new and confirmed findings. It should not generally exceed 12 typed pages including tables/figures etc. A paper has the following features.

Title followed by Author(s) and Affiliation: Address of the institution(s) where the research was undertaken.

**Abstract:** A concise summary (200 to 300 words) of the entire work done along with the highlights of the findings.

**Key words:** Maximum five keywords to be indicated.

**Introduction:** A short introduction of the research problem followed by a brief review of literature and objective of the research.

**Materials and Methods:** Describe the materials used in the experiments, year of experimentation, site etc. Describe the methods employed for collection of data in short.

**Results and Discussion:** This segment should focus on the fulfillment of stated objectives as given in the introduction. It should contain the findings presented in the form of tables, figures and photographs. As far as possible, the data should be statistically analyzed following a suitable experimental design. Same data should be presented in the table and figure form. Avoid use of numerical values in findings, rather mention the trends and discuss with the available literatures. At the end give short conclusion. Insertion of coloured figures, as photograph(s) will be charged from the author(s) as applicable and suggested by the printer.

**Acknowledgements** (where applicable).

**References:** Reference to literature should be arranged alphabetically and numbered according to author's  names, should be placed at the end of the article. Each reference should contain the names of the author with initials,  the year of the publication, title of the article, the abbreviated title of the publication according to the World List of Scientific Periodicals, Volume and page(s). In the text the reference should be indicated by the author's name, followed by the year in brackets.

Brown, M.E. and Carr, G.R.(1984) : Interaction between Azotobacter chroococcum and VAM and their effects plant growth.J.App.

Bact., **56:** 429-437.

Gloveannetti, M. and Mosse, B. (1980).An evaluation of techniques for measuring VAM infection in roots. New Phytol., 84 : 489-500.

Sharma, S.D. and Bhutani, V.P.(1999). Influence of VA-mycorrhizae and P fertilization on growth leaf area and root colonization of apple seedlings. Haryana. J.Hort.Sci, 28(3): 134-137.

**Short communication:** The text including table(s) and figure(s) should not exceed five pages. It should have a short title; followed by name of author(s) and affiliation and references. There should be no subheadings,i.e. Introduction, Materials and Methods etc. The manuscript should be in paragraphs mentioning the brief introduction of the topic and relevance of the work, followed by a short description of the material and methods employed, results and discussion based on the data presented in 1or 2 table(s)/figure(s) and a short conclusion at the end. References, should be maximum of seven in Nos. at the end.

## General Instructions

- Author must submit the following materials to the Editor:
  - (a) Three computer printed copies of the manuscript, printed on one side of A4 size paper with proper margin.
  - (b) CD of the printed research paper should be in only one file in M.S.Word.
  - (c) Short running title.
- Processing fee Rs.100 per research paper/article.
- **MEMBERSHIP FEE FOR EACH AUTHOR.**
- At the bottom of the first page, address of the corresponding author and co-author(s) Tel.No.Fax No. and E-mail ID etc. must be specified.
- Article forwarded to the Editor for publication is understood to be offered to **THE ASIAN JOURNAL OF HORTICULTURE** exclusively. It is also understood that the authors have obtained a prior approval of their Department, Faculty or Institute in case where such approval is a necessary.
- Acceptance of a manuscript for publication in **THE ASIAN JOURNAL OF HORTICULTURE** shall automatically mean transfer of copyright to the Hind Agri-Horticultural Society. The Editorial Board takes no responsibility for the fact or the opinion expressed in the journal, which rests entirely with the author(s) thereof.
- **Author Page Charges:** Because of the high cost of publishing articles in the journal payment of page charges is mandatory and charges are subject to change without notice. Current charges are Rs.200/- ($8.00) per printed page (A4 size, approx.800 words) For further details please contact on our society's website www.hindagrihorticulturalsociety.com

# CHAPTER - 118

# The Cashew

## Attention Authors

Articles intended for publication in "The Cashew" may kindly be sent in duplicate. The material for publication should be typed in double space on one side only without correction and overwriting. If possible Electronic copy in the CD prepared in MS Word/Page maker. (Font – Times New Roman, 12 Point size) along with the hard copy may please be sent. The authors are requested to intimate their address clearly in their letters, for correspondence.

The Director

Directorate of Cashewnut & Cocoa Development

Government of India

Ministry of Agriculture (Department of Agriculture & Co-operation)

Cochin – 682 011, Kerala.

Phone: 2377239, 2377151

Fax: 0484-2377239

Website: dacnet.nic.in/cashewcocoa

E-mail: dccd@hub.nic.in

# CHAPTER - 119

# The Indian Forester

## Rules for Contributions of Articles

1. Contributions should be sent to the Honorary Editor, Indian Forester, P.O. New Forest, Dehra Dun – 248 006 (Uttaranchal). Email : indfor@icfre.org.

2. The Central Editorial Board does not hold itself responsible for the opinions expressed by the authors.

3. The Central Editorial Board reserves the right of accepting or rejecting any contribution and of making suitable alterations and amendments, without assigning any reason.

4. Contributors will be supplied 8 extra copies of their contribution, free of charge.

## Instructions to Authors

**Title:** This should be brief, it should be inserted on the top of the first page in block capitals.

**Summary :** Each paper must commence with an accurate, informative summary (abstract) in one paragraph, that is complete in itself and intelligible without reference to text or figures. It should not exceed 200 words.

**Introduction :** An introduction should present the general object of the work

**Keywords :** Three or four keywords.

**Presentation :** The contribution should be submitted in the following format : (1) Summary, (2) Introduction, (3) Material and Methods, (4) Results and Discussion, (5) Conclusion and (6) References. **The matter should be typed in double spacing on one side only of A4 size paper with a  margin of 2.5 cm. Contributions should be sent in duplicate.**

**Electronic manuscript :** For the initial submission of manuscripts for consideration, hardcopies are sufficient. **For the processing of accepted papers, electronic versions are preferred. After final acceptance, your CD plus two final and exactly matching printed versions should be submitted together on CD. Label the CD with the word processing package used, your name, and the name of the file on the disk.**

**Illustrations :** The number of illustrations should be kept at minimum as their reproduction is costly. Photographs and drawings should, if possible, be made to the correct proportion of the printed page of the magazine. Nothing should be written in pencil or ink or typewritten on the back of illustration. Photographs and illustrations should not be pinned or sewn with thread. The former should be in black and white, with pronounced contrasts and glossy finish: the letter should be neatly drawn in Indian ink on white thick paper or drawing paper. Graph paper should preferably have black co-ordinate lines on white background. Captions should be typed or written on separate strips of paper and attached with colourless adhesives to the back of the prints. The approximate place where the illustrations are to be inserted should be indicated in the text, and the illustrations should be serially numbered. When mailing, illustrations should be placed between cardboards held together by thin thread or rubber bands. **In case originals are not being sent, please send line drawings as 'bmp' files and photographs as 'jpeg' or 'tif' files.**

**Reference:** The References should be given at the end of the article on separate sheet of paper. Reference to a citation in the text should be made by means if the author's name followed by the year of publication in parenthesis. The reference will consist of the serial number, the author's name and initials, the year of publication, the title of the article, the title of the book, paper or periodical, volume number and date of issue and the page numbers, thus:

Hamilton, A.P.F. (1947). Farm Forestry in India, Indian Forester, 74(3): 84-88.

**Tables :** Tables should be reduced to the simplest form and present only essential data. They should be submitted on separate sheets at the end of the article and must fit into single column. Full width or landscape (If absolutely necessary) format. The use of vertical rules should be avoided. Numerical results should be presented as means with standard errors and/or statistically significant differences, quoting probability levels.

# CHAPTER - 120

# The Indian Journal of Agricultural Science

## Guidelines to Authors

*The Indian Journal of Agricultural Sciences* is published every month. The following types of material are considered for publication on meeting the style and requirements of the journal.

ARTICLES ON ORGINAL RESEARCH COMPLETED, not exceeding 4000 words (up to 15 typed pages, including references, tables, etc) should be exclusive for the journal. They should present a connected picture of the investigation and should not be split into parts. Complete information of Ph D thesis should preferably be given in 1 article.

SHORT RESEARCH NOTES, not more than 1300 words (total 5 typed pages), which deal with (i) research results that are complete but do not warrant comprehensive treatment, (ii) descriptions of new material or improved techniques or equipment, with supporting data, and (iii) a part of thesis or study. Such notes require no headed sections.

CRITICAL RESEARCH REVIEW ARTICLES, showing lacunae in research and suggesting possible lines of future work. These are mostly invited from eminent scientists.

The research article or note submitted for publication should have a direct bearing on agricultural production or open up new grounds for productive research. Articles on agricultural engineering, agricultural economics and Home Science are also considered. Basic type of articles and notes relating to investigation in a narrow specialized branch of a discipline may not form an appropriate material for this journal, nor do the articles of theoretical nature, or those of local importance, repetitive, based on old data, with no positive significance, or on extension education.

Author should note: (a) period (years) of conducting the experiment must be indicated, (b) article should preferably be submitted soon after completing of experiment, (c) articles on genetics and plant breeding and on plant crops should be based on data of minimum 2 years, (d) contribution involving a former or present student must clarify that it is not based / based on complete M Sc the thesis, or complete or a part of the Ph D thesis, indicating thesis year of submission and (e) Article Certificate must be signed by all the authors and must contain subscription numbers of authors.

TITLE  should be short, specific and informative. It should be phrased to identify the  content of the article and include the nature of the study and the technical approach, essential for key-word indexing and information retrieval.

A SHORT TITLE not exceeding 35 letters should also be provided for running headlines.

BY-LINE should contain, in addition to the names and initials of the authors, the place(organization) where research was  conducted. Change of address should be given as a footnote and correspondence address separately.

ABSTRACT, written in complete sentences, should have more than 150 words. It should contain a very brief account of the materials, methods, results, discussion and conclusion so that the reader need not refer to the article except for details. It should not have references to literature, illustrations and tables.

INTRODUCATION part of should be brief and limited to the statement of the problem or the aim and scope of the experiment. The review of recent literature should be pertinent to the problem. Key words of the article should be given in the beginning.

Relevant details should be given of the METERIALS AND METHODS including experimental design and the techniques used. Where the methods are well known, citation of the standard work is sufficient. Mean results with the relevant standard errors should be presented rather than detailed data.  The statistical methods used should be clearly indicated.

RESULTS AND DISCUSSION should be combined, to avoid repetition.

The results should be supported by brief but adequate tables or graphics or pictorial materials wherever necessary. Self-explanatory tables should be typed on separate sheets, with appropriate titles.

The tables should fit in the normal lay-out of the page. All weights and measurements must be in SI (metric) unit. Tables and illustrations (up to 20% of text) should not reproduce the same data.

The discussion should relate to the limitations or advantages of the author's experiment in comparison with the work of others. All relevant literature should be discussed critically.

Line-drawings should be clearly drawn (8.5 or 17 cm width) in black waterproof ink on smooth, tough paper. Minor points of style should be noted carefully. Photographs should be large, unmounted, glossy prints of good quality. They should be clear and relevant to the subject. Colour photographs may be sent for better identification or differentiation of different parts of the object. Line drawings and photographs should have legends (typed). Original artwork should accompany 2 copies. Repetition in graphic and tabular matter should be avoided.

1.  For citing REFERENCES a recent issue of the journal should be consulted, ensuring that the names and dates in the text and at the end correspond. Each citation should have the names of the authors, initials, year of publication, full title of the article, name of the journal (without abbreviation), volume, preferably the issue( within parentheses), and complete page-range (not merely the first page). Complete name of publisher and place of publication of books should be given. For proceedings or other publications complete details should be given.

2.  All articles are sent to referees for scrutiny, and authors should meet criticism by improving the article, indicating the modifications made (in separate sheet, 2 copies).

3.  Articles should be TYPEWRITTEN, and double spaced throughout (including by-line, abstract, references and tables) on white, durable A-4 size paper, with a 4 cm margin at the top, bottom and left. Articles should be sent in triplicate after checking typographical errors.

4.  FOR WRITING, authors are requested to consult The Council of Biology Editors Style Manual, edn 4 (or later edition), American Institute of Biological Sciences, Washington DC. The language and spelling are followed as per British style, not American.

5.  PROOF-CORRECTION should be in ink, in the margin. All queries marked in the article should be answered. Proofs are supplied for a check-up of the correctness of typesetting and facts. Excessive alterations may be charged to the author. Proofs must be returned in time.

6.  Contributors will receive one free copy of the issue.

# CHAPTER - 121

# The Indian Journal of Genetics and Plant Breeding

## Guidelines to Contributors

**The Indian journal of Genetics and Plant Breeding** is a periodical for the publication of records of original research in all branches of genetics, plant breeding and cytology, including human genetics, molecular biology and biotechnology, and other cognate sciences of sufficient importance and of such a character as to be of primary interest to the geneticist and plant breeders.

**Preparation of manuscript :** Manuscript should be submitted in triplicate to the Editor, The Indian Society of Genetics and Plant breeding, P.B.11312, IARI, New Delhi 110 012.  On receipt of an article an acknowledgement giving the registration number of the paper is sent to the author. This number should always be quoted while makings further enquiry about its publication.

Each full-length research paper must not exceed **3000 words** including tables and illustrations. **Short communications** should be restricted to about **1000** words but not exceeding two printed pages including tables, figures and references.  Manuscripts should be typed in *double space* on one side of bond paper (A4 size), and thoroughly revised before submission. No editing or material changes at the proof stage will be permitted. While the short communication will have only title, author's name and address, followed by text and references, the full length paper should have the following headings.

**Short Title : A** short title, not exceeding 35 letters, should be typed at the top of the first page in first letters capital and underlined.

**Title :** The title should be short, specific and informative, typed in first letters capitals, Latin names in italics underlined.

**Authors :** Names of authors to be typed in first letters capitals unaccompanied by degrees, titles etc.

**Address :** Address of the institution where the work was carried out. Present address of correspondence should be given as footnote.

**Abstract :** A brief abstract, not exceeding **150 words,** of the principal points and important conclusions should be typed after address in double spacing entirely indented on both sides, without additional paragraph indentation.

**Key words :** The Abstract should be followed by not more than five key words indicating the contents of the research paper.

**Main headings :** Each full-length research paper will be divided into the following main headings which are to be typed only the first letter capitals in left of the page: Abstract, Introduction, Materials and methods, Results and discussion, Acknowledgements, References.

**Sub-headings :** Typed flush left in first letters capital, underlined/italicized.

**Paragraph heading :** Typed indented, only the first letter capital, ending with a period underlined.

**Introduction :** Should be brief and limited to the statement of the problem and aim of the experiment. The literature reviewed should be pertinent to the problem under study.

**Materials and Methods :** Should include relevant details on the nature of material, experimental design, the techniques employed, and the statistical methods used. For well-known methods, citation of reference will suffice.

**Results and Discussion :** Should preferably be combined to avoid repetition. Statistically analysed data, which are essential for drawing main conclusion from the study, and not presented in tables and figures form only to be given.

**Nomenclature :** Generic and species names should be italicized/underlined and the first reference to the letter should be accompanied by the authority.

**Tables :** Tables should be typed on separate sheets, each with a heading. Tables should be typed with the first letter (T) only capital, table No. in Arabic numerals, followed by a period. All measurements should be in metric units.

**Figures/Line drawings :** Only good quality figures that are essential shall be accepted. The illustrations should not repeat the data presented in tables and vice versa. Text figures should be used in preference to plates. Should be clearly drawn in black Indian ink on good tracing paper and should be of the size: width 12 cm, height 9 cm to a maximum of 18 cm. Figures; meant for reduction should have multiple dimensions of the above size. Letters, numbers, dots, lines etc. in the drawing should be large enough to permit reduction without loss of details. Text figures should be numbered in Arabic numerals in order of their reference. Captions and legends to illustrations should be typed on a separate sheet of paper.

**Photographs :** Should be of high contrast on glossy paper. Photographs number and title of the article with author's names should be given on the back of each photograph. Half-tone/coloured plates will be accepted only, if they are paid for by the authors.  Coloured plates processing charges are Rs.4000/-per page to be paid in advance.

**Acknowledgements :** Should mention only assistance received in real terms, and financial grant provided by an agency.

## References

To be cited as specified below:

(a)  Text citation by numbers in square brackets

(b)  Listing of references by S.No.in accordance with their order of text citations.

(c)  Author's names to be written in normal sequence, with surnames first, followed by initials (in double space) a period, and year (followed by a period)

(d)  Journal names in standard abbreviations and book titles in expanded form in first capitals

(e)  Volume of Journal underlined, followed by No. in parentheses (not underlined), colon and page numbers

(f)  Volume or part of serialized books to be printed as vol.2 or pt. III (not underlined), followed by a colon, and page numbers.  Name of publishers not to be abbreviated

(g)  Only S.No.in the list of references projecting out on the left.  Remaining text in same alignment (without further indentation of second line onward)

## Examples

1.  Miller P.A.and Marani A.  1963. Heterosis and combining ability in diallel crosses of upland cotton Gossypium hirsutum.  Crop Sci., 3: 441-444.

2.  Weber D.F. 1967. On the Interaction of Nonhomologous Segments of Chromosomes in Zea mays.  Unpubl.Ph.D. Thesis.  Indiana University, Bloomington, Indiana, USA. (For Indian universities, country name not to be indicated).

3.  Sunderland N. 1977.Nuclear cytology In: Plant Cell and Tissue Culture, vol. II (ed.H.E.Street). University of California press, Berkeley, California, USA: 171-206. (For Indian publishers, country name not to be indicated.)

4.  Chase S.S. 1974. Utilization of haploids in plant breeding: breeding diploid species.  In: Haploids in Higher Plants: Advances and Potential.  Proc. I Intern. Symp., 10-14 June, 1974 University of Guelphs (ed k. J. Kasha). University of Guelph, Canada: 211-230.

5.  Falconer D.S. 1960 Introduction to Quantitative Genetics.  The Ronald Press Co., New York, USA 365.

# CHAPTER - 122

# The Indian Journal of Microbiology

## Instruction to Authors

The Indian Journal of Microbiology publishes peer reviewed papers of original research work in the areas of agricultural, food, environmental, industrial, medical, pharmaceutical, veterinary and molecular microbiology from the members of the Association of Microbiologists of India as well as from the non-members. Review articles from specialists in the above areas are also considered for publication.

**General :** Submission of paper will imply that it contains unpublished original work, and that it is not submitted elsewhere even part for publication. If any related paper is published or is in press, a copy of the same must be enclosed alongwith the manuscript.

Three types of papers are considered.

1. *Full length :* The paper should describe new and confirmed findings. Experimental procedures should give sufficient details for others to verify the work. The paper should have a 10-12 typed pages comprising (a) an abstract, summarizing the findings, (b) Key words, not more than 5, (c) introduction, precise to introduce the subject with citations of relevant literature, (d) Material and Methods, (e) Results and Discussion, (f) Acknowledgement, and (g) References. No heading for abstract and introduction to be given.

2. *Short communication :* A short communication should be a record of completed short investigation giving details of new methods or findings. It should not exceed 4 to 5 typed pages with an abstract followed by Key words. Body of the text will not have any title, like Abstract, Material Methods, Results and Discussion except the Acknowledgement and References.

3. *Review :* It should be comprehensive, up to date and critical on a recent topic of importance. It should cite latest references and identity the gaps for future research. It should also contain an abstract (without heading), Key words, Acknowledgement and References.

4. ***Submission of manuscript:*** Submit the manuscript (original and one photo copy) on A4 size bond paper typed in 12 font in Microsoft Version 6 (Windows 98) with double spacing and 0.9" margin on both sides, to the Editor in Chief, Indian Journal of Microbiology, Division of Microbiology, Indian Agricultural Research Institute, New Delhi 110012, India. Cover page should have title of the article, author(s) names, address of the place of work and also for correspondence, if deferent, with E-mail address, telephone and Fax no. Also provide a short running title containing 4-5 words and the area of microbiology for indexing. Wherever there are more than one author, a certificate, signed by all the authors and stating that neither this paper nor any part of this paper has been published or sent for publication in this or any other journal, must be enclosed. Authors are advised to provide names and addresses of five experts in the specific area of the article and also 8 gummed labels containing self-address for  correspondence. Authors should see latest issue of IJM to ensure that the MS is in the format of IJM.

**References :** Arrange all the citations under References, in the order as they appear in the text and serially number them. Quote these numbers as superscript in the text. Use standard abbreviation of the journals. The examples of citations are as follows:

Conrad R (1999) Soil microorganisms oxidizing atmospheric trace gases ($CH_4$ , CO, $H_2$, NO). Indian J Microbiol 39:193-203.

Rand MC, Greenberg AE, Taras MJ & Franson MA (1975) Standard Methods for the Examination of Water and Wastewater. Amer Public Health Assoc, Washington DC, USA, 1193 p:

Kaspar Hf & Tiedje JM (1982) Anaerobic bacteria and processes. In: Methods of Soil Ananlysis, Part 2 (Page AL, Miller RH & Keeney Dr eds). Soil Sci Soc Amer Inc, Madison, Wisconsin, USA, pp 987-1009.

**Tables :** Tables should be numbered with a brief title. Type each table on separate page using tab.

**Illustrations/figures/graphs :** Each graph, figure, photo or diagram should be provided on a separate sheet. Author's name and Fig no. should be written on the back of the figure with pencil. Diagrams should be computer drawn laser print. Line work should be sharp in not more than 0.3 mm thickness. One copy of figure should be on a high quality art paper also. Legends for all figures must be typed on a separate page. Indicate approximate position of each table and figure in the text on the margin with pencil.

Colour photographs can be published at the cost of the authors.

**Abbreviations :** Use standard abbreviations. A few are: mL (milliliter), g (gram), t (ton), mM(milli mole), d (days), min (minutes), h (hours), sec (seconds), ha (hectare), SD (standard deviation), kD (kilo Dalton), SE (standard earror), wt (weight), DAS (days after sowing), mol wt (molecular weight), cfu (colony forming unit) and standard symbols. The Latin names of microorganisms, plants and animals should be italicized.

**Revised manuscript :** Carefully revised manuscript should be submitted on high quality HD diskette (3.25") or CD, free from computer virus, along with one hard copy on A4 size bond paper. Mention manuscript NO, date of revision and Microsoft version on the diskette. All letters in final manuscript must be of 12 font size in Times New Roman. Revision received after 8 weeks of dispatch date with be considered a new submission. Improperly revised MS, will not be considered.

**Galley proof :** Galley proof is sent to the corresponding author. It must be checked carefully and returned to the Editor in Chief within 48 h of receipt, by spead post. No major change at this state should be made. However, if any essential change is made, that will be charged from the authors.

**Reprints :** Order for reprints should be placed while returning the galley proof. The charges per page for 25 reprints are Rs. 150/- for articles from India and US$ 10 from abroad.

Acceptance of manuscript for publication in IJM shall automatically mean transfer of copyright to AMI. AMI shall not be responsible for authenticity of the results and conclusions drawn by the authors.

# CHAPTER - 123

# The Journal of AOAC International

## Instructions to Authors

### Scope of Articles and Review Process

The Journal of AOAC INTERNATIONAL publishes articles that present, within the fields of interest of the Association: unpublished original research; new methods; further studies of previously published methods; background work leading to development of methods; compilations of authentic data of composition; technical communications; cautionary notes; comments on techniques, apparatus, and reagents; and invited reviews and features. Emphasis is on research and development of precise, accurate, sensitive methods for analysis of foods, food additives, supplements contaminants, cosmetics, drugs, toxins, hazardous substances, pesticides, feeds, fertilizers, and the environment. The usual review process is as follows: (I) Author submits manuscript online; (2) AOAC editorial office transmits each submitted paper to appropriate subject matter editor, who solicits peer reviews; (3) editor sends e-mail to author for revision in response to reviewers' comments; editor accepts or rejects revision; and informs author and AOAC editorial office; (4) AOAC editorial staff edits accepted papers following the journal of AOAC INTERNATIONAL style guide and transmits approved manuscripts to desktop publisher; (5) desktop sends page proofs to author for final approval.

### General Information

The Journal of AOAC INTERNATIONAL now has an online submission process. Please see **http://www.aoac.rog/pubs/joural.htm** to submit. **Papers will NOT be accepted through mail or fax. Additional hardcopies are no longer necessary.** Please follow these instructions closely; doing so will save time and revision. For all questions of format and style addressed in these instructions, consult recent issue of Journal or current edition of *Council of Science Editors Style Manual* (http//:www.councilscienceeditors.org/ publications/ style.cfm). For describing collaborative studies, contact Robert Rathbone (+1-301-924-70077 x105, rrathbone@aoac.org) or Yinka Ladeji (+1-301-924-7077 x 33, oladeji@aoca.org) for guidelines.

1. Write in clear, grammatical English

2. Submit all the following mandatory information: (1) title; (2) all author's full names (full first middle initial if any, full last), addresses including mail codes, and corresponding author's e-mail address; (3) abstract; (4) text (introduction, method or experimental, results and/or discussion, acknowledgments, references); (5) figure captions; (6) foot notes; (7) tables with captions, 1 per page; and (8) figures (see instructions below).

3. Suggest at least 4 qualified reviewers, i,e., individuals engaged in or versed in research of the type reported.

4. Double space all materials. Do not right justify or use proportional spacing; avoid hyphenation.

5. If the Section Editor sends notification to author that changes are necessary, it is the author's responsibility to upload updated files per Section Editor's comments before approval is granted.

## Format and Style

1. *Abstract :* 200 words. Provide specific information, not generalized statements.

2. *Text :* Upload document or copy and paste into appropriate field. **Please send manuscript text as Word, WordPerfect, or txt file. Do NOT embed figures or tables in the document. Please include all figures with captions and tables at the end of the manuscript, following the references.**

*Introduction :* Include information on why work was done, previous work done, use of compound, or process being studied.

*Methods or Experimental* Consult recent issue of *Journal* for proper format. Separate special reagents/apparatus from details of procedure and list in sections with appropriate headings; avoid use of brand names. (Common reagents/ apparatus or those that require no special treatment need not be listed separately.) Place detailed operations in  separate sections with appropriate heading (e.g., *Preparation of Sample, Extraction, and Cleanup*). Include necessary calculations; number of significant figures must reflect accuracy of method. Use metric units for measurements of quantities wherever possible. Note hazardous and/or carcinogenic chemicals.

**Results/Discussion :** Cite tables and figures consecutively in text with Arabic numerals. Do not intersperse tables and figures in text.

**Acknowledgments :** Give brief thanks (no social or academic titles) or acknowledge financial aid in this section.

**References :** Submitted papers or unpublished oral presentations may  not be listed as references; cite them in text as published data or personal communications. Cite all references to previously published papers or papers in press in numerical order in text with number in parentheses on line (*not superscript*). List references numerically in *References* in *exactly* (arrangement, punctuation, capitalization, use of ampersand, etc.) the same styles of examples shown below or *see* resent issue of *Journal* of less often used types of entries. Follow *Chemical Abstract* for abbreviations of journal titles.

Journal Article Reference

   (1)  Esgstrom, G.W Richard J.L., &

       Cysewski, S.J. (1977) *J. Agric. Food Chem,* **25** 833-836

Book Chapter Reference

   (1)  Hurn, B.A.L., & Chantler, S.M.(1980) *Methods in Enzymology*, Vol 70 H. Van Vunakis & J.J. Langone (Eds), Academic Press, New York, NY, pp 104-142

Book Reference

   (1)  Siegel, S. (1956) *Nonparametric Statistics for the behavioral Sciences,* McGraw-Book Co., New York, NY

Official methods Reference

   (1)  *Official methods of Analysis (2000)* 17th Ed., AOAC INTERNATIONAL, Gaithersburg, MD., Method **2000.01**

  **3.**  *Tables :* Please submit tables, preferably in Word or WordPerfect, in table format. Each "entry" must be in a cell of its own.

  **4.**  *Figures :* Figures must be submitted in a graphs format, preferably .tif, however, .jpg and .gif are acceptable. Usually a minimum of 300 dots per inch is required for decent print quality. Each figure should be submitted as its own file.

  **5.**  *Footnotes :* Avoid use of footnotes to text. Include location/date of presentation, if appropriate; present address(es) of authors(s); identification of corresponding author, if not senior author, proprietary disclaimers; institution journal series numbers.

  **6.**  *Copyright :* Authors must print out, sign, and fax or mail the copyright form to AOAC INTERNATIONAL. This form can be found online during submission.

  **7.**  *Miscellaneous :* Abbreviation for liter is L; abbreviation for micron is l. Do not italicize common Latin expressions such as et al. and in vitro; for nomenclature of spectrometry, gas chromatography, and liquid chromatography, follow practice of American Society for Testing and Materials (in particular, do not use "high performance," "high pressure," or the abbreviation "HP" with "liquid chromatography")

**ANY FILES SUBMITTED INCORRECTLY WILL BE SENT BACK TO AUTHOR AND MAY DELAY PUBLICATION.**

# CHAPTER - 124

# The Journal of Plant Biochemistry and Biotechnology

## Instructions to Authors

The Journal publishes research papers in the areas of plant biochemistry, plant molecular biology, microbial and molecular genetics. DNA finger printing, micro propagation, and plant biotechnology including plant genetic engineering & genomics. Review articles in the above mentioned areas would also be accepted. However, for this prior permission of the Chief Editor may be obtained. Chief Editor on the advice of Editorial Board members may also invite review articles from scientists. The length of the review article must not exceed 5000 words including tables and illustrations. Page charge for authors including tables and illustrations. Page charge for authors who do not order reprints would be @ Rs.100 (US $ 10) per page. All the manuscripts will be refereed. The responsibility for statements in the article rests entirely with the authors thereof. The authors are advised to follow the following style while preparing the manuscript.

**General :** Each full length research paper must not exceed 3,000 words including tables and illustrations.

Short communications are also accepted but their length should be restricted to about 1000 words. Manuscripts should be typed on one side of the bond paper of 22 × 28 cm, double-spaced. While the short communication will have only title, authors' name and address, abstract, key words, followed by text and references without headings, the full length paper should have the following headings.

**Short Title :** A short title, not exceeding 35 letters, should be typed at the top of first page in first letters capitals and underlined.

**Title :** The title should be short and informative, typed in first letters capital, Latin names underlined.

**Authors :** Names of authors to be typed in first letter capital.

**Address :** Address of the institution where the work was carried out. Footnote should indicate the author for correspondence.

**Main Headings :** Each full-length research paper should be divided into the following main headings which are to be typed in first letters capitals on the left hand side of the page: abstract, Introduction, Materials and Methods, Results, Discussion, References.

**Sub-headings :** Typed flush left in first letter capital, underlined.

**Abstract :** A brief abstract, not exceeding 200 words should be typed after address.

**Key words :** The Abstract should be followed by about five key words indicating the contents of the manuscript.

**Introduction :** Introduction should be brief and pertinent.

**Materials and Methods :** Materials and Methods should include the source and nature of material, experimental design and the techniques employed.

**Results :** Results should contain data, which are essential for drawing main conclusion from the study. Wherever needed the data should be statistically analysed. Same data should not be presented in tables and figure form.

**Discussion :** The discussion should deal with the interpretation of the results. Wherever possible, Results and Discussion can be combined.

**Tables :** Tables should be typed on separate sheets, each with a heading. Tables should be typed with the first letter capital, table number in Arabic numerals, followed by a period.

**Figures :** For each figure submit a glossy print or original drawing. Photomicrographs should have a scale bar. Line drawings should be roughly twice the final printed single column size of 7.5 cm width. Text figures should be numbered in Arabic numerals. Captions and legends to illustrations should be typed on a separate sheet of paper. Line drawings and photograph should contain figure number and authors' name, on the back with soft lead pencil. Colour photographs will be printed if author(s) pay Rs 2000/- (or US $ 50) as processing charges.

**References :** References should be cited in the text by number in brackets. References should be numbered in the order in which they appear in the text. Use conventional abbreviations for names of well known Journals. Each reference should provide: name(s) of the author(s) followed by Journal Name (italicized or underlined), Volume No., Year in bracket and page No. The style of references should be:


1.  **Schubert KR & Evans HJ,** Proc *Natl.Acad Sci,* USA, 73 (1976) 1207.

2.  **Renaud CS, Pasternak JJ & Glick BR,** Arch Microbiol, 152 (1989) 437.

3.  **Vincent JM,** *A manual for the practical study of root nodule bacteria,* Blackwell scientific Publications, Oxford (1970).

4. **Larkins BA,** In the biochemistry of plants, vol 6 (PK Studmpf, EE Conn, Editors), Academic Press, New York (1981) pp 449-489.

**Abbreviations :** Use standard abbreviations. Some of the common abbreviations are given below:

A (absorbance); h (hour); min (minutes); sec (seconds); cpm (counts per minute); Ci. (Curie); mol wt (molecular weight); KD (kilo Dalton); kb (kilo base); sp act (specific activity); wt (weight); SD (standard deviation); SE (standard error); DAF (days after flowering)

**Reprints :** Only 50 reprints will be available on payment and should be ordered while returning the galley proof.

# CHAPTER - 125

# The Journal of Research ANGRAU

## Guidelines to Contributors

*Submission of Manuscripts :* All the authors and co-authors of a research paper should essentially be the subscribers for the Journal of Research ANGRAU. The papers relating to agriculture, veterinary, fisheries or home science are invited to this journal. The papers should by typed neatly with "Times New Roman" letter type in double space on A4 size bond paper leaving a margin of 5 cm on all sides. Three copies of the paper have to be submitted along with self-addressed envelope with sufficient stamps. fulfilling the subscription requirements to the **Managing Editor. Agricultural Information and Communication Centre, Rajendranagar, Hyderabad – 500 030.** The authors are requested to follow the latest guidelines of the journal as furnished here. The length of the paper as furnished here. The length of the paper shall be restricted to a maximum of 6 typed pages for research papers and 3 for a research note including tables. The paper/note will be published after the approval by the Editorial Board. Each article has to be organized as follows.

**Title :** It is should be specific, short, unique and informative and reflect the content of the paper. The starting letter or each word should be in capitals, set in bold with a font size of 12 pt. The common names should be followed by scientific names in brackets and be in italics. The end of the title should be marked with asterisk, the particulars of which are given in footnote at the bottom of the page.

**Byline :** The name of authors should be prefixed with initials in capitals with appropriate punctuations, font size being 10 pt. Followed by 'and' in lower case letters before the name of the last author. The name of the authors should be consecutively numbered by plain Arabic superscripts, the details of which have to be given in footnote. The place where research work was conducted has to be given after names.

**Foornote :** It should contain details of an asterisk marked at the end of the title indicating if it is a part of M.Sc.(Ag.) or Ph.D. thesis or scheme work. For the plain Arabic superscripts, details of present addresses of authors be given for correspondence. Footnote has to be typed in 8 pt font size.

**Abstract :** The heading should be in capitals, bold, (10 pt.) not exceeding 100 words and the content be typed in single space (not exceeding 5 per cent the length of the paper). It shall include the principal findings, the names of organisms, effects of major treatments and valid conclusions.

The entire text has to be typed in 10 pt font size including headings. Introduction (without heading) should be brief, stating the background relevance, main objectives and importance of the study. It must end with proper justification for undertaking the work reported in the paper.

**Materials and Methods :** The starting letter of each work in the heading be given in capitals, in bold. Relevant details have to be given for the materials used and methods followed including experimental design, treatments, techniques and statistical designs employed. Any new or modified technique, if followed be given in detail. Whenever the methods of other authors (old) are followed, mere reference to their paper be made instead of repeating the procedure.

**Results and Discussion :** The starting letter of each word in heading be given in capitals in bold. The results and discussion must be combined to avoid repetition. The results have to be supported by brief but adequate number of tables or graphic material wherever necessary. The number of tables shall  not be too many and unwieldy. Data presented in tables shall not be duplicated in figures and vice-versa. The salient results may be highlighted and discussed in relation to reported findings of related works. Discussion has to be oriented towards explaining the significance and implication of results, supported with previous works with logical interpretations and end with conclusions.

Only the references mentioned in the text are to be cited under references. Metric system of units have to be used for standard measurement of physical quantities. The latest expression of units viz., kg ha-1, t ha-1 etc. be used.

**Tables and Figures :** Tables and figures shall be consecutively arranged, numbered with Arabic numerals and order in which they are discussed in the text. Eg. Table 1, Fig.1 The tabulated mater should not exceed 20% of the text and be arranged as per the dimensions of the printed page. The headings of tables and figures (heading abbreviated) have to start with capital letter being capitalized and in bold. Each table should be typed in double spacing and have only vertical lines between the headings and horizontal lines one above the column heading and the other below the column heading and the other at the table foot. The first word of each column and row headings has to begin with capital letter and bold.  The figures in tables, which need footnote, have to be denoted by asterisk. The figures shall fit in the normal layout of the page and be drawn on white art card paper in black Indian ink and all letters may be and symbols shall be defined when they are used for the first time in the paper at the foot of the figure/table.

Use figures to express all numbers 10 and above. Use words to express the numbers lower than 10 and common fractions that begin a sentence/title. The seasons namely *kharif, rabi* and vernacular names shall be mentioned in italics.

In the text, citations have to be made according to the name and year system. Depending upon the construction of the sentence, the citation will appear thus: Rao (1970) or (Rao, 1970) or Rao and Murthy (1995) or Rao and Murthy, (1995). Etc. When there are more than two authors, it should be given as Rao *et al.* (1995) if given in brackets and the works *et al.* be given in italics.

**References :** The starting letter of the heading shall be given in capital and bold. References shall be cited in single space and with double in between successive references. References shall be furnished at the end of the paper. They have to include and correspond with the names mentioned in the text. In references the names of the authors be arranged alphabetically and mentioned in full followed by corresponding initials in capitals. The last but one author shall be prefixed with 'and' in lower case letters followed by year of publication, title of article/thesis/journal name with volume number in 'bold' page numbers denoted by first and last pages separated by hyphen in that order and end with a full stop. The name of the journals, periodicals, and reports shall be cited in full. While listing references, the following  examples may be followed.

**Journals and bulletins :** PANEERASALVAM K, LOGISWARAN G and GOPAL S 1990 Time of insecticidal application for control of groundout leaf miner *Aproaerema modicella.* Indian Journal of Applied Sciences 60 : 86-87.

**Books :** EDWARDS A C 1957 Techniques of Attitude Scale Construction. Yakis Eelter and Simons Private Limited. Haque Buildings, 9 sports road, Ballard Estate, Bombay, pp: 149-171.

**Thesis :** Rao V S N 1997 Job satisfaction of veterinary assistant surgeons of Vijayawada and Hyderabad areas. Unpublished M.Sc., (Dairy Extension) Thesis submitted to National Dairy Research Institute, Karnal.

**Papers presented in symposium/seminar/workshop and annual meeting :** KUNJAMMA K K  1996 Flowering studies in three sorghum hybrids. Paper resented at the All India Symposium of Modern Concepts in Plant Protection Udaipur 26-28 March, 1976.

**Newsletter & Proceedings of seminar/symposia/workshop(published) :** BOYCE H R 1961 Insecticidal activity of manual formulations against greenhouse whitefly. Proceedings of Entomological Society of Ontario 92: 196-200.

**SUBMISSION OF REVISED SCRIPTS :** Two revised scripts typed on paper of size on 14 ½ x  18cm (text and tables) along with CD and the corrected scripts by the referees have to be sent to the Managing Editor within a month.

**Reprints :**  20 copies will be sent on payment of Rs. 50/- which is compulsory

# CHAPTER - 126

# The Journal of Vegetation Science

## Publishes Instructions to Authors
## Scope

The Journal of Vegetation Science publishes papers on all aspects of vegetation science, with particular emphasis on papers that develop new concepts or methods, test theory, identify general patterns, or that are otherwise likely to interest a broad readership.

Papers may focus on any aspect of vegetation science including theory, methodology, spatial patterns (including plant geography and landscape ecology), temporal changes (including palaeoecology and demography), processes (including ecophysiology), and description of ecological communities (by phytosociological or other methods), provided the focus is on increasing our understanding of plant communities. Papers with a more applied focus should be directed to our sister journal, Applied Vegetation Science.

Applied Vegetation Science includes any community-level topic relevant to human impact on vegetation, including global change, nature conservation, nature management, restoration of plant communities and of the habitats of threatened plant species, and the planning of semi-natural and urban landscapes. Vegetation modelling and remote-sensing applications are especially welcome.

Applied fields covered by the journal include human impact on vegetation, particularly eutrophication and global change, nature conservation, nature management, restoration of plant communities and habitats of threatened plant species, and the planning of semi-natural and urban landscapes.

## Acceptance criteria

To be acceptable, a paper must have something that will interest an international readership, even if its immediate scope is local (Other papers, even if competently executed, would be better published in a local or regional journal). A paper can be new/interesting by doing one of several things:

- Develop new concepts in understanding vegetation, or
- Test concepts applicable to all plant communities, or

- Add particularly well executed empirical examples that are part of a growing literature on a general conceptual issue, or

- Represent a particular interesting combination of models, observational data, and experiments, or

- Demonstrate new and generally useful methods, or

- Present a particularly exemplary or thorough analysis, i.e. be a definitive paper, even if concepts and methods are not novel. A definitive paper is one that represents the state of the art (methods and statistics) that presents a critical and final (definitive) test for an interesting hypothesis, even if that hypothesis is not new and even if many other papers have collected data that bears, in a less definitive way, on the hypothesis. It might be regional in its scope, but be a model of how to write a type of paper or how to apply methods, or

- Describe the vegetation of an area, whether large or small, when that description will be of interest to readers worldwide because that habitat/vegetation will be of such interest, or when it attains the exemplary qualities described above, or

- Demonstrate how vital vegetation science is to social questions of the day (species invasion, global warming, nitrogen deposition).

The questions in the paper can be addressed by many means, including description, experiments, simulations, meta-analysis, inference, extrapolation, etc. We do not limit the nature of the approach, as long as the work is sound.

As a very rough rule of thumb, we would accept a paper if in our judgement we think that 66% of vegetation scientists would regard it as having some interest, or 10% would regard it as being very interesting.

## Manuscripts

The language must be in English (either British or American throughout). We see the norm as eight journal pages. Shorter articles are welcome, and may be published sooner. There is no maximum length, but longer papers should contain greater content of interest, in proportion to their length. The Editorial Office (edit@opuluspress.se) can provide an advance length estimate of a manuscript you are preparing: preferably send an MS-Word copy of the ms including tables and figures.

One of the Chief Editors or Associate Editors will be selected as Co-ordinating Editor for each submitted ms, and will make the final decision on acceptance.

There are four categories of contributed paper: Ordinary article (including reviews), Forum, Letter, and Report.

## Forum

Forum papers are essays with original ideas / speculation / well-sustained arguments. They will usually contribute to free debate of current and often controversial ideas in vegetation science. There may be criticism of papers published in JVS/AVS, or (if interesting to our readers) of papers published elsewhere. An Abstract is required, but otherwise the sectional format is flexible. Maximum length is normally four pages. Forum papers have high priority in publication.

## Letter

A letter is a very short but refereed paper (= 0.5 pp.) with one scientific idea. It will have the highest priority in publication. The section is intended for positive ideas, not for papers primarily intended to criticise.

## Report

This includes items that are not scientific papers: e.g. news items, obituaries (normally commissioned), the existence of databases and technical information.

A report cannot comprise the announcement and/or description of a new computer program. We can accept paid advertisements for computer programs. Moreover, we carry reviews of computer programs, and authors of new programs are very welcome to submit them for review. [Papers that, whilst mentioning a particular program, are basically descriptions of a new method, can be submitted as ordinary articles.]

## Submission

Submit manuscripts either to the web page www.OpulusPress.se ? 'For Authors', or to the Editorial Office, edit@opuluspress.se, in either case as:

An MS-Word file, preferably with figures included/embedded (each with its caption below it) at the end of the Word document. If you are unable to provide MS-word, we can accept almost any other word-processing format. [We ask for a word-processor file because it is convenient for referees and editors who want to make corrections/suggestions directly in the text.]

Preferably plus also a PDF file, containing the entire ms as above, including any tables, figures and/or or appendices. [We ask for a PDF file in addition because we can be more sure of your intended format from a PDF file.]

When submitting a manuscript, indicate in a covering letter:

1.  Whether the manuscript is a revision of a manuscript previously rejected by J. Veg. Sci. or Appl. Veg. Sci. If so, state how previous concerns have been addressed.

2.  That it comprises work that has not been published, and is not being considered for publication elsewhere. If there be substantial overlap between the submitted manuscript and manuscripts under consideration elsewhere or with papers already published, explain the situation in detail.

A manuscript will not be entered into the editing process until a covering letter addressing the above two points has been received.

## Format

Double-space manuscripts (including tables and references). Number all pages, and number the lines on each page.

Use scientific names of taxa, and avoid vernacular names.

Units of measurement must follow the International System of Units: e.g. $mg.m^{-2}.yr^{-1}$.

Numbers with units of measurement must be in digits, 3.5 g . Numbers in the text of up to ten items (i.e. integers) should be in words, e.g. "ten quadrats", "five sampling times"; otherwise in digits, e.g. "11 sampling times". Use '.' for a decimal point. Thousands in large numbers should be indicated by a space, e.g. 10 000 for ten thousand.

For an example of format, see www.OpulusPress.se ? Sample papers.

## Technical issues

If your paper is eventually accepted, there are several technical issues that will need to be checked (listed on page 2). You can check these when you receive the Co-ordinating Editor's response and make necessary modifications (the Co-ordinating Editor may give you directions on such issues, or may not, depending on how busy they are at the time). If your paper is accepted, it will be passed via the Editorial Office to the Final Editors (Eddy van der Maarel and Val Cochrane). If only minor technical issues remain, they may make the changes themselves, perhaps checking with you first, or asking you by a note on the proofs to check the changes. For more major change (e.g. if there are many language problems), the Final Editors will be unable to correct your paper for you, and you will be given the choice of doing this work yourself, even at this late stage, or having us do it for you and charging you for the time it takes (The journal operates on a low-cost basis, to make information on Vegetation Science available to all, including students and those in developing countries, so it cannot afford numerous staff. Exceptions to these charges can be made only for ecologists from the very poorest countries). It is quite possible that none of this will apply to your paper, but we warn all authors at the submission stage just in case it turns out that it does.

In-press papers are available for anyone to download from the Opulus site, and you might find other interesting papers there too.

## Technical Issues

**Abstract:** Up to 250 words (fewer for a Forum paper). No references. The abstract for ordinary papers should have **named sections,** normally: Question, Location, Methods, Results, and Conclusions. Vary this structure when necessary: e.g. for reviews; omit Location for theoretical papers; make other small changes when expression is clearly enhanced

**Title :** Catch the reader's attention with topical issues or an interesting hypothesis

**Logical structure :** The Introduction should state what topics will be addressed, and those topics should be addressed by the Methods, Results and Discussion

**Author list :** e.g.:

**Bush, George W.[1,2]; Smith, A.B.[1,3] & Coxon, E. Fred[4*]**

[1]*Ecology Department, Little Marsh University, 11 Main St., Little Marsh, Berkshire, UK;*

[2]*E-mail  gw_bush@lmu.ac.uk;* [3]*E-mail  ab_smith @lmu.ac.uk;*

[4]*Botany Department, Herbicide Manufacturers, P.O. Box 2002, Southend-on-Sea, UK;*

*Corresponding    author;    Fax    +4412345678;    E-mail doughnut@herbicide.co.uk; Url: www.herbicide.co.uk/efcoxon*

**Keywords :** e.g. Aardvark; Copper mobilty; New Zealand; Zebra tail [don't duplicate words from the title]

**Nomenclature source :** Give if relevant

**Introduction :** Ending with questions or hypotheses

**Results :** The claims in the Results section text should match what's in the figures and tables

**Citation format in text,** e.g.: (Smith et al. 1999; Tan & Sim 1970)

**Table and Figure captions :** Understandable without reading the text.

**Tables :** Concise, with row and column labels as self-explanatory as possible

**Figures**

   a.  Not too many of them, and compact

   b.  Readable at the size they will be printed

   c.  No superfluous lines (e.g. across a graph, or to the top and right of a graph).

      d Lines and symbols explained in direct language, e.g. * = Litter removed

      not: * = LRT  or  * = Treatment LR  or  * = Treatment 3

**Reference format** [take especial care with book chapters]

   Greig-Smith, P. 1983. *Quantitative plant ecology.* 3rd ed. Blackwell, Oxford.

   Lane, D.R., Coffin, D.P. & Lauenroth, W.K. 2000. Changes in grassland canopy structure across a precipitation gradient. *J. Veg. Sci.* 11: 359-368.

Levin, S.A. 2001. Immune systems and ecosystems. *Conserv. Ecol.* 5(1): article 17. URL: http://www.consecol.org/vol5/iss1/art17 [Ecological Society of America].

Whittaker, R.H. 1969. Evolution of diversity in plant communities. In: Woodwell, G.M. & Smith, H.N. (eds.) *Stability and diversity in ecological systems*, pp. 178-196. Brookhaven National Laboratory, Brookhaven, NY.

There should not be an unnecessary number of references. A reference that is cited only once, and that time along with several others, may be redundant.

**Citation of JVS/AVS papers :** Are any directly-relevant recent JVS or AVS papers cited?

**Electronic appendices :** All appendices (except mathematical ones), arge figures & tables, extra photographs and raw data, go here.

## Figures and tables

Figures in the submitted manuscript should be reproduced at the size at which they are intended to be printed: either one-column or full-page width. Place each figure/table caption on the same page as the figure/table to which it refers. The caption should contain sufficient information for the figure/table to be understood without reference to the text of the paper, even if its significance cannot be fully appreciated. Figures and tables may optionally be embedded in the text.

## Sections

**Title :** This should be strongly directed towards attracting the interest of potential readers.

**Author names and addresses :** Follow exactly the format in an issue of the journal. Give an E-mail address for all authors. Fax and web addresses may be given for the corresponding author.

**Abbreviations :** List any that will be frequently used in the text.

Nomenclatural reference (as a source for the authors of scientific names).

**Abstract :** Up to 250 words (fewer for a Forum paper). Include no references.

The abstract for ordinary papers should have named sections, normally: Question, Location, Methods, Results, and Conclusions. Vary this structure when necessary: e.g. for reviews use whatever structure is appropriate; for theoretical papers Location is not needed; for Forum papers a different and compressed structure will probably be appropriate.

**Keywords :** These should not duplicate the title.

**Main text :** Start this on a new page. Indicate new paragraphs by indentation. Avoid footnotes.

Variation from the usual Introduction - Methods - Results - Discussion structure is acceptable when appropriate.

If coloured figures are required, a financial contribution for the extra cost will be negotiated.

**Acknowledgements :** Keep them brief. References to research projects/funds and institutional publication numbers go here.

**References :** see below.

## References

**Citations in the text :** Use forms such as: Smith (2005) or (Smith 2005) or Smith et al. (2005) or (Smith 2005 a, b; Jones 2006).

**References section :** Use the formats below. Journal names; always give the full name.

Lane, D.R., Coffin, D.P. & Lauenroth, W.K. 2000. Changes in grassland canopy structure across a precipitation gradient. *Journal of Vegetation Science* 11: 359-368.

Greig-Smith, P. 1983. *Quantitative plant ecology*. 3rd ed. Blackwell, Oxford, UK

Whittaker, R.H. 1969. Evolution of diversity in plant communities. In: Woodwell, G.M. & Smith, H.N. (eds.) *Stability and diversity in ecological systems*, pp. 178-196. Brookhaven National Laboratory, Brookhaven, NY, US.

Levin, S.A. 2001. Immune systems and ecosystems. *Conserv. Ecol.* 5(1): article 17. URL: http://www.consecol.org/vol5/iss1/art17 [Ecological Society of America].

**Unpublished material :** The References section can contain only material that is published, in press or is a thesis. Indicate all other material as "unpubl." or "pers. comm." (the latter with date and description of the type of knowledge, e.g. "local farmer"); "submitted" may be used only if the cited item is in some journal's editorial process, and the reference will have to be removed if the item has not been firmly accepted by that journal by the time proofs are corrected for citing paper.

References to web pages that cannot be guaranteed to remain valid for say 25 years should be avoided. It is preferable to include such material as an electronic appendix to the paper. When essential, reference to such 'ephemeral' web pages should be given in the text, and the citation not listed under References.

## Photographs

Photographs (referred to in the text as Plates) of the vegetation or landscapes studied are encouraged. Supply high-contrast glossy prints. They may be reproduced in colour if this enhances the reader's understanding appreciation of the results presented, and if an arrangement can be made regarding the extra costs. Extra, non-essential photographs that help the reader appreciate the landscape/vegetation/plants are welcome; they will be printed if there is spare space in the article, or otherwise put in the electronic archive.

When a paper is accepted, photographs will be solicited as candidates for the cover of the issue in which the paper will appear.

## Appendices, and the journal's electronic archive

Appendices (except mathematical ones) are not printed, but are placed in the journal's electronic archive, and a web address to the archive is given in the printed paper. This facility can be used for raw data, extra tables, extra figures or maps (including coloured ones), etc. The archive is held at several geographic locations, to ensure permanence.

## Criticism and responses

If a paper (Forum or otherwise) has a major element criticising a particular paper or body of work of (an)other scientist(s), the latter will be invited to comment on the paper (doing this does not prevent the criticised scientist(s) from writing a reply). However, those comments will be taken in context, and there will in addition be one or two referees who are outside the controversy.

The author who has been criticised will be offered a right of reply in the same journal issue, so long as the reply is received before an indicated deadline, typically four weeks from acceptance of the criticism, especially since the criticised author will already have seen the criticism (see above). The reply will be refereed. It will be sent to the author of the original criticism, to check there is nothing unreasonable or offensive.

The criticising authors have no automatic right of further responses, but the editor may allow this. The sequence will normally finish with the author(s) of the originally criticised paper, or when the participant next due to submit does not do so, or submits an article that says nothing that is both new and valid as judged by referees. Such further responses will normally be published in a later issue.

The editor will ensure that the process is fair to all concerned, and that the readers of the journal can evaluate both sides and make their own decision.

Correction of simple errors

If a paper is submitted that contains only a correction for a simple error in a paper published by another author, but one that is likely to mislead readers, if the error does not appear to affect the results (e.g. it is an error in a formula, though the correct formula was used in the original calculations), the original author will be offered the opportunity to submit an erratum with an acknowledgement to the author who pointed it out, which would be published in place of the submitted critical paper.

An author may of their own initiative submit an erratum note for an error that is likely to mislead readers.

## Page charges and subscriptions

There are no page charges. However, please consider taking a subscription to *J. Veg. Sci.* and/or *Appl. Veg. Sci.*: they carry important articles in your field. Subscriptions help us to avoid charges. The personal subscription rates are very reasonable and include membership of IAVS.

For those in poor countries, assistance may be available through the International Association for Vegetation Science: contact the Secretary General (Joop.Schaminee@wur.nl).

In addition, both *J. Veg. Sci.* and *Appl. Veg. Sci.*, as well as many other journals of ecological interest, are available online in the BioOne package and they offer schemes by which institutions in less-developed countries can obtain access at reduced rates, or in cases of extreme hardship possibly free. Please contact the following programs for eligibility and subscription rates:

HINARI (World Health Organization initiative) www.who.int/hinari/eligibility/e/

e-IFL (Soros Foundation initiative) www.eifl.net/

ALUKA (Andrew W. Mellon Foundation initiative) www.ithaka.org/aluka/index.htm

## Offprints

Authors receive 25 free offprints, plus a. PDF file of their paper from which they are free to make paper or electronic copies. Further offprints may be ordered from the Publishers (an order form accompanies the proofs).

# CHAPTER - 127

# The Karnataka Journal of Horticulture

## INSTRUCTIONS TO AUTHORS

**General :** Research papers for publication must contribute substantially to the advancement of knowledge in horticulture. The Editorial Committee has right to accept or reject a paper and the committee does not shoulder any responsibility for the opinion(s) expressed in the paper by the author(s). Once a paper is accepted for publication, it should not be published elsewhere either in the same or abridged form or in any other language, without the written permission of the Society. They should also specify that they are not sending the paper to any other journal simultaneously. The manuscripts are published on cost basis. Therefore the authors are requested to cooperate in bearing the printing cost if the manuscript is accepted for publication. All correspondence should be addressed to The Chief Editor, The Karnataka Journal of Horticulture, K.R.C. College of Horticulture; Arabhavi-591 310, Taluk: Gokak, Dist: Belgaum, Karnataka. The senior author of manuscript should be the member of the Society.

**Manuscript and its Arrangement :** Manuscript should be submitted in duplicate (Hard Copy) along with soft copy. Manuscript, not more than ten and 3-4 pages inclusive of tables, diagrams and references for research paper and research notes, respectively, should be typed in double line space on A4 size paper on one side of the page only, with atleast 5 cm margin on the left and 2 cm margin on the right. The language should be simple (UK English, not US English) and care should be taken to check up the spellings, punctuations, etc. All tables should be serially numbered and should be typed on a separate sheet.

The contents of research paper should be organized as, Title, Abstract, Introduction, Material and Methods, Results and Discussion and References. Acknowledgement, if any, may be included in the last paragraph of the text before the References part. The contents for research note should be organized without mentioning the sub-headings as above, but with the Title and References.

**Cover Page :** The cover page of the manuscript should carry the title of the research paper, name(s) of author(s) and their address. The authorship and its order furnished at the time of registration of the manuscript is final. If the paper forms a part of M.Sc./Ph.D. thesis, a foot-note should indicate the same. No manuscript should carry author(s) name(s) and address anywhere on the body of the manuscript other than on cover page.

**Title :** This should be informative; but concise. While typing the title of the paper/note, only first letter of each word must be in capital. The title must be typed just before the commencement of abstract.

**Abstract :** The abstract must be brief and informative and should not be more than 300 words.

**Introduction :** The introduction should be brief and state the objective of the experiment. The review of literature should be pertinent to the problem.

**Material and Methods :** This should be precise and include experimental design, treatments and techniques employed. Whenever the methods of other authors are followed, it would suffice if references to their paper are made instead of repeating the procedure.

**Results and Discussion :** This should govern the presentation and interpretation of experimental data only and each of the experiments should be properly titled. Common names of plant species, microorganisms, insects, etc. should be supported with authentic, latest Latin names, which should be underlined in the typescript or italiced  in the computer typing. When such names are first mentioned, the full generic name and the species name with the authority should be mentioned and in subsequent reference, the generic name be abbreviated and the authority be deleted.

All headings must be typed in lower case from the left hand margin. The sub-headings must be typed in lower order capitals and must start from the left hand margin and underlined (italiced in case of computer typing). The Para under a sub-heading must start from a line below the sub-heading.

**Tables :** Every table should be on a separate sheet with bold letter titles and clear with proper reference in the text. The units of the data should be defined. Appropriate statistical tests should be applied to the data presented. Footnote should be seldom used. Lengthy tables should be avoided.

**Illustrations :** The illustrations and text-figures should be clear and capable of reproduction in print. They should be drawn neatly on tracing paper in Indian ink with serial numbers in clear, uniformly large letters. Photo prints should be on glassed paper with proper contrast and at least of half plate size pasted on thick mount. If there are many photo prints, they are to be grouped compactly. Whenever photo prints are included, the magnification should be indicated.

Photostat copies of any illustrations are not accepted and in all cases, originals should be sent. Legends to figures and photos should be typed on the page on the figure/photo.

Abbreviations should be used sparingly if advantageous to the reader. All new or unusual abbreviations should be defined when they are used for the first time in the paper. Ordinarily, the sentences should not begin with abbreviations or numbers.

**References :** This should be typed in double line space along with the body of the paper but starting on a fresh page. These should be listed in the alphabetical and chronological order. In the text, the references should be cited as Patil (2002) or (Patil, 2002), Patil and Swamy (2004), Patil et al. (2004) or (Patil et al. 2004), when there are more than two authors.

**While listing the references, the following examples should be followed.**

While citing an article from:

## Journal
Conn HJ, 1942, Validity of the genus alcaligens. *Journal of Bacteriology,* 44(1): 353-360.

## Papers presented in Symposium / Seminar / Workshop
Lal B, Mishra, Dand Tripathi S, 2003, Effect of cowdung paste on rejuvenated mango trees. Paper presented at the *National Symposium on Organic Farming in Horticulture for Sustainable Production,* CISH, Lucknow, 29-30 August 2003.

## Proceedings of Seminar / Symposia / Workshop (published)
Lal B, Mishra D and Tripathi S, 2004, Effect of cowdung paste on rejuvenated mango trees. Proceedings of National Symposium on Organic Farming in Horticulture for Sustainable Production, pp.270-271.

## Edited Book
Pandey SN, 1998, Mango cultivars. In *Mango Cultivation.* Ed. Srivastava RP, International Book Distributing co., Lucknow, UP, pp.39-100.

## Bulletin
Gray p, 1914, The compatability of insecticides and fungicides. *Monthly Bulletin of California,* July 1914.

## Annual meetings
Schenck NC, 1965, Compatability of fungicides with insecticides and foliar nutrients. 57th *Annual Meeting of American Phytopathological Society,* 3-7 October 1965.

## Books

Panse VG and Sukhatme PV, 1967, *Statistical Methods for Agricultural Workers,* ICAR, New Delhi, pp. 70-75.

## Reports

Anonymous, 1971, Investigations of insect pests of sorghum and millets. *Final Technical Report,* 1965-70, IARI, New Delhi, pp.157.

## Annual Report

Anonymous, 2003, *Annual Report for 2002-03.* University of Agricultural Sciences, Dharwad, pp. 90-95.

## Newsletter

Katyal JC, 2003, Need for reforms in farm education. ***CSK HPKV Newsletter,*** 21(1-2): 3.

## Thesis

Adiga JD, 1996, Clonal propagation of Singapore jack through tissue culture. *Ph.D. Thesis,* University of Agricultural Sciences, Bangalore.

The name of the source (periodicals/reports/seminar/symposia) should be spelt out completely.

# CHAPTER - 128

# The Maharashtra Journal of Extension Education

**Notice to Authors**

Send Research Papers to the Editor by name. Each manuscript in duplicate typed on A-4 size paper should be accompanied by the declaration that the research paper has neither been published not sent for publication in any other journal duly signed by authors.

All correspondence and remittance relating to subscription, printing charges, sales and advertisement etc. should be addressed to the Secretary, Maharashtra Society of Extension Education, Department of Extension Education, Konkan Krishi Vidyapeeth, Dapoli. Dist. Ratnagiri –415712, Maharashtra, India.

# Chapter - 129

# The Nucleus

**An International Journal of Cytology and Allied Topics**
**The nucleus edited and published by**

**AK Sharma and A Sharma,** Centre for Advanced Study in Cell and Chromosome Research, Department of Botany, University of Calcutta, Kolkata 700 0019, India and UC Lavania, CIMAP, Lucknow 226 015, India (e-mail: lavania@cimap.res.in)

**Postal Address**

The Nucleus, C/o a Sharma, 2F2, Meghmallar,

18/3 Gariahat Road, Kolkata 700 019, India

## Rules and Regulations

The nucleus, a journal of cytology and allied topics, started in 1958, is published in three issues, in April, August and December of every year. Original articles and reviews in English on cytology, cytogenetics, cytochemistry, genomics and related subjects are accepted for publication. Manuscripts intended for publication in the journal may be sent directly to the editors by registered post (hard copy and diskette) and e-mail. The *authors are requested to include their fax and e-mail numbers.* Reference and citations should be given according to the standard form followed in the recent issues of the journal. Photographs and line drawings in original should be sent so as to fit the printing space on reduction.

Subscriptions and orders for the journal should be sent to the editors. Price of an yearly volume of the journal is Rs.600.00 (inland) or US $ 60.00 (for foreign subscribers). Supplements and airmail postage are charged extra. Cheque/ Draft should be drawn in favour of "THE NUCLEUS"

# CHAPTER - 130

# The Orrissa Journal of Horticulture

## Instructions to Contributors

Publication in THE ORISSA JOURNAL OF HORTICULTURE is generally open to the members of the Horticultural Society of Orissa except when the articles are called for by special invitation.

**One copy of the print out and the floppy disk of the article should be sent to the Editor and the floppy should contain the text of the article in MS word or pagemaker package only for publication.** In exceptional cases article may be typed doubled spaced on one side of the page only and should not generally exceed 10 typed pages. Clearness, brevity and conciseness are essential in form, style, punctuation, spelling and use of italics. Manuscripts should conform to the best usage in the leading journals published in the India and abroad. Numerical data and calculations should be carefully checked.

The names of the authors, accompanied by their degrees, titles, should be written on the next line after the title of the paper followed by the name and address of the institution where the work was done.

Generic and specific names should be underlined throughout the manuscript and the first reference to the latter should be followed by the name of the author. When a plant is comparatively little known, the name of the family should be given in brackets. Specific name should always be written with a small initial letter. First reference to common name should be followed by Latin name in parentheses.

Reference to literature, arranged alphabetically according to author's names, should be placed into a list of 'References' at the end of the article, the various references to each author being arranged alphabetically. Each reference should contain the names of the author (with initials), the year of the publication, title of the article, the abbreviated title of the publication according to the world List of Scientific Periodicals, volume and page. In the text, the reference should be indicated by the author's name, followed by the serial number only in brackets.

Each table should have a heading stating its content clearly and concisely. Places at which tables are inserted should be indicated in the text. Research articles must conclude with a Summary drawing attention to the main facts and conclusions.

Only illustrations of good quality on best quality glossy contrast paper, unmounted and unfolded, which are essential to a clear understanding of the paper will be accepted. Each illustration must be specifically referred to in the text and Roman numerals should be used in numbering the plates and also in the text. Individual illustrations within a plate should be designated by Arabic numerals. Line drawings will be preferred both because of the lower cost of production and their relative clearness.

Articles forwarded to the Editor for publication are understood to be offered to the Orissa Journal of Horticulture exclusively. It is also understood that the authors have obtained the approval of their Department, Faculty, or Institute in case where such approval is necessary.

The Editorial Board takes no responsibility for the fact or the opinion expressed in this journal, which rests entirely with the authors thereof.

# CHAPTER - 131

# The World Weeds

## Notice to Contributors

**The World Weeds** is an international quarterly journal in which are published papers, concerned with the study of weeds and parasitic flowering plants viz. Taxonomy, External and Development Morphology, Anatomy, Ecology, Physiology, Cytology, Damages to Crop and Forest Plants, Control Methods and other related topics.

Papers on work carried out in any part of the world are welcomed. Reviews of technical conferences and symposia and news items are also welcomed.

Manuscripts should be typed double-spaced on one side of the paper, preferably on A4 white paper, leaving wide margins on both sides.

Authors are asked to study the recent issue of **The World Weeds** and to observe the style of the journal in respect of layout, abbreviation and reference citation. Metric units must be used.

Page one of the manuscript should contain title of the paper (in capital letters), name(s) of the author(s), name(s) of the organization (s) where the work has been carried out and an **ABSTRACT** not exceeding 250 words with all research papers, regardless of length.

The text of the paper should be divided into INTRODUCTION, MATERIALS and METHODS, OBSERVAIONS / RESULTS, DISCUSSION, ACKNOWLEDGEMENTS and REFERENCES.

Tables should be typed on separate sheets, each bearing a short title.

Photographs, line drawings and graphs should be about the width of a single or of two columns of the text. Authors should aim at a height/width proportion of either 3:2 or 2:3.  The drawings or graphs should be drawn in Indian ink on white card, tracing paper or on a graph paper with feint blue ruling. All illustrations must be sent in duplicate (One original and one Photostat copy). Legends for Figures and Plants should be typed in consecutive paragraphs on separate sheet(s).

Manuscripts must be submitted in duplicate (One original and one carbon or Photostat copy).  To : Chief Editor, Dr. Y.P.S. Pundir, Head, Botany Department, D.B.S. (P.G.) College, Dehradun – 248001, India . Other correspondence should be addressed to the publisher : Internal Book Distributors, Post Box No. 4, 9/3 Rajpur Road, Dehradun – 248001 (India).

# CHAPTER - 132

# Trends in Plant Science

## Guide for Authors

*Trends in Plant Science* is now review only, and is divided into three main sections: Update, Opinion and Review. Articles for *Trends in Plant Science* are usually commissioned by the Editor, but proposals for articles are welcome, especially for the Update section. Prospective authors should send a brief summary, citing key references, to the Editor, who will supply guidelines on manuscript preparation if the proposal is accepted. The submission of completed manuscripts without prior consultation with the Editor is strongly discouraged. Authors should note that all major articles in *Trends in Plant Science* are peer-reviewed and publication cannot be guaranteed.

Reviews and Opinion articles form the foundation of each monthly issue. Reviews are invited from leading researchers in a specific field and objectively chronicle recent and important developments. Opinion articles present a personal, authoritative viewpoint of a field or research-related subject. They can cover timely controversial topics or debates, provide a new framework for, or interpretation of, an old problem or current issue, or speculate in depth on the implications of some recently published research or data.

The Update section includes Research Focus articles and Letters to the Editor. Research Focus articles highlight recent research papers of particular note and report on groundbreaking technical advances. The decision to publish Letters rests with the Editor, and the author(s) of any article discussed in a Letter will normally be invited to reply. Letters can address topics raised in recent issues of the journal, or other matters of general interest to plant scientists. Letters should be no more than 700 words long with a maximum of 12 references and one figure. An electronic version of the text should be e-mailed to

plants@current-trends.com

(or post a hardcopy plus an electronic copy on disc to

The Editor

*Trends in Plant Science*

Elsevier
84 Theobald's Road
London
UK WC1X 8RR

**Abstracting/Indexing**

Bibliography of Agriculture (AGRICOLA)

CAB Abstracts International

Current Contents (Life Science)

Food Science & Technology Abstracts

Index Medicus

Life Sciences Collection

Medline/MEDLANS Online

Reference Update

SciSearch/Science Citation Index Expanded

## Statement on publishing ethics

The Editor(s) and Publisher of this Journal believe that there are fundamental principles underlying scholarly or professional publishing. While this may not amount to a formal 'code of conduct', these fundamental principles with respect to the authors' paper are that the paper should: (i) be the authors' own original work, which has not been previously published elsewhere, (ii) reflect the authors' own research and analysis and do so in a truthful and complete manner, (iii) properly credit the meaningful contributions of co-authors and co-researchers, (iv) not be submitted to more than one journal for consideration, and (v) be appropriately placed in the context of prior and existing research.

Of equal importance are ethical guidelines dealing with research methods and research funding, including issues dealing with informed consent, research subject privacy rights, conflicts of interest, and sources of funding. While it may not be possible to draft a 'code' that applies adequately to all instances and circumstances, we believe it useful to outline our expectations of authors and procedures that the Journal will employ in the event of questions concerning author conduct.

## Conflict of Interest Policy

The Publisher now requires authors to declare any conflicts of interest that relate to papers accepted for publication in this Journal. A conflict of interest may exist when an author or the author's institution has a financial or other relationship with other people or organizations that may inappropriately influence the author's work. A conflict can be actual or potential and full disclosure to the Journal is the safest course. All submissions to the Journal must include disclosure of all relationships that could be viewed as presenting a potential conflict of interest. The Journal may use such information as a basis for editorial decisions and may publish such disclosures if they are believed to be important to readers in judging the manuscript. A decision may be made by the Journal not to publish on the basis of the declared conflict. For more information, please refer to: http://www.elsevier.com/wps/find/authorshome.authors/conflictsofinterest

## Copyright statement



## Photocopying

Tel: +44 (0) 20 7631 5555

Fax: +44 (0) 20 7436 5500

Other countries may have a local reprographic rights agency for payments.

## Derivative Works

Subscribers may reproduce tables of contents or prepare lists of articles including abstracts for internal circulation within their institutions. Permission of the publisher is required for resale or distribution outside the institution.

Permission of the publisher is required for all other derivative works, including compilations and translations.

## Electronic Storage or Usage

Permission of the publisher is required to store or use electronically any material contained in this journal, including any article or part of an article. Contact the publisher at the address indicated. Except as outlined above, no part of this publication may be reproduced, stored in a retrieval system or transmitted in any form or by any means, electronic, mechanical, photocopying, recording or otherwise, without prior written permission of the publisher. Address permissions requests to: Elsevier Science Rights & Permissions Department, at the mail, fax and e-mail addresses noted above.

## Notice

No responsibility is assumed by the Publisher for any injury and/or damage to persons or property as a matter of products liability, negligence or otherwise, or from any use or operation of any methods, products, instructions or ideas contained in the material herein. Because of rapid advances in the medical sciences, in particular, independent verification of diagnoses and drug dosages should be made.

Although all advertising material is expected to conform to ethical (medical) standards, inclusion in this publication does not constitute a guarantee or endorsement of the quality or value of such product or of the claims made of it by its manufacturer.

# Section – III

# Appendices

# Directory of Agricultural Journals

1.  **African Journal of Food Agriculture, Nutrition and Development**
    The Editor-in-Chief
    *African Journal of Food Agriculture, Nutrition and Development*
    (AJFAND)
    P.O. Box No. 29086
    Nairobi, Kenya
    E-mail : oniango@iconnect.co.ke

    Three issues for year

2.  **Agribusiness**
    Food Marketing Policy Center,
    Agricultural and Resource Economics,
    Units 4021, 1376  Storrs Road,
    Storrs, CT  06269-4021
    E-mail: fmpc@uconn.edu;
    www.econlit.org/subjectdescriptors.html

    Quarterly

3.  **Agricultural and Forest Meteorology**
    Editor-in-Chief
    University of California
    Atmospheric Sciences
    One Shields Avenue
    Davis, CA 95616-8627
    http://ees.elsevier.com/agrformet/

    Half Yearly

4.  **Agricultural_Economics Research Review**
    The *Editor, Agricultural Economics Research Review,*
    Division of Agricultural Economics,
    IARI, New Delhi – 110 012.
    E-mail : aera@rediffmail.com

5.  **Agricultural Situation in India**

>   The Economic and Statistical Adviser,
>   Directorate of Economics and Statistics,
>   Room No.152, Krishi Bhawan,
>   New Delhi-110001.
>
>   Monthly

6.  **Agricultural Systems**

>   International Research Institute for climate Prediction
>   Palisades, NY, USA
>   http://www.elsevier.com/locate/agsy
>
>   Monthly

7.  **Agricultural Water Management**

>   Water Management Research Lab.
>   USDA-ARS, P.O. Drawer 10
>   Bushland, TX 79012, USA.
>   http://ees.elsevier.com/agwat
>
>   Half Yearly

8.  **Agriculture, Ecosystems & Environment**

>   Agriculture Canada
>   Central Experimental Farm
>   Ottowa, Ontario K1A 0P6, Canada
>   E-mail: agee@elsevier.com
>   http://www.elsevier.com/locate/agee
>
>   Half Yearly

9.  **Agronomy Journal**

>   The Associate Production Editor,
>   Agronomy Journal,
>   American Society of Agronomy,
>   677 South Segoe Road, Madison, WI, USA 53711
>   E-mail: mnilsson @ agronomy.org
>   www.asa-cssa-ssa.org/
>
>   Bi-Monthly

10. **Agronomy Research**
    Agronomy Research
    Estonian Grassland Society
    St. Teaduse 4, 75501 Saku, Estonia
    E-mail: rein.viiralt@emu.ee
    http://www.eau.ee/~agronomy/

    Half yearly

11. **Agropedology**
    The Honorary Secretary
    Indian Society of Soil Survey and Land Use Planning,
    NBSS&LUP, Amravati Road,
    Nagpur – 440 010, India

12. **Allelopathy Journal International**
    Prof. S.S. Narwal,
    8/515, Haryana Agricultural University,
    Hisar-125 004, India.
    allelopathy1974@yahoo.com

    Half Yearly

13. **American Journal of Agricultural Economics**
    Black well Publishing Ltd.
    Customer Service Department
    P.O.Box 805, 108 Cowley Road
    Oxford Ox4,IFH,UK
    http://ajae.aem.cornell.edu

    Five issues for Year (February, May, August, November, and
    December)

14. **Annals of Plant Physiology**
    The Editor-in-Chief
    Annals of Plant Physiology
    Saint Tukaram Hospital Chowk,
    01, Kamala Nagar,  Akola-444 004 (M.A) India.

    Half Yearly

15. **Applied Soil Ecology**
    University College Dublin
    School of Biological and Environmental Science

Agriculture and Food Science Centre

Belfield, Dublin 4, Ireland

Nine issues for year

**16.  Arid Land Research and Management**

Research Institute for Soil Science and Agricultural Chemistry,
Hungarian Academy of Sciences, (MTA TAKI-RISSAC),
P.O. Box 35, Budapest, 1525, Hungary.
E-mail: tibor@rissac.hu.

Quarterly

**17.  Artha Vijnana**

Editor

Gokhale Institute of Politics and Economics

Pune-411004

Quarterly

**18.  Asian Journal of Microbiology, Biotechnology and Environmental Sciences**

The Executive Editor, AJMBES

C/O GLOBAL SCIENCE PUBLICATIONS

23, Maharshi Dayanand Nagar, Surendra Nagar,

ALIGARH 202 001 (U.P). Ph : (0571) 404271.

**19.  Australian Journal of Plant Physiology**

www.publish.csiro.au/journals/ajpp

CSIRO PUBLISHING

PO Box 1139 (150 Oxford St)

Collingwood VIC 3066, Australia

E-mail: Jennifer.mccutchan@publish.csiro.au

Monthly

**20.  Bhagirath**

The Editor, Bhagirath

Central Water Commission,

Ministry of Water Resources,

425 (N), Sewa Bhavan, R.K.Puram

New Delhi –110066

Quarterly

**21. Biology and Fertility of Soils**

Editorial Office of the Breeding Science,

C/o Dr. Masumi Katsuta

National Institute of Crop Science

2-1-18 Kannondai, Tsukuba, Ibaraki 305-8518, Japan

Bi Monthly

**22. Breeding Science**

Editorial Office of the Breeding Science,

C/o Dr. Masumi Katsuta

National Institute of Crop Science

2-1-18 Kannondai, Tsukuba, Ibaraki 305-8518, Japan

Quarterly

**23. Calcutta Statistical Association Bulletin**

The Editor, Calcutta Statistical Association Bulletin,

C/o Department of Statistics, University of Calcutta,

35, Ballygunge Circular Road, Kolkata – 700 019

Quarterly

**24. Clay Research**

The Officer Incharge, Publication Office,

The Clay Minerals Society of India,

Division of Soil Resource Studies,

National Bureau of Soil Survey and Land Use Planning,

Amaravathi Road, Nagpur –440 010, India

Half Yearly (June and December)

**25. Communicator**

The Editor, Communicator,

Indian Institute of Mass Communication,

Aruna Asaf Ali Marg, New Delhi 110 067, India

E-mail: abhilashakumari@hotmail.com

Half Yearly (January and June)

**26. Crop Protection**

Elsevier Japan, 4F Higashi-Azabu,

1-chome Bldg, 1-9-15, Higashi-Azabu,

Minato-ku, Tokyo 106-0044; Japan;

E-mail: jp.info@elsevier.com.
http://www.elsevier.com/locate/cropro

Monthly

27. **Crop Science**

The Managing Editor,
Crop Science,
677 S. Segoe Road,
Madison, WI 53711–1086
www.asa-cssa-sssa.org
www.crops.org

Bi Monthly

28. **Economic and Political Weekly**

Economic and Political Weekly,
Hitkari House,
284 Shahid Bhagatsingh Road,
Mumbai 400 001, India.
Email: edit@epw.org.in,
epw.Mumbai@gmail.com

Weekly

29. **European Journal of Agronomy**

Editor-in-Chief
Instituto de Agricultura Sostenible (CSIC)
Apartado 4084,14080 Cordoba, Spain
http://ees.elsevier.com/euragr
Eight issues for year

30. **European Journal of Soil Science**

Editor-in-Chief
Rothamsted Research
Harpenden, Hertfordshire
AL5 2JQ,UK
E-mail: ejss@bbsrc.ac.uk
www.blackwellpublishng.com/ejs

Bi Monthly

31. **Everyman's Science**

The Hony, Editor, Everyman's Science,
The Indian Science Congress Association

14, Dr. Biresh Guha Street, kolkata-700 017
Email : iscacal@vsnl.net.

## 32.  Experimental Agriculture

Editor, Experimental Agriculture
Crop and Water Management Systems  (International) Ltd.,
Pear Tree Cottage, Frog Lane, Ilmington, Shipston-on-STOUR,
Warwickshire CV36 4LG, UK.
Email: mikecarr.rtcs@freeUK.com

Quarterly

## 33.  Farming Systems

The Director,
Institute for Studies on Agriculture and Rural Development,
Belgaum Road, Dharwad – 580 008, India

Quarterly

## 34.  Field Crops Research

Field Crops Research, Publications Expediting, Inc.,
200 Meacham Ave, Elmont, NY 11003, USA

Eighteen Issues for year

## 35.  Financing Agriculture

The Editor,
Agricultural Finance Corporation Ltd.,
Dhanraj Mahal, 1$^{st}$ Floor,
Chhatrapati Shivaji Maharaj Marg,
Mumbai – 400 001, India
E-mail  : afc@vsnl.com
          piplapure@afcmails.com

Quarterly

## 36.  Forage Research

The Editor, Forage Research,
CCS Haryana Agricultural University,
Hisar – 125 004, India

Quarterly

## 37.  Genetics

GENETICS Editorial Office
Mellon Institute, Box I
4400 fifth Avenue

Pittsburgh, AP 51213-2683, USA

E-mail: genetics-gsa@andrew.cmu.edu

Monthly

**38.  Genome**

NRC Research Press,

National Research Council of Canada, Ottawa,

ONKIA 0R6, .Canada (fax: 613-925-7656;

E-mail: pubs@nrc-cnrc.gc.ca; http://pubs.nrc-crnc.gc.ca

Monthly

**39.  Indian Agriculturist**

The Editor,

The Agricultural Society of India 35,

Ballygunge Circular Road, Kolkata – 700 019 (India)

Quarterly

**40.  Indian Coconut Journal**

Coconut Development Board

Ministry of Agriculture,

Department of Agriculture and Co-operation,

Government of India

Kera Bhavan, Kochi – 683 001, India

E-mail          :  enk_cdrkochi@sancharnet.in

www.coconutboard.nic.in

Monthly

**41.  Indian Farmers' Digest**

The Editor

Indian Farmers' Digest,

Department of Agricultural Communication,

College of Agriculture,

G.B. Pant University of Agriculture & Technology,

Pantnagar-263145 (Uttaranchal)

Monthly

**42.  Indian Farming**

Editor

Indian Council of Agricultual Research

Krishi Anusandhan Bhavan, Pusa

New Delhi-110012

Monthly

43.  **Indian Food Packer**
     Executive Editor
     All India Food Processor's Association
     206 Aurobinda Place, Houzkhas
     New Delhi – 110016

     Bi Monthly

44.  **Indian Journal of Agricultural Economics**
     Editor
     The Indian Society of Agricultural Economics
     46-48, Esplanade Mansion's, M.G.Road
     Mumbai – 400 001

     Quarterly

45.  **Indian Journal of Agricultural Marketing**
     Editor
     Indian Society of Agricultural Marketing
     57,Type IV, CPWD quarters, Seminary Hills
     Nagpur – 440006

     3 Issues for year

46.  **Indian Journal of Agronomy**
     Editor
     Indian Society of Agronomy
     Division of Agronomy
     Indian Agricultural Research Institute
     New Delhi – 110012

     Quarterly

47.  **Indian Journal of Arecanut, Spices and Medicinal Plants**
     The Director
     Directorate of Arecanut and Spices Development
     Ministry of Agricultural, Government of India
     Department of Agriculture & Co-operation
     West Hill P.O., Calicut – 673 005, Kerala
     E-mail: spicedte@nic.in

     Bi Monthly

**48.  Indian Journal of Biochemistry and Biophysics**

The Editor, Indian Journal of Biochemistry and Biophysics,
National Institute of Science Communication and Information
Resources (NISCAIR),  Dr K S Krishnan Marg, New Delhi 110012

Bi Monthly

**49.  Indian Journal of Biotechnology**

Editor
National Institute of Science Communication and Information
Resources
CSIR, Dr. K.S. Krishnan Marg, New Delhi – 1100012

Quarterly

**50.  Indian Journal of Horticulture**

The Editor, the Horticultural Society of India,
C/O Division of Fruits and Horticultural Technology,
I.A.R.I. Pusa, New Delhi-110 012, India.

Quarterly

**51.  Indian Journal of Marketing**

The Editor
Indian Journal of Marketing
Y-21, Hauzkhas, New Delhi-110 016

Monthly

**52.  Indian Journal of Nematology**

The Chief Editor, Indian Journal of Nematology,
Division of Nematology. Indian Agricultural Research Institute,
New Delhi – 110012, India

Half Yearly

**53.  Indian Journal of Plant Physiology**

The Editor-in-Chief, Indian Journal of plant Physiology,
Division of Plant Physiology, Indian Agricultural Research Institute,
New Delhi 110 012, India.

Quarterly

**54.  Indian Journal of Plant Protection**

The Chief Editor
Indian Journal of Plant Protection
National Bureau of Plant Genetic Resources – Regional Station

Rajendranagar, Hyderabad – 500 030

Andhra Pradesh, India.

E-mail: ijpp@epatra.com

Half Yearly

**55.   Indian Journal of Social Research**

Editor

M/s. Academic and Law Serials

F22 B/3, Laxminagar, Delhi – 110092

Quarterly

**56.   Indian Journal of Weed Science**

The Chief Editor

Indian Journal of Weed Science

Department of Agronomy

G.B. Pant University of Agriculture & Technology

Pant Nagar – 263 145, India

Quarterly

**57.   Indian Perfumer**

The Editor, Indian Perfumer

Essential Oil Association of India

301, 4832/24 Ansari Road, Daryaganj

New Delhi 110 002, India

Quarterly

**58.   Irrigation Science**

Management Director

Springer, Heidelberger Platz 3

14197 Berlin, Germany

E-mail: anzeigen@springer.com

www.springer.de/wikom

Quarterly

**59.   Journal of Agribusiness**

The editor:

Journal of Agribusiness

Department of Agricultural & Applied Economics

313-E Conner Hall Judy Harrison

The University of Georgia, Athens, GA 30602-7509
E-mail: jab@agecon.uga.edu

Half Yearly

**60.    Journal of Agricultural Economics**
Journal of Agricultural Economics Office,
Centre for Rural Economy, School of Agriculture,
Food and Rural Development,
University of Newcastle upon Tyne, NE1 7RU, UK
E-mail: jae.editor@ncl.ac.uk; http://www.aes.ac.uk

**61.    Journal of Agricultural Engineering**
Editor
Indian Society of Agricultural Engineering
New Delhi

**62.    Journal of Applied Horticulture**
Chief Editor
Horticulture and Food Processing
Deputy Director General (Horticulture)
UPCAR,UP,Lucknow-226016, India

Half Yearly

**63.    Journal of Biotechnology**
The Chief Editor:
Universitat Bielefeld, Universitatstraâe 25,
D-33615 Bielefeld, Germany
E-mail: jbiotech@genetik.uni-bielefeld.de

Half Yearly

**64.    Journal of Central European Agriculture**
The Main Editor
Prof. Dr. Nikola Keziæ; Faculty of Agriculture,
Svetošimunska 25, 10000 Zagreb, Croatia
E-mail: nkezic@agr.hr
http://www.agr.hr/jcea

Quarterly

**65.    Journal of Cereal Science**
The Editorial Office
Journal of Cereal Science
Elsevier Editorial Services Office

The Boulevard, Langford Lane
Kidlington, Oxford OX5 1GB, UK
E-mail: jcs@elsevier.com
www.elseveir.com/locate/jcs
Bi Monthly

**66. Journal of Ecobiology**

Palani Paramount Publications
57, Anna Nagar, Palani 624 602 India.

**67. Journal of Eco-friendly Agriculture**

Dr. R.P. Srivastava, Chairman,
*Doctor's Krishi Evam Bagwani Vikas Sanstha,*
A-601, Sector-4, Indira Nagar,
Lucknow-226 016 (U.P.) India,
E-mail: vrsystem@sify.com.
www.ecoagrijournal.com

Half Yearly

**68. The Journal of Economic Entomology**

Entomological Society of America
930/Annapolis Road, Scute 300
Lamham, MDZ0706, USA
E-mail: *pubs@entsoc.org*
www.entsoc.org

Bi Monthly ( Feb.,April,June, August,Oct.,and December)

**69. Journal of Education and Psychology**

The Editor,
Journal of Education and Psychology,
Sardar Patel University,
Vallabh Vidyanagar – 388 120, Gujarat, India

Quarterly

**70. Journal of Entomological Research**

The Editor, Journal of Entomological Research,
C/o Mlahotra publishing House, B-6 DSIDC Complex,
Kirti Nagar, New Delhi-110 015, India

Quarterly

**71.  Journal of Food Science and Technology**
The Editor-in-chief,
Journal of Food Science and Technology,
AFST(I), CFTRI Campus, Mysore-570 020, India

Bi Monthly

**72.  Journal of General Plant pathology**
The Editor-in-Chief
Dr. Ichiro Uyeda
Graduate School of Agriculture,
Hokkaido University, Sapporo 060-8589, Japan
e-mail: uyeda@res.agr.hokudai.ac.jp

Quarterly

**73.  Journal of Information Technologies in Agriculture**
Department of Agricultural Economics
Prof. Dr. Gerhard Schiefer
Meckenheimer Allee 174
D-53115 Bonn, Germany
E-mail: info@infita.org
www.jitag.org

**74.  Journal of Maharashtra Agricultural Universities**
The Editor-in-Chief,
Journal of Maharashtra Agricultural Universities,
College of Agriculture, Pune – 411 005, INDIA

**75.  Journal of Nuclear Agriculture and Biology**
The Secretary
Indian Society for Nuclear Techniques in Agriculture & Biology,
Nuclear Research Laboratory, I.A.R.I.,
New Delhi – 110 012, India

Quarterly

**76.  Journal of Oilseeds Research**
The Editor,
Journal of Oilseeds Research
Directorate of oilseeds Research
Rajendranagar
Hyderabad-500 030, A.P., India

Half Yearly

**77.  Journal of Ornamental Horticulture**
The Editor
Indian Society of Ornamental Horticulture
Division of Floriculture and Landscaping
Indian Agriculture Research Institute
New Delhi – 110 012

Quarterly

**78.  Journal of Plant Growth Regulation**
JPGR Editorial Office
1002 Stonebriar Drive
Verona, WI 53593, USA
Email: jpgr@tds.net

Quarterly

**79.  Journal of Plant Nutrition**
Dr. Harry A. Mills,
183 Paradise Boulevard,
Suite 104, Athens,
Georgia 30607 USA.

Monthly

**80.  Journal of Plantation Crops**
The Editor, Journal of Plantation Crops,
Central Plantation Crops Research Institute,
Kasaragod – 671 124, Kerala, India.

3 issues for year

**81.  Journal of Potassium Research**
The Editor, *Journal of Potassium Research,*
Potash Research Institute of India, Sector 19,
Dundahera, Gurgaon-122 001 (Haryana)

**82.  Journal of Spices and Aromatic Crops**
Chief Editor
Journal of Spices & Aromatic Crops (JOSAC)
C/o National Research Center for Spices
P.O. Box. 1701, Marikunnu P.O.
Calicut – 673 012, Kerala, India

Half Yearly

83.    **The Journal of the Indian Society for Cotton Improvement**
Secretary
Editorial Board
Journal of the Indian Society for Cotton Improvement
C/o. Central Institute for Research on Cotton Technology
Adenwala Road,
Matunga Mumbai – 400 019

3 issues for year (April, August and December)

84.    **Journal of the Indian Society of Agricultural Statistics**
The Secretary,
Indian Society of Agricultural Statistics,
IASRI campus, Library Avenue,
New Delhi-110 012 (INDIA)
E-Mail: isas@iasri.res.in

3 issues for year

85.    **Journal of the Indian Society of Coastal Agricultural Research**
Editor
Indian Society of Coastal Agricultural Research
Central Soil Salinity Research Institute
Regional Research Station
Canningtown 743329, West Bangal
E-mail : iscar@rediffmail.com,   cssri@wb.nic.in
arijit_sen@vsnl.net

Half Yearly

86.    **Journal of the Indian Society of Soil Science**
The Secretary
Indian Society of Soil Science
Division of Soil Science and Agricultural Chemistry
Indian Agricultural Research Institute
New Delhi – 110 012 (India)
Email: isss@vsnl.com
www.isss-india.org

Quarterly

87.    **Journal of the South Carolina Academy of Science**
Editor in Chief
Division of Natural Sciences

USC Spartanburg
800 University Way
Spartanburg, SC 29303
http://www.uscupstate.edu/~dkferris/scas.html

Three issues for year

**88.   Journal of Tropical Agriculture**
The Managing Editor,
Journal of Tropical Agriculture.
College of Forestry, KAU-PO,
Thrissur 680 656, Kerala, India
e-mail: bmkumar53@yahoo.co.uk
ww.kau.edu

Half Yearly

**89.   Karnataka Journal of Agricultural Sciences**
The Editor,
Karnataka Journal of Agricultural Science,
University of Agricultural Sciences,
Dharwad – 580 005, Karanataka, India

**90.   Kisan World**
Managing Editor
101, Mount Road
Guindy, Chennai – 600 032
E-Mail: chennai @ sakthi sugars.com

Monthly

**91.   Legume Research**
The Managing Editor,
Agricultural Research Communication Centre,
1130, Sadar Bazar, Near Post office,
KARNAL-132 001, Haryana (INDIA)

Quarterly

**92.   Molecular Breeding New Strategies in Plant Improvement**
Editor-in-Chief
Universitat de Lleida
Lleida, Spain
http://www.springeronline.com/jounral/11032

Quarterly

**93.  Paddy and Water Environment**

The Editor-in-Chief
Department of Bio-Business Management and Information
Faculty of International Agricultural and Food Studies
Tokyo University of Agriculture
1-1-1  Sankuragaoka, Setagayaku
Tokyo 156-8502, Japan
E-mail: ynakano@ggr.kyushu-u.ac.jp
          sjhwang@konkuk.ac.kr

**94.  Pest Management and Economic Zoology**

The Chief Editor,
Pest Management and Economic Zoology,
Department of Entomology and Apiculture,
Dr Y S Parmar University of Horticulture and Forestry,
Solan, H P 173 230, India

Haf Yearly

**95.  Physiological and Molecular Plant Pathology**

Physiological and Molecular Plant Pathology
Elsevier Editorial Services Office
Bampfylde Street, Exeter,
Devon, EX1 2AH, UK
E-mail: pmpp@elsevie.com
http://www.nejm.org

Monthly

**96.  Physiological Entomology**

Blackwell Publishing Ltd.
P.O.No.1354, 9600 Garsington Road
Oxford, U.K.

Quarterly

**97.  Plant Biotechnology Journal**

Blackwell Publishing Ltd.
Customer Services Department
P.O.Box – 805, 108 Cowley Road
Oxford, OX4, IFG UK
http://plantbiotechjournal.manuscrptcentral.com/

Bi Monthly

**98.  Pest Management in Horticultural Ecosystems**
>The Editor
>AAPMHE Division of Entomology and Nematology
>Indian Institute of Horticultural Research
>Hessaraghatta Lake (Post)
>Bangalore- 560 089
>
>Half Yearly (June & Dec.)

**99.  Productivity**
>The Editor, PRODUCTIVITY,
>National Productivity Council,
>Utpadakta Bhawan, Lodi Road,
>New Delhi-110 003
>
>Quarterly

**100.  Rural Sociology**
>Rural Sociological Society
>104 Gentry Hall,
>University of Missouri
>Collumbia MO 65211 -7040 USA
>rslsu@isu.edu
>
>Quarterly

**101.  Sankhya**
>*Sankhya,*
>Indian Statistical Institute,
>203 B.T. Road, Calcutta – 700035
>E-mail : sankhya@isical.ac.in
>Series A  Three issues for Year(Feb.,June & Oct.)
>Series B  Two issues for year (April, August)

**102.  Sarvekshana**
>The Chief Executive Officer,
>National Sample Survey Organization
>Sardar Patel Bhavan, Parliament Street,
>New Delhi – 110001
>
>Half Yearly

**103.  Scientia Horticulturae**
>Department of Horticulture and Landscape Architecture
>College of Agriculture
>N 318 Ag Science Bldg North

Lexinton, K Y 40546 – 0019, USA
http://www.elsevier.com/locte/scihorti

Sixteen  issues for year

**104.    Seed Science and Technology**
The Chief Editor, Dr.Anne Bulow-Olsen,
Furesf Parkvej 23,DK-2830 Virum, Denmark
E-mail : sst@ista.ch
www.seedtest.org

Three issues for year

**105.    Seed Research**
The Chief Editor, Seed Research,
Division of Seed Science and Technology,
Indian Agricultural Research Institute,
New Delhi 110012, India
E-mail: cd_seedresearch@yahoo.com

Half Yearly

**106.    Soil and Tillage Research**
The Editorial Office of Soil & Tillage Research
Elsevier Ireland Ltd., Brookvale Plaza
East Park, Shannon Co. Clare, Ireland

Ten issues for year

**107.    Soil Biology and Biochemistry**
University of Exeter
Biological Sciences
Hatherly Laboratories
Prince of Wales Road
Exeter Ex4 4PS, UK.
http://authors.elsevier.com/journal/soilbio

Monthly

**108.    Soil Science Society of America Journal**
Soil Science Society of America Journal
677 South Segoe Road Madison, WI, USA 53711
E-mail: rfunck@agronomy.org

Bi Monthly

**109.  The Andhra Agricultural Journal**
Secretary
Andhara Agricultural Journal
Agricultural College
Bapatla – 522 101, Guntur Dist.

Quarterly

**110. The Asian Journal of Horticulture**
The Editor, The Asian Journal of Horticulture
Hind Agri-Horticultural Society, Ashram
418/4-South civil lines ( Numaish camp)
MUZAFFARANGAR-251001(U.P) .INDIA

**111.  The Cashew**
The Director
Directorate of Cashewnut & Cocoa Development
Government of India
Ministry of Agriculture (Department of Agriculture & Co-operation)
Cochin – 682 011, Kerala.
Website: dacnet.nic.in/cashewcocoa
E-mail: dccd@hub.nic.in

**112.  The Indian Forester**
The Honorary Editor,
Indian Forester, P.O. New Forest,
Dehra Dun – 248 006 (Uttaranchal)
E-mail : indfor@icfre.org

Monthly

**113.  The Indian Journal of Agricultural Sciences**
Editor
Director of Information and Publication of Agriculture
Indian Council of Agricultural Research
Krishi Anusandhan Bhavan, Pusa, New Delhi – 110012

Monthly

**114.  The Indian journal of Genetics and Plant Breeding**
The Editor,
The Indian Society of Genetics and Plant breeding,
P.B.11312, IARI, New Delhi -110 012

Quarterly

**115.   The Indian Journal of Microbiology**

The Editor in Chief, Indian Journal of Microbiology,
Division of Microbiology, Indian Agricultural Research Institute,
New Delhi 110012, India

Quarterly

**116.   The Journal of AOAC INTERNATIONAL**

481, Northrederic K Avenue
Suite 500, Gaithersburg
Maryland 20877-2417
http://www.aoac.rog/pubs/joural.htm

Bi Monthly

**117.   The Journal of Plant Biochemistry and Biotechnology**

Chief Editor
Society for Plant Biochemistry and Biotechnology
Division of Biochemistry
I.A.R.I. New Delhi – 110 012

**118.   The Journal of Research ANGRAU**

The Managing Editor
Agricultural Information and Communication Centre,
Rajendranagar, Hyderabad – 500 030, India

Quarterly

**119.   The Journal of Vegetation Science**

Chief Editor
Opulus Press
Gamla Vagen 40
77013 Grangarde
Sweden
E-mail edit @opuluspress.se

Bi Monthly

**120.   The Karnataka Journal of Horticulture**

The Chief Editor,
The Karnataka Journal of Horticulture,
K.R.C.College of Horticulture; Arabhavi-591 310,
Taluk: Gokak, Dist: Belgaum, Karnataka

**121.  The Maharashtra Journal of Extension Education**
The Secretary,
Maharashtra Society of Extension Education
Department of Extension Education,
Konkan Krishi Vidyapeeth, Dapoli.  Dist.
Ratnagiri –415712, Maharashtra, India.

Annual

**122.  The Nucleus**
Editor
Md.Publications Pvt.Ltd.
Md.House, 11 Daryaganj
New Delhi - 110002
Email: lavania@cimap.res.in
Three issues for year (April, August and December)

**123.  The Orissa Journal of Horticulture**
Editor
The Orissa Horticultural Society
Department of Horticulture
College of Agriculture, OUAT
Bhubaneswar – 751 003

Half Yearly

**124.  The World Weeds**
The Chief Editor,
Dr. Y.P.S. Pundir, Head,
Botany Department, D.B.S. (P.G.) College,
Dehradun – 248001, India

Quarterly

**125.  Trends in Plant Science**
The Editor
*Trends in Plant Science*
Elsevier
84 Theobald's Road
London, UK WC1X 8RR
E-mail: plants@current-trends.com

Monthly

# Directory of Open Access Publishers

Open Access (OA) is the free online availability of digital contents. It is best-known and most feasible for peer-reviewed scientific and scholarly journal articles, which scholars publish without expectations of payment. The author (usually the author's research funds or institution/ pays the publication costs in Open Access publishing which has been proposed as an alternative to a subscription-based cost recovery model. The growth of this alternative cost recovery model, however, has been slow sofar. The list of major open access publishers is given below with their websites :

| Sl. No. | Open Access publisher | URL |
|---|---|---|
| 1. | Directory of Open Access Journals | www.doaj.orj |
| 2. | BioMed Central Journals | http://www.biomedicentral.com/info/librariesoajpurnals |
| 3. | Oxford Open Journals | http://www.oxfordjournals.org |
| 4. | Ingenta Open Access Journals | www.ingentacontract.com |
| 5. | University of Toronto Research Repository | http://tspace.library.utorontoca/ |
| 6. | Access to Global Online Research in Agriculture | http://www.aginternetwork.org |
| 7. | Indian Journals | www.indianjournals.com |
| 8. | Ivyspring International Publishers | www.ivysring.com/ |
| 9. | Molecular Diversity Preservation International | www.mdpi.org/ |
| 10. | Public Library of Science | /wiki/public_library_of_science |
| 11. | Hindawi Publishing Corporation | www.hindawi.com/ |
| 12. | Scholarly Exchange | www.scholarlyexchange.org/ |
| 13. | Springer | www.springerlink.com |
| 14. | Medknow Publications | www.medknow.com/ |
| 15. | Copernicus open-access journals | www.open-access-publishing.net/ |

# Rating of Scientific Journals

According to National Academy of Agricultural Sciences (NAAS), scientists are identified to ensure excellence in different subject matter areas is agricultural sciences, as the main purpose of Academy is to recognize and support excellence in scientific research. For the purpose of giving weightage to the  academy has done an exercise to identify journals of standing relevant to agricultural sciences and the final outcome is given hereunder. Each journal is assessed for a maximum of 4.0 marks and its standing is indicated in appropriate numeral against each Journal (NAAS - 2004).

| Sl. No. | International Journal Code | NAME OF THE JOURNAL | MARKS |
|---|---|---|---|
| 1. | (A001) | ANGRAU Journal of Research | 0.5 |
| 2. | (A028) | Advances in Agronomy | 4.0 |
| 3. | (A036) | Advances in Genetics | 4.0 |
| 4. | (A037) | Advances in Insect Physiology | 4.0 |
| 5. | (A039) | Advances in Plant Breeding | 1.5 |
| 6. | (A040) | Advances in Plant Pathology | 1.5 |
| 7. | (A041) | Advances in Plant Physiology | 1.5 |
| 8. | (A042) | Advances in Plant Science | 1.5 |
| 9. | (A045) | Advances in Rice Blast Research | 1.5 |
| 10. | (A046) | Advances in Soil Science | 1.5 |
| 11. | (A049) | Advances in Virus Research | 4.0 |
| 12. | (A050) | Advances in Water Resources | 3.0 |
| 13. | (A051) | Agricultural Economics | 2.5 |
| 14. | (A052) | Agricultural Economics Research Review | 3.0 |
| 15. | (A053) | Agricultural Engineering | 1.5 |

*Contd...*

| Sl. No. | International Journal Code | NAME OF THE JOURNAL | MARKS |
|---|---|---|---|
| 16. | (A054) | Agricultural Engineering Journal | 1.5 |
| 17. | (A055) | Agricultural Extension Review | 1.5 |
| 18. | (A056) | Agricultural History | 2.5 |
| 19. | (A057) | Agricultural Mechanization in Asia | 1.0 |
| 20. | (A058) | Agricultural Research Journal of Kerala | 0.5 |
| 21. | (A059) | Agricultural Situation in India | 2.0 |
| 22. | (A060) | Agricultural Systems | 3.0 |
| 23. | (A061) | Agricultural Wastes | 3.0 |
| 24. | (A062) | Agricultural Water Management | 3.0 |
| 25. | (A063) | Agricultural and Biological Chemistry, Tokyo | 3.0 |
| 26. | (A064) | Agriculture Ecosystem and Management | 1.5 |
| 27. | (A065) | Agriculture and Forest Entomology, UK | 1.5 |
| 28. | (A066) | Agriculture and Forest Meteorology | 3.5 |
| 29. | (A067) | Agriculture, Ecosystem and Environment | 3.5 |
| 30. | (A068) | Agrobiology | 1.5 |
| 31. | (A069) | Agrochimica | 2.5 |
| 32. | (A070) | Agroforestry Systems | 3.5 |
| 33. | (A071) | Agron. J. Fert. Research | 1.5 |
| 34. | (A072) | Agronomie | 2.5 |
| 35. | (A073) | Agronomy Journal (Journal of American Society of Agronomy) | 3.0 |
| 36. | (A074) | Agropedology | 1.5 |
| 37. | (A078) | American Bee Journal | 2.5 |
| 38. | (A079) | American Entomologist | 2.0 |
| 39. | (A080) | American Journal of Agricultural Economics | 3.0 |
| 40. | (A081) | American Journal of Alternative Agriculture | 2.0 |
| 41. | (A082) | American Journal of Botany | 4.0 |
| 42. | (A087) | American Journal of Potato Research | 3.0 |
| 43. | (A092) | American Potato Journal | 3.0 |
| 44. | (A093) | American Reviews of Phytopathology | 1.5 |
| 45. | (A098) | Analytical Biochemistry | 4.0 |

*Contd...*

| Sl. No. | International Journal Code | NAME OF THE JOURNAL | MARKS |
|---|---|---|---|
| 46. | (A101) | Andhra Agricultural Journal | 0.5 |
| 47. | (A117) | Annals of Applied Microbiology | 1.5 |
| 48. | (A118) | Annals of Arid Zone | 2.5 |
| 49. | (A121) | Annals of Entomology | 1.5 |
| 50. | (A122) | Annals of Forestry Research | 1.0 |
| 51. | (A124) | Annals of Phytopathological Society of Japan | 1.5 |
| 52. | (A125) | Annals of Plant Protection Sciences | 0.5 |
| 53. | (A133) | Annual Review of Entomology | 4.0 |
| 54. | (A134) | Annual Review of Genetics | 4.0 |
| 55. | (A135) | Annual Review of Microbiology | 4.0 |
| 56. | (A139) | Annual Review of Phytopathology | 4.0 |
| 57. | (A140) | Annual Review of Plant Physiology & Plant Molecular Biology | 4.0 |
| 58. | (A142) | Applied Biochemistry and Biotechnology | 3.0 |
| 59. | (A146) | Applied Engineering in Agriculture | 2.5 |
| 60. | (A147) | Applied Entomology and Phytopathology | 1.5 |
| 61. | (A148) | Applied Entomology and Zoology | 3.0 |
| 62. | (A154) | Applied Soil Ecology | 3.5 |
| 63. | (A155) | Applied and Environmental Microbiology | 4.0 |
| 64. | (A181) | Arid Land Research and Management | 2.0 |
| 65. | (A182) | Arid Soil Research and Rehabilitation | 2.5 |
| 66. | (A187) | Asian Economic Review | 0.5 |
| 67. | (A189) | Asian Journal of Geoinformatics | 2.0 |
| 68. | (A192) | Australasian Plant Pathology | 2.5 |
| 69. | (A193) | Australian Dairy Foods | 2.0 |
| 70. | (A194) | Australian Entomologist | 1.5 |
| 71. | (A195) | Australian Forestry | 1.5 |
| 72. | (A196) | Australian Journal of Agricultural Research | 3.0 |
| 73. | (A197) | Australian Journal of Agricultural and Resource Economics | 3.0 |
| 74. | (A200) | Australian Journal of Botany | 3.0 |
| 75. | (A202) | Australian Journal of Ecology | 3.5 |

*Contd...*

| Sl. No. | International Journal Code | NAME OF THE JOURNAL | MARKS |
|---|---|---|---|
| 76. | (A203) | Australian Journal of Entomology | 2.5 |
| 77. | (A204) | Australian Journal of Experimental Agriculture | 3.0 |
| 78. | (A205) | Australian Journal of Marine and Freshwater Research | 3.0 |
| 79. | (A206) | Australian Journal of Plant Pathology | 1.5 |
| 80. | (A207) | Australian Journal of Plant Physiology (Functional Plant Biology) | 3.5 |
| 81. | (A208) | Australian Journal of Soil Research | 3.5 |
| 82. | (A209) | Australian Journal of Soil Science | 2.5 |
| 83. | (A210) | Australian Journal of Soil and Water Conservation | 2.5 |
| 84. | (B003) | Bangladesh Horticulture | 1.0 |
| 85. | (B011) | Bihar Journal of Agricultural Marketing | 0.5 |
| 86. | (B028) | Biodiversity and Food Security | 1.5 |
| 87. | (B035) | Biological Agriculture and Horticulture | 2.5 |
| 88. | (B036) | Biological Conservation | 3.5 |
| 89. | (B038) | Biological Control of Insect Pests | 1.5 |
| 90. | (B044) | Biology and Fertility of Soils | 3.5 |
| 91. | (B081) | British Journal of Entomology and Natural History | 2.5 |
| 92. | (B093) | Bulletin of Entomological Research | 3.0 |
| 93. | (B094) | Bulletin of Entomology (New Delhi) | 0.5 |
| 94. | (B096) | Bulletin of Grain Technology | 0.5 |
| 95. | (B097) | Bulletin of Marine Science | 3.0 |
| 96. | (C001) | Canadian Agricultural Engineering | 2.5 |
| 97. | (C002) | Canadian Entomologist | 3.0 |
| 98. | (C009) | Canadian Journal of Plant Pathology | 3.0 |
| 99. | (C010) | Canadian Journal of Plant Science | 2.5 |
| 100. | (C011) | Canadian Journal of Soil Science | 3.0 |
| 101. | (C030) | Cereal Food World | 3.0 |
| 102. | (C031) | Cereal Research Communication | 2.5 |
| 103. | (C046) | Communication in Soil Science and Plant Analysis | 2.5 |
| 104. | (C053) | Computers and Electronics in Agriculture | 3.0 |

*Contd...*

| Sl. No. | International Journal Code | NAME OF THE JOURNAL | MARKS |
|---|---|---|---|
| 105. | (C054) | Comunication in Soil Science and Plant Analysis | 2.5 |
| 106. | (C060) | Critical Reviews in Plant Sciences | 4.0 |
| 107. | (C062) | Critical Reviews of Food Science and Nutrition | 4.0 |
| 108. | (C064) | Crop Improvement | 0.5 |
| 109. | (C065) | Crop Protection | 3.0 |
| 110. | (C066) | Crop Science | 3.0 |
| 111. | (C085) | Cytogenetics and Cell Genetics | 3.5 |
| 112. | (E015) | Economical and Political Weekly | 3.0 |
| 113. | (E026) | Energy for Sustainable Development | 1.0 |
| 114. | (E027) | Energy in Agriculture | 2.0 |
| 115. | (E038) | Entomologist | 1.5 |
| 116. | (E039) | Entomon | 2.5 |
| 117. | (E040) | Entomotaxonomia | 1.0 |
| 118. | (E041) | Environment | 3.5 |
| 119. | (E042) | Environment and Development Economics | 2.0 |
| 120. | (E043) | Environment and Ecology | 1.5 |
| 121. | (E045) | Environmental Conservation | 3.0 |
| 122. | (E047) | Environmental Management | 3.0 |
| 123. | (E049) | Environmental Modeling and Softwares | 3.0 |
| 124. | (E050) | Environmental Monitoring and Assessment | 2.5 |
| 125. | (E053) | Environmental Research | 3.5 |
| 126. | (E060) | Environmental and Resource Economics | 2.0 |
| 127. | (E069) | European Journal of Agricultural Education and Extension | 2.0 |
| 128. | (E070) | European Journal of Agronomy | 3.0 |
| 129. | (E073) | European Journal of Entomology | 3.0 |
| 130. | (E077) | European Journal of Plant Pathology | 3.5 |
| 131. | (E079) | European Journal of Soil Biology | 2.5 |
| 132. | (E080) | European Journal of Soil Science (Journal of Soil Science) | 3.5 |

*Contd...*

| Sl. No. | International Journal Code | NAME OF THE JOURNAL | MARKS |
|---|---|---|---|
| 133. | (E082) | European Review of Agricultural Economics | 2.5 |
| 134. | (E088) | Experimental Agriculture | 2.5 |
| 135. | (F006) | Field Crops Research | 3.5 |
| 136. | (F036) | Food Processing | 2.0 |
| 137. | (F038) | Food Research International | 3.0 |
| 138. | (F059) | Functional Analysis and Its Applications | 2.5 |
| 139. | (G012) | Genetic Resources and Crop Evolution | 3.0 |
| 140. | (G016) | Genetics | 4.0 |
| 141. | (H001) | Haryana Journal of Horticultural Science | 1.0 |
| 142. | (H010) | Horticultural Reviews | 2.0 |
| 143. | (H011) | Hortscience | 3.0 |
| 144. | (I012) | Indian Economic Review | 2.0 |
| 145. | (I014) | Indian Journal of Agricultural Biochemistry | 0.5 |
| 146. | (I015) | Indian Journal of Agricultural Chemistry | 0.5 |
| 147. | (I016) | Indian Journal of Agricultural Economics | 3.0 |
| 148. | (I017) | Indian Journal of Agricultural Marketing | 2.0 |
| 149. | (I018) | Indian Journal of Agricultural Sciences | 2.5 |
| 150. | (I019) | Indian Journal of Agroforestry | 1.0 |
| 151. | (I020) | Indian Journal of Agronomy | 2.5 |
| 152. | (I045) | Indian Journal of Extension Education | 2.5 |
| 153. | (I047) | Indian Journal of Forestry | 1.0 |
| 154. | (I048) | Indian Journal of Genetics and Plant Breeding | 2.5 |
| 155. | (I049) | Indian Journal of Heredity | 0.5 |
| 156. | (I050) | Indian Journal of Hill Farming | 0.5 |
| 157. | (I051) | Indian Journal of Horticulture | 2.5 |
| 158. | (I052) | Indian Journal of Labour Economics | 1.5 |
| 159. | (I066) | Indian Journal of Plant Genetic Resources | 0.5 |
| 160. | (I067) | Indian Journal of Plant Pathology | 1.0 |
| 161. | (I068) | Indian Journal of Plant Physiology | 0.5 |
| 162. | (I069) | Indian Journal of Plant Protection | 1.0 |

*Contd...*

| SI. No. | International Journal Code | NAME OF THE JOURNAL | MARKS |
|---|---|---|---|
| 188. | (J017) | Journal of Agrometeorology | 1.5 |
| 189. | (J018) | Journal of Agronomy and Crop Science | 2.5 |
| 190. | (J022) | Journal of American Society for Horticultural Science | 3.0 |
| 191. | (J038) | Journal of Applied Engineering in Agriculture | 1.0 |
| 192. | (J039) | Journal of Applied Entomology | 2.5 |
| 193. | (J072) | Journal of Cell Biology | 4.0 |
| 194. | (J073) | Journal of Cell Science | 4.0 |
| 195. | (J086) | Journal of Coffee Research | 0.5 |
| 196. | (J091) | Journal of Consumer Studies and Home Economics | 1.5 |
| 197. | (J098) | Journal of Economic Entomology | 3.0 |
| 198. | (J104) | Journal of Entomological Research | 0.5 |
| 199. | (J105) | Journal of Entomological Science | 2.5 |
| 200. | (J119) | Journal of Extension Education | 0.5 |
| 201. | (J120) | Journal of Extension Systems | 1.0 |
| 202. | (J121) | Journal of Farming System Research | 0.5 |
| 203. | (J129) | Journal of Food Processing and Preservation | 2.5 |
| 204. | (J132) | Journal of Food Quality | 2.5 |
| 205. | (J144) | Journal of Herbs, Spices and Medicinal Plants | 2.0 |
| 206. | (J146) | Journal of Horticultural Science and Biotechnology | 3.0 |
| 207. | (J151) | Journal of Indian Academy of Wood Science | 2.0 |
| 208. | (J155) | Journal of Indian Potato Association | 2.0 |
| 209. | (J157) | Journal of Indian Society of Agricultural Statistics | 1.5 |
| 210. | (J158) | Journal of Indian Society of Coastal Agricultural Research | 0.5 |
| 211. | (J160) | Journal of Indian Society of Soil Science | 2.5 |
| 212. | (J172) | Journal of Intellectual Property Rights | 0.5 |
| 213. | (J174) | Journal of Irrigation and Drainage | 2.5 |
| 214. | (J177) | Journal of Japanese Society for Horticultural Science | 2.5 |

*Contd...*

| Sl. No. | International Journal Code | NAME OF THE JOURNAL | MARKS |
|---|---|---|---|
| 215. | (J178) | Journal of Kansas Entomological Society | 2.5 |
| 216. | (J181) | Journal of Maharashtra Agricultural University | 0.5 |
| 217. | (J206) | Journal of Nuclear Agriculture and Biology | 0.5 |
| 218. | (J208) | Journal of Nutrient Cycling in Agro-ecosystem | 1.0 |
| 219. | (J215) | Journal of Oilseeds Research | 0.5 |
| 220. | (J216) | Journal of Orchid Society of India | 0.5 |
| 221. | (J217) | Journal of Organic Chemistry | 4.0 |
| 222. | (J218) | Journal of Ornamental Horticulture | 1.5 |
| 223. | (J221) | Journal of Pathology | 4.0 |
| 224. | (J224) | Journal of Pesticide Science | 2.5 |
| 225. | (J233) | Journal of Plant Biochemistry and Biotechnology | 2.5 |
| 226. | (J234) | Journal of Plant Biology | 0.5 |
| 227. | (J235) | Journal of Plant Disease and Protection | 0.5 |
| 228. | (J236) | Journal of Plant Growth Regulation | 3.0 |
| 229. | (J237) | Journal of Plant Nutrition | 3.0 |
| 230. | (J238) | Journal of Plant Nutrition and Soil Science | 3.0 |
| 231. | (J239) | Journal of Plant Physiology (Biochem. Physiol. Pflanzen.) | 3.5 |
| 232. | (J240) | Journal of Plant Research | 3.0 |
| 233. | (J241) | Journal of Plantation Crops | 1.5 |
| 234. | (J243) | Journal of Potassium Research | 0.5 |
| 235. | (J245) | Journal of Protozoology Research | 3.5 |
| 236. | (J246) | Journal of Quantitative Economics | 1.0 |
| 237. | (J250) | Journal of Research, AAU | 0.5 |
| 238. | (J251) | Journal of Research, HAU | 0.5 |
| 239. | (J252) | Journal of Research, PAU | 0.5 |
| 240. | (J253) | Journal of Research, RAU Bihar | 0.5 |
| 241. | (J255) | Journal of Root Crops | 0.5 |
| 242. | (J257) | Journal of Rural Development | 1.5 |
| 243. | (J261) | Journal of Social and Economic Development | 1.0 |

*Contd...*

| Sl. No. | International Journal Code | NAME OF THE JOURNAL | MARKS |
|---|---|---|---|
| 244. | (J262) | Journal of Soil Biology and Ecology | 2.0 |
| 245. | (J263) | Journal of Soil Conservation | 2.0 |
| 246. | (J264) | Journal of Soil Contamination | 3.0 |
| 247. | (J265) | Journal of Soil and Water Conservation | 3.0 |
| 248. | (J266) | Journal of Soil and Water Conservation, India | 0.5 |
| 249. | (J267) | Journal of Solar Energy | 0.5 |
| 250. | (J268) | Journal of Spices Arom. Crops | 1.5 |
| 251. | (J272) | Journal of Sustainable Agriculture | 2.5 |
| 252. | (J273) | Journal of Sustainable Forestry | 1.0 |
| 253. | (J292) | Journal of Water Management | 0.5 |
| 254. | (J293) | Journal of Water Resources, Planning and Management | 3.0 |
| 255. | (M001) | Madras Agricultural Journal | 0.5 |
| 256. | (M003) | Maharashtra Journal of Extension Education | 0.5 |
| 257. | (M004) | Maharastra Journal of Horticulture | 0.5 |
| 258. | (M040) | Molecular Plant Pathology | 2.5 |
| 259. | (M041) | Molecular Plant-Microbe Interactions | 4.0 |
| 260. | (N012) | Netherlands Journal of Agricultural Sciences | 3.0 |
| 261. | (N018) | New Zealand Journal of Agricultural Research | 2.5 |
| 262. | (N020) | New Zealand Journal of Crop and Horticultural Science | 2.5 |
| 263. | (N029) | Nucleus | 2.5 |
| 264. | (N030) | Nutrient Cycling in Agroecosystems (Fertilizer Research) | 3.0 |
| 265. | (O008) | Orissa Journal of Agricultural Research | 0.5 |
| 266. | (O009) | Orissa Journal of Horticulture | 0.5 |
| 267. | (O010) | Oryza | 0.5 |
| 268. | (O011) | Outlook on Agriculture | 3.0 |
| 269. | (P001) | PKV Research Journal | 0.5 |
| 270. | (P005) | Pan-Pacific Entomologist | 2.5 |
| 271. | (P011) | Pest Management Science | 3.0 |

*Contd...*

| Sl. No. | International Journal Code | NAME OF THE JOURNAL | MARKS |
|---|---|---|---|
| 272. | (P012) | Pest Mgmt. Hort. Ecosystem | 0.5 |
| 273. | (P016) | Pesticide Research Journal | 2.5 |
| 274. | (P017) | Pesticide Science | 3.5 |
| 275. | (P029) | Physiological Entomology | 3.0 |
| 276. | (P040) | Phytopathology | 4.0 |
| 277. | (P045) | Plant Breeding (Zeitschrift fur pflanzenzuchtung) | 3.0 |
| 278. | (P046) | Plant Cell | 4.0 |
| 279. | (P047) | Plant Cell Reports | 3.5 |
| 280. | (P048) | Plant Cell Tissue and Organ Culture | 3.0 |
| 281. | (P049) | Plant Cell and Environment | 4.0 |
| 282. | (P050) | Plant Diseases (Plant Disease Reporter) | 3.5 |
| 283. | (P051) | Plant Ecology | 3.5 |
| 284. | (P053) | Plant Growth Regulation | 3.0 |
| 285. | (P057) | Plant Pathology | 3.5 |
| 286. | (P058) | Plant Physiology | 4.0 |
| 287. | (P059) | Plant Physiology and Biochemistry | 3.5 |
| 288. | (P060) | Plant Protection | 1.5 |
| 289. | (P061) | Plant Science (Plant Science Letters) | 3.5 |
| 290. | (P062) | Plant Systematics and Evolution | 3.5 |
| 291. | (P063) | Plant Varieties and Seeds | 2.5 |
| 292. | (P064) | Plant and Cell Physiology | 4.0 |
| 293. | (P065) | Plant and Soil | 3.5 |
| 294. | (P071) | Postharvest Biology and Technology | 3.5 |
| 295. | (P072) | Potato Research | 2.5 |
| 296. | (P086) | Progressive Horticulture | 0.5 |
| 297. | (R014) | Review of Agricultural Economics | 2.0 |
| 298. | (R015) | Review of Plant Pathology, UK (Review of Applied Mycology) | 2.5 |
| 299. | (R031) | Rural Sociology | 2.0 |
| 300. | (R032) | Rural Women | 0.5 |
| 301. | (S011) | Seed Research | 0.5 |
| 302. | (S012) | Seed Science Research | 3.5 |

*Contd...*

| Sl. No. | International Journal Code | NAME OF THE JOURNAL | MARKS |
|---|---|---|---|
| 303. | (S013) | Seed Science and Technology | 2.5 |
| 304. | (S022) | Soil Science | 3.5 |
| 305. | (S023) | Soil Science Society of America Journal (Proceedings of Soil Science Society of America) | 3.5 |
| 306. | (S024) | Soil Science and Plant Nutrition | 3.0 |
| 307. | (S025) | Soil Use and Management | 3.5 |
| 308. | (S026) | Soil and Tillage Research | 3.0 |
| 309. | (S028) | South Indian Horticulture | 0.5 |
| 310. | (S037) | Systematic Entomology | 3.5 |
| 311. | (S038) | Systematic Parasitology | 3.0 |
| 312. | (T006) | The American Economic Review | 1.5 |
| 313. | (T009) | The Horticulture Journal | 0.5 |
| 314. | (T010) | The Indian Economic Journal | 1.5 |
| 315. | (T024) | Transactions of American Society of Agricultural Engineering | 2.5 |
| 316. | (T028) | Transactions of The American Entomological Society | 2.5 |
| 317. | (T040) | Trends in Plant Science | 4.0 |
| 318. | (T041) | Tropical Agriculture | 2.5 |
| 319. | (T045) | Tropical Pest Management | 2.5 |
| 320. | (V002) | Vegetable Science | 2.5 |
| 321. | (W004) | Water Resource Journal | 1.5 |
| 322. | (W005) | Water Resource Management | 2.5 |
| 323. | (W006) | Water Resources Bulletin | 3.0 |
| 324. | (W007) | Water Resources Research | 3.5 |
| 325. | (W008) | Water Science and Technology | 3.0 |
| 326. | (W010) | Water, Air and Soil Pollution | 3.0 |
| 327. | (W011) | Weed Biology and Management | 1.5 |
| 328. | (W012) | Weed Research | 3.0 |
| 329. | (W013) | Weed Science | 2.5 |
| 330. | (W014) | Weed Technology | 2.5 |